T0212353

Lecture Notes in Physics

Volume 1005

The series Lecture Notes in Physics (LNP), founded in 1969, reports new developments in physics research and teaching - quickly and informally, but with a high quality and the explicit aim to summarize and communicate current knowledge in an accessible way. Books published in this series are conceived as bridging material between advanced graduate textbooks and the forefront of research and to serve three purposes:

- to be a compact and modern up-to-date source of reference on a well-defined topic;
- to serve as an accessible introduction to the field to postgraduate students and non-specialist researchers from related areas;
- to be a source of advanced teaching material for specialized seminars, courses and schools.

Both monographs and multi-author volumes will be considered for publication. Edited volumes should however consist of a very limited number of contributions only. Proceedings will not be considered for LNP.

Volumes published in LNP are disseminated both in print and in electronic formats, the electronic archive being available at springerlink.com. The series content is indexed, abstracted and referenced by many abstracting and information services, bibliographic networks, subscription agencies, library networks, and consortia.

Proposals should be sent to a member of the Editorial Board, or directly to the responsible editor at Springer:

Dr Lisa Scalone
Springer Nature
Physics
Tiergartenstrasse 17
69121 Heidelberg, Germany
lisa.scalone@springernature.com

Silvia M. Lenzi • Dolores Cortina-Gil
Editors

The Euroschool on Exotic Beams, Vol. VI

Springer

Editors
Silvia M. Lenzi
Department of Physics and Astronomy
University of Padova and INFN
Padova, Italy

Dolores Cortina-Gil
Instituto Gallego de Física de Altas
Energías (IGFAE)
University of Santiago de Compostela
Santiago de Compostela, Spain

ISSN 0075-8450   ISSN 1616-6361   (electronic)
Lecture Notes in Physics
ISBN 978-3-031-10750-4   ISBN 978-3-031-10751-1   (eBook)
https://doi.org/10.1007/978-3-031-10751-1

Preface

This book is the sixth volume of the series Lecture Notes in Physics, which collects a selection of review articles on nuclear physics subjects, based on lectures given at the Euroschool on Exotic Beams. The Euroschool is an annual event, addressed to PhD students and young researchers, and consists of several lecture courses given by specialists in fields related to the physics of radioactive ion beams, starting from a basic level. Initially funded by the EU, it is now supported by several funding agencies and large research facilities in Europe. With one exception (in 1999), the Euroschool on Exotic Beams has been held every year, first in Leuven from 1993 to 2000, and then, starting in 2001, it travelled to various places in Europe: Jÿvaskÿla (2001, 2011), Les Houches (2002), Valencia (2003), Surrey (2004), Mainz (2005, 2016), Trento (2006), Houlgate (2007), Piaski (2008), Leuven (2009, 2018), Santiago de Compostela (2010), Athens (2012), Dubna (2013), Padova (2014), Dubrovnik (2015), Cabourg (2017), and Knebel (2019). The 2020 edition had to be cancelled due to the pandemic, and in 2021 we organized an online Euroschool edition for the first time. This year, 2022, the Euroschool will be held in Spain, in collaboration with the International Scientific Meeting on Nuclear Physics in La Rábida.

In the last few years, we have experienced a substantial progress in the understanding of the structure of atomic nuclei and the forces at play in the nuclear medium. This has been possible due to the important theoretical and experimental developments achieved. The implementation of ab-initio calculations, the link between quantum chromodynamics and the development of realistic interactions used in various models, and the progress in shell model calculations with large-scale configuration mixing are some examples illustrating this progress. On the experimental side, the availability of exotic nuclear beams combined with novel instrumentation has been essential to improve our knowledge of the nuclear structure properties, allowing for the discovery of new phenomena and constraining and guiding the theoretical predictions. Moreover, our discipline has a very promising outlook with the construction and start-up of new facilities capable of providing intense beams of nuclei far from the valley of stability, as well as more intense and new stable beams. Training and education of our young researchers is the best guarantee we have to take advantage of the enormous opportunities and the great challenges that await us.

The Euroschool Lecture Notes aim to provide graduate and PhD students, as well as young researchers, with a comprehensive and pedagogical introduction to exciting research topics in nuclear physics with radioactive ion beams and their applications. They have been part of the program of the Euroschool lectures in recent years and are here enriched with recent and novel developments. Our goal is to fill the gap between the academic education and the research activities. Indeed, the selected topics are not treated in textbooks but mainly in scientific journals. For this sixth volume, the Board of Directors of the Euroschool on exotic beams has selected topics that have not been treated in the previous volumes of this collection as well as those that deserve an updated review. Two chapters of this sixth volume are devoted to theoretical nuclear physics. They present recent developments in ab initio approaches and large-scale shell model calculations for the interpretation and prediction of nuclear structure properties. From an experimental approach, two different methods to study nuclear structure have been selected: low-energy Coulomb excitation and nuclear reactions using active targets. Interesting nuclear astrophysics features are discussed from a theoretical and experimental point of view. This volume also includes a chapter on the recent developments in gamma-ray spectrometers. Finally, the application of gamma-ray emission imaging to medical and safeguards fields completes the book. It is our pleasure to thank the sponsors for their support over many years; this support makes the Euroschool events possible and contributes to the education of next-generation scientists. The sponsors, to whom we are indebted, are:

- ADS, Arenberg Doctoral School (Belgium)
- CEA, Commissariat à l'energie atomique et aux énergies alternatives (France)
- CNRS, Centre national de la recherche scientifique (France)
- Demokritos, National Center for Scientific Research, Athens (Greece)
- ECT*, European Centre for Theoretical Studies in Nuclear Physics and Related Areas, Trento (Italy)
- GANIL, Grand Accelerateur National d'Ions Lourds, Caen (France)
- Gobierno de España, Ministerio de Economia y Competitividad FANUC Network and CPAN Ingenio 2010, Madrid (Spain)
- GSI, Helmholtz Centre for Heavy Ion Research, Darmstadt (Germany)
- HIC-4-FAIR, Helmholtz International Center for FAIR, Darmstadt (Germany)
- IAEA, International Atomic Energy Agency
- IFIC-CSIC, Instituto de Fisica Corpuscular, Consejo Superior de Investigaciones Cientificas, Madrid (Spain)
- INFN, Istituto Nazionale di Fisica Nucleare (Italy)
- INFN-LNL, Laboratori Nazionali di Legnaro (Italy)
- IRB, Institut Ruder Boskovic (Croatia)
- ISOLDE-CERN, Geneva (Switzerland)
- JINR, Joint Institute for Nuclear Research, Dubna (Russia)
- JYFL, University of Jÿvaskÿla (Finland)
- KU Leuven, Instituut voor Kern- en Stralingsfysica, Leuven (Belgium)

- KVI-CART, Center for Advanced Radiation Technology, Groningen (the Netherlands)
- UCL, Centre de Ressources du Cyclotron, Louvain-la-Neuve (Belgium)
- Università degli Studi di Padova (Italy)
- University of Warsaw (Poland)
- University of Zagreb (Croatia)
- University of Santiago de Compostela (Spain).

We would like to thank all those who have contributed to this volume. First and foremost, we are very grateful to the authors, world class experts in their domains and lecturers in past editions of the Euroschool, for their excellent work in preparing their lectures and these contributions with a very pedagogical and, at the same time, thorough and rigorous approach. We are grateful to the members of the Board of Directors of the Euroschool on Exotic Beams, who have helped us select the topics. Finally, we would like to thank Dr. Lisa Scalone and her colleagues at Springer-Verlag for their support and encouragement in this adventure. Enjoy the lectures!

Padova, Italy
Santiago de Compostela, Spain
May 2022

Silvia M. Lenzi
Dolores Cortina-Gil

Contents

Shell Model Approaches: From $N = Z$ Towards the Neutron Drip Line

Alfredo Poves and Frédéric Nowacki

Abstract

The shell model with large-scale configuration mixing (SM-CI) is the theoretical tool of choice in nuclear spectroscopy. In this chapter, we introduce its basic concepts and discuss our present understanding of the model in terms of the competition between the spherical mean field and the nuclear correlations. A key aspect we shall treat is the choice of the valence spaces and effective interactions. We shall discuss as well the main collective modes of the nucleus—superfluidity, associated with the pairing interaction and vibrations and rotations originating in the multipole-multipole terms—using simple models. The emergence of permanent quadrupole deformation and rotational bands brings us to study Elliott's model and some of its variants. These models make it possible to give a physically intuitive interpretation of the full-fledged SM-CI calculations. First, we examine the cases of shape coexistence in two paradigms of doubly magic nuclei, ^{40}Ca and ^{56}Ni. We then move into the neutron-rich regime, to study the mechanisms that lead to the appearance of islands of inversion (IoI) at $N = 40$ and $N = 50$ and its relationship with the phenomenon of shape coexistence in ^{68}Ni and ^{78}Ni.

A. Poves (✉)
Departamento de Física Teórica and IFT UAM-CSIC, Universidad Autónoma de Madrid, Madrid, Spain
e-mail: alfredo.poves@uam.es

F. Nowacki
Université de Strasbourg, CNRS, IPHC UMR 7178, Strasbourg, France
e-mail: frederic.nowacki@iphc.cnrs.fr

© The Author(s), under exclusive license to Springer Nature Switzerland AG 2022
S. M. Lenzi, D. Cortina-Gil (eds.), *The Euroschool on Exotic Beams, Vol. VI*,
Lecture Notes in Physics 1005, https://doi.org/10.1007/978-3-031-10751-1_1

1.1   Basic Concepts

In the standard model of nuclear structure, the elementary components are nucleons
(N neutrons and Z protons, N+Z = A). The mesonic and quark degrees of freedom
are integrated out. In most cases non-relativistic kinematics is used. The bare
nucleon-nucleon (or nucleon-nucleon-nucleon) interactions are inspired by meson
exchange theories or more recently by chiral effective field theory (χ-EFT) and
must reproduce the nucleon-nucleon phase shifts and the properties of the deuteron
and other few-body systems. The challenge is to find $\Psi(\mathbf{r}_1, \mathbf{r}_2, \mathbf{r}_3, \ldots \mathbf{r}_A)$ such that
$H\Psi = E\Psi$, with

$$H = \sum_i^A T_i + \sum_{i,j}^A V_{2b}(\mathbf{r}_i, \mathbf{r}_j) + \sum_{i,j,k}^A V_{3b}(\mathbf{r}_i, \mathbf{r}_j, \mathbf{r}_k) \qquad (1.1)$$

The knowledge of the eigenvectors Ψ and the eigenvalues E makes it possible
to obtain electromagnetic moments, transition rates, weak decays, cross sections,
spectroscopic factors, etc. The task is indeed formidable. Only recently and only
for very light nuclei $A \leq 10$ the problem has been solved "exactly" thanks to
the pioneer work of Pandharipande, Wiringa and Pieper [1], who used variational
methods (Green function) solved by Monte Carlo (GFMC) techniques. The perturbative approach has been implemented in the framework of the no-core shell model
(NCSM) by Barrett, Navratil and Vary [2]. And even more recently, the techniques
of lattice gauge theory together with χ-EFT interactions have been used with very
promising results in very light nuclei [3]. A mixed approach between the "ab initio"
program based upon effective interactions obtained by χ-EFT, and the shell model
with (large-scale) configuration mixing (SM-CI), is the valence-space in-medium
similarity renormalization group (VS-IMSRG9 approach [4].

A very important outcome of these calculations is the compulsory need to
include three-body forces in order to get correct solutions of the nuclear many-body
problem. The GFMC and the NCSM approaches are severely limited by the huge
size of the calculations when A becomes larger than 12. For the rest of the chart of
nuclides, approximate methods have to be used. Except for the semi-classical ones
(liquid drop) and the α-cluster models, all are based on the independent particle
model (IPM). Beyond the limits of applicability of the fully "ab initio" descriptions,
the methods of choice are the SM-CI and the beyond mean field (BFM) approaches
using energy density functionals (a.k.a. density-dependent effective interactions,
like the Gogny force). There are nowadays renewed efforts to connect rigorously
these two global methods and the bare two- and three-body nuclear interactions by
means of the full palette of the many-body perturbation methods. If this is achieved,
they will deserve the "ab initio" label as well.

1.1.1 The Independent Particle Model

The basic idea of the independent particle model is to assume that, at the zeroth order, the result of the complicated two-body interactions among the nucleons is to produce an average self-binding potential. Mayer and Jensen (1949) proposed a spherical mean field consisting in the isotropic harmonic oscillator plus a strongly attractive spin-orbit potential and an orbit-orbit term:

$$H = \sum_i h(\mathbf{r}_i) \tag{1.2}$$

$$h(r) = -V_0 + \mathcal{T} + \frac{1}{2}m\omega^2 r^2 - V_{so}\mathbf{l} \cdot \mathbf{s} - V_B \mathbf{l}^2 \tag{1.3}$$

where we have used $\mathcal{T}$ for the kinetic energy to avoid confusion with the isospin quantum number T.

Later, other functional forms which follow better the form of the nuclear density and have a more realistic asymptotic behaviour, e.g. the Woods-Saxon (WS) well, were adopted:

$$V(r) = V_0 \left(1 + e^{\frac{r-R}{a}}\right)^{-1} \tag{1.4}$$

with

$$V_0 = \left(-51 + 33\frac{N-Z}{A}\right) \text{MeV} \tag{1.5}$$

and

$$V_{ls}(r) = \frac{V_0^{ls}}{V_0}(\mathbf{l} \cdot \mathbf{s})\frac{r_0^2}{r}\frac{dV(r)}{dr} \; ; \; V_0^{ls} = -0.44V_0 \tag{1.6}$$

The eigenvectors of the IPM are characterized by the radial quantum number n, the orbital angular momentum l, the total angular momentum j and its z projection m. With the choice of the harmonic oscillator, the eigenvalues are

$$\epsilon_{nljm} = -V_0 + \hbar\omega(2n + l + 3/2)$$

$$- V_{so}\frac{\hbar^2}{2}(j(j+1) - l(l+1) - 3/4) - V_B\hbar^2 l(l+1) \tag{1.7}$$

In order to reproduce the nuclear radii,

$$\hbar\omega = 45A^{-1/3} - 25A^{-2/3} \tag{1.8}$$

With a suitable choice of the parameters, it explains the magic numbers and, in the large A limit, the volume, the surface and (half) the symmetry terms of the semi-empirical mass formula as well (more on that later).

The wave functions of the isotropic HO can be written as

$$\Psi_{nlm}(r, \theta, \phi) = \frac{1}{r} R_{nl}(r) \, Y_{lm}(\theta, \phi) \tag{1.9}$$

By convention the ns start at zero; therefore the eigen-energies read:

$$E_{nl} = (2n + l + 3/2) \, \hbar\omega = (p + 3/2)\hbar\omega \tag{1.10}$$

$Y_{lm}(\theta, \phi)$ are the spherical harmonics and

$$R_{nl}(r) = (-1)^l \left(\frac{2 \, (2\nu)^{l+3/2} \, n!}{\Gamma(n + l + 3/2)} \right)^{1/2} r^{l+1} \, e^{-\nu r^2} \, L_n^{l+1/2}(2\nu r^2) \tag{1.11}$$

The parameter ν is defined as $\dfrac{m\omega}{2\hbar}$, and it is related to the length parameter of the HO $2\nu = \dfrac{1}{b^2}$. The degeneracy of each shell is $(p + 1)(p + 2)$, and the functions L are the Laguerre (associated) polynomials.

When the spin-orbit coupling is taken into account, we must include explicitly the spin part of the wave function and change the coupling scheme from [L S] to [J J]:

- VOCABULARY
    - STATE: a solution of the Schrödinger equation with a one-body potential, e.g. HO or WS. It is characterized by the quantum numbers $nljm$ and by its neutron or proton nature (or equivalently by the projection of the isospin t_z).
    - ORBIT: the ensemble of states with the same nlj, e.g. the $0d_{5/2}$ orbit. Its degeneracy is $(2j + 1)$.
    - SHELL: an ensemble of orbits quasi-degenerated in energy, e.g. the sd-shell, that includes the orbits $0d_{5/2}$, $1s_{1/2}$ and $0d_{3/2}$.
    - MAGIC NUMBERS: the numbers of protons or neutrons that fill orderly a certain number of shells. For instance, 28 corresponds to the filling of the $s(2)$, $p(6)$ and $sd(12)$ shells plus the orbit $0f_{7/2}(8)$ and 50 to the filling of the s, p, sd, and $pf(20)$ shells plus the orbit $0g_{9/2}(10)$.
    - GAP: the energy difference between two shells.
    - SPE, single-particle energies: the eigenvalues of the IPM Hamiltonian.
    - ESPE, effective single-particle energies: the eigenvalues of the monopole Hamiltonian to be introduced in Sect. 1.4.

1.1.2 The Independent Particle Model and the Liquid-Drop Mass Formula

The IPM explains the magic numbers, spins and parities of the ground states and some excited states of doubly magic nuclei plus or minus one nucleon, their magnetic moments, etc. With the addition of a schematic pairing term between like particles, it can go a bit further in semi-magic nuclei (Schmidt lines). What is less well known is that in the large A limit, the IPM can reproduce the volume, the surface and the symmetry terms of the semi-empirical mass formula as well.

Let's take the IPM with an HO potential and neglect the spin-orbit term. Then

$$H = \sum_i \mathcal{T}_i - V_0 + \frac{1}{2}m\omega^2 r_i^2 \tag{1.12}$$

The single-particle energies are $\epsilon_i = -V_0 + \hbar\omega(p_i + 3/2)$ and $< r_i^2 > = b^2(p_i + 3/2)$ with $b^2 = \dfrac{\hbar}{m\omega}$.

Assuming $N = Z$, to accommodate $\frac{A}{2}$ identical particles, we need to fill all the shells up to a maximum value of $p = p_F$. Experimentally, the radius of the nucleus is given by $< r^2 > = \frac{3}{5}R^2 = \frac{3}{5}(1.2A^{1/3})^2$ and in the IPM by

$$< r^2 > = \frac{2}{A}\sum_i^{A/2} < r_i^2 > = \frac{2}{A}\sum_{p=0}^{p_F} b^2(p + 3/2)(p + 1)(p + 2) \tag{1.13}$$

From

$$\frac{A}{2} = \sum^{p_F}(p + 1)(p + 2) \tag{1.14}$$

it obtains at leading order

$$\frac{A}{2} = \frac{1}{3}p_F^3 \tag{1.15}$$

Hence, $p_F = (\frac{3}{2}A)^{1/3}$. Inserting this value in Eq. (1.13) it is easy to find that at leading order in p_F, $b^2 = A^{1/3}$ and $\hbar\omega = 41 \cdot A^{-1/3}$. We can now compute the total binding energy as

$$B = \sum_{i=1}^{A}(-V_0 + \hbar\omega(p_i + 3/2)) \tag{1.16}$$

that gives at leading order:

$$\frac{B}{A} + V_0 = \hbar\omega \cdot \frac{p_F^4}{4} \cdot \frac{2}{A} = \hbar\omega \left(\frac{3A}{2}\right)^{4/3} \frac{1}{2A} = \hbar\omega A^{1/3} \frac{1}{2} \left(\frac{3}{2}\right)^{4/3} \tag{1.17}$$

Finally we have

$$\frac{B}{A} = -V_0 + 41 \times 0.86 \tag{1.18}$$

and we recover the volume term of the semi-empirical mass formula for $V_0 \sim 50$ MeV.

If we go to next to leading order, keeping the terms in p_F^3, we recover the surface term and with a coefficient that agrees with the empirically determined one. We can repeat the calculation at leading order but with $N \neq Z$ and obtain

$$B = -AV_0 + \frac{\hbar\omega}{4}((p_F^\nu)^4 + (p_F^\pi)^4) = -AV_0 + \frac{\hbar\omega}{4}((3N)^{4/3} + (3Z)^{4/3}) \tag{1.19}$$

Making a Taylor expansion around the minimum at $N = Z$ and using the previously determined values, we find an extra term of the form $(N - Z)^2/A$ with a coefficient which does not agree with the one resulting from the fit of the semi-empirical mass formula to the experimental binding energies ($a_{sym} = 23$ MeV). This reflects the fact that the nuclear two-body neutron-proton interaction is in average more attractive than the neutron-neutron and the proton-proton ones, and it is related as well to the experimental evidence of the near equality of the neutron and proton radii for $N \neq Z$. Therefore we should use different values of $\hbar\omega$ and V_0s for protons and neutrons in the derivation, which complicates a lot the calculation because both effects go in opposite directions.

1.2   The Meaning of the Independent Particle Model

The usual procedure to generate a mean field in a system of N interacting fermions, starting from their free interaction, is the Hartree–Fock (HF) approximation, extremely successful in atomic physics. Whatever the origin of the mean field, the eigenstates of the N-body problem are Slater determinants, i.e. antisymmetrized products of N single-particle wave functions. In the nucleus, there is a catch, because the very strong short-range repulsion and the tensor force make the HF approximation based upon the bare nucleon-nucleon force impracticable. However, at low energy, the nucleus does manifest itself as a system of independent particles in many cases, and when it does not, it is due to the medium-range correlations that produce strong configuration mixing and not to the short-range repulsion. Does the success of the shell model really "prove" that nucleons move independently in a fully occupied Fermi sea as assumed in HF approaches? In fact, the single-particle

Table 1.1 The parameters of the Brink and Boeker interaction

i	μ_i (fm)	v_i (MeV)	m_i
1	0.7	471.1	−0.43
2	1.4	−163.8	0.51

motion can persist at low energies in fermion systems due to the suppression of collisions by the Pauli exclusion principle (see Pandharipande et al. [5]). Brueckner theory takes advantage of the Pauli blocking to regularize the bare nucleon-nucleon interaction, in the form of density-dependent effective interactions of use in HF calculations or G-matrices for large-scale shell model calculations.

An example of regularized interaction is the one proposed by Brink and Boeker [6], whose central part is

$$V_c(|\mathbf{r}_1 - \mathbf{r}_2|) = \sum_{i=1}^{2} [1 - m_i(1 + P_\sigma P_\tau)]\, v_i\, e^{-|\mathbf{r}_1 - \mathbf{r}_2|^2/\mu_i^2} \tag{1.20}$$

where μ are the widths of the Gaussians and P the spin and isospin projectors. The values of m are fitted to produce the attraction in the S=0, T=1 and S=1, T=0 channels and repulsion in the others, whereas the vs give the energy scale of the two Gaussians. For the spin-orbit term, they took a one-body approximation (Table 1.1):

$$V_{ls} = \frac{-12\,\text{MeV}}{\hbar^2 \sqrt{A}}\, \mathbf{l} \cdot \mathbf{s} \tag{1.21}$$

To be more realistic, one should refine the channel dependence of the central terms, include a two-body spin-orbit interaction and more importantly add a term which depends on the density. After this re-vamping, the Brink and Boeker interaction becomes the Gogny interaction [7] extremely successful in numerous mean field applications (and beyond).

The wave function of the ground state of a nucleus in the IPM is the product of a Slater determinant for the Z protons that occupy the Z lowest states in the mean field and another Slater determinant for the N neutrons in the N lowest states of the mean field. In the second quantization, this state can be written as

$$|N\rangle \cdot |Z\rangle \tag{1.22}$$

with

$$|N\rangle = n_1^\dagger n_2^\dagger \ldots n_N^\dagger |0\rangle \tag{1.23}$$

$$|Z\rangle = z_1^\dagger z_2^\dagger \ldots z_Z^\dagger |0\rangle \tag{1.24}$$

In a system of noninteracting fermions, the occupied states have occupation number 1, and the empty ones occupation number 0. In reality there is a dilution of the strength leading to a nonzero value above the Fermi level. In spite of that, the nuclear quasi-particles resemble extraordinarily to the mean field solutions of the IPM. This was demonstrated by the beautiful electron scattering experiment of Cavedon et al. [8] in which they extracted the charge density difference between ^{206}Pb and ^{205}Tl that in the IPM limit is just the square of the $2s_{1/2}$ orbit wave function. The shape of the $2s_{1/2}$ orbit is very well given by a mean field calculation with the Gogny functional. To make the agreement quantitative, the calculated density had to be scaled down with the occupation number. This is a first example of the necessity of using effective transition operators consistent with the regularized interactions that provide the natural basis for the many-body description of nuclei. For a very pedagogical discussion of the basis of the IPM, see Ref. [5].

1.3   Beyond the Independent Particle Model

It is quite obvious that the IPM cannot encompass the extreme variety of manifestations of the nuclear dynamics. In fact, even in the most favourable cases, as at the doubly magic nuclei, its limitations are dramatically evident. Let's take ^{40}Ca as an example. In the IPM limit, we expect a 0^+ ground state (no problem) and a gap of about $\hbar\omega$ (9 MeV) before finding a bunch of quasi-degenerate levels of particle-hole type and negative parity. In fact, the first excited state lies at 3.5 MeV and is again a 0^+, which, upon experimental and theoretical scrutiny, turns out to be the band head of a rotational band of 4p-4h nature. Even more exotic is another 0^+ at 5.1 MeV, which is the band head of a superdeformed band of 8p-8h structure. Going beyond the mean field is compulsory because the nuclear dynamics is dominated in most cases by the correlations. We shall show in what follows how these coexisting structures can be reproduced by large-scale shell model calculations and interpreted using analytic models.

To go beyond the IPM, there are two main routes: In the mean field way, the correlations are taken into account by explicitly breaking the symmetries of the mean field HF wave functions and employ density-dependent interactions of different sorts: Skyrme, Gogny or relativistic mean field parametrizations. They are often referred to as "intrinsic" descriptions. Projections before (VAP) or after (PAV) variation are enforced to restore the conserved quantum numbers. Ideally, configuration mixing is also implemented through the generator coordinate method. The other route pertains to the SM-CI which can be seen as an approximation to the exact solution of the nuclear A-body problem using effective interactions in restricted spaces. The SM-CI wave functions respect the symmetries of the Hamiltonian, and these approaches are sometimes called "laboratory frame" descriptions.

Let's proceed through a kind of formal solution to the A-body problem. The single-particle states $(i, j, k, \ldots)$, which are the solutions of the IPM, provide as well a basis in the space of the occupation numbers (Fock space). The many-body wave functions are Slater determinants:

$$\Phi = a_{i_1}^\dagger, a_{i_2}^\dagger, a_{i_3}^\dagger, \ldots a_{i_A}^\dagger |0\rangle \tag{1.25}$$

We can distribute the A particles in all the possible ways in the available single-particle states. This provides a complete basis in the Fock space. The number of Slater determinants will be huge but not infinite because the theory is no longer valid beyond a certain energy cut-off. Therefore, the "exact" solution can be expressed as a linear combination of the basis states:

$$\Psi = \sum_\alpha c_\alpha \Phi_\alpha \tag{1.26}$$

and the solution of the many-body Schödinger equation

$$H\Psi = E\Psi \tag{1.27}$$

is transformed in the diagonalization of the matrix:

$$H_{\alpha,\beta} = \langle \Phi_\alpha | H | \Phi_\beta \rangle \tag{1.28}$$

whose eigenvalues and eigenvectors provide the "physical" energies and wave functions. A shell model calculation thus amounts to diagonalizing the effective nuclear Hamiltonian in the basis of all the Slater determinants that can be built distributing the valence particles in a set of orbits which is called "valence space". The orbits that are always full form the "core". If we could include all the orbits in the valence space (a full no-core calculation), we should get the "exact" solution. The effective interactions for SM-CI calculations are obtained from the bare nucleon-nucleon interaction by means of a regularization procedure aimed to soften the short-range repulsion. In other words, using effective interactions we can treat the A-nucleon system in a basis of independent quasi-particles. As we reduce the valence space, the interaction has to be renormalized again using many-body perturbation theory. Up to this point, these calculations can be labelled as "ab initio". In fact, the realistic NN interactions seem to be correct except for its simplest part, the monopole Hamiltonian responsible for the evolution of the spherical mean field. Therefore, we surmise that the three-body forces will mainly contribute to the monopole Hamiltonian.

The three basic ingredients of the SM-CI approach are then the effective interactions, the valence spaces and the algorithms and codes put at work to solve the huge computational challenges posed by the solution of this secular problem. See, for instance, Ref. [9] for a full-fledged presentation of our approach.

1.4   The Effective Interactions in Fock Space

Using the creation and annihilation operators of particles in the states of the underlying spherical mean field in the coupled representation, we can write the Hamiltonian as

$$\mathcal{H} = \sum_{rr'} \epsilon_{rr'} (a_r^+ a_{r'})^0 + \sum_{r \leq s, t \leq u, \Gamma} W_{rstu}^\Gamma Z_{rs\Gamma}^+ \cdot Z_{tu\Gamma} \tag{1.29}$$

where Z_Γ^+ (Z_Γ) is the coupled product of two creation (annihilation) operators:

$$Z_{rs\Gamma}^+ = [a_r^\dagger a_s^\dagger]^\Gamma \tag{1.30}$$

Γ is a shorthand for (J,T), r, s ... run over the orbits of the valence space, $\epsilon_{rr'}$ are the single-particle energies (or the kinetic energies in the no-core calculations) and W_{rstu}^Γ the antisymmetrized two-body matrix elements:

$$W_{rstu}^\Gamma = \langle j_r j_s (JT) | V | j_t j_u (JT) \rangle \tag{1.31}$$

In the occupation number representation (Fock space), all the information about the interaction is contained in its two-body matrix elements. The many-body problem then reduces to the manipulation of the creation and annihilation operators using the Wick theorem and techniques alike.

The most general method to compute the two-body matrix elements is due to Slater and carries his name. When the independent particle wave functions are those of the harmonic oscillator or if they can be represented by linear combination of a few harmonic oscillator states, the method of choice is that of Brody and Moshinsky [10]. Both methods are described in detail in Ref. [11].

1.4.1   Monopole and Multipole Components of the Interaction

Without losing the simplicity of the Fock space representation, we can recast the two-body matrix elements of any effective interaction in a way full of physical insight, following Dufour–Zuker rules [12].

"Any effective interaction can be split in two parts:

$$\mathcal{H} = \mathcal{H}_m (monopole) + \mathcal{H}_M (multipole) \tag{1.32}$$

where $\mathcal{H}_m$ contains all the terms that are affected by a spherical Hartree–Fock variation; hence it is responsible for the global saturation properties and for the evolution of the spherical single-particle energies".

Considering two-body interactions only, we can write

$$\mathcal{H}_m = \sum \epsilon_i n_i + \sum \left[\frac{1}{(1 + \delta_{ij})} a_{ij} n_i (n_j - \delta_{ij}) \right.$$
$$\left. + \frac{1}{2} b_{ij} \left(T_i \cdot T_j - \frac{3n_i}{4} \delta_{ij} \right) \right] \tag{1.33}$$

where n_i and T_i are the number and isospin operators for the orbit i. The coefficients a and b are defined in terms of the centroïds (angular averages):

$$V_{ij}^T = \frac{\sum_J W_{ijij}^{JT}[J]}{\sum_J [J]} \tag{1.34}$$

as $a_{ij} = \frac{1}{4}(3V_{ij}^1 + V_{ij}^0)$, $b_{ij} = V_{ij}^1 - V_{ij}^0$, the sums running over Pauli allowed values $[J]$ is a shorthand for $(2j + 1)$.

It is easy to verify that the expectation value of the full Hamiltonian in a Slater determinant for closed shells has the same expression than the Hartree–Fock energy:

$$\langle H \rangle = \sum_i \langle i | \mathcal{T} | i \rangle + \sum_{ij} \langle ij | V | ij \rangle \tag{1.35}$$

where i and j run over the occupied states. $\mathcal{T}$ is the kinetic energy and V the effective interaction. If the two-body matrix elements are written in coupled formalism and we denote the orbits by α, β, ..., the expression reads

$$\langle H \rangle = \sum_\alpha (2j_\alpha + 1) \langle \alpha | \mathcal{T} | \alpha \rangle + \sum_{\alpha \leq \beta} \sum_{J,T} (2J + 1)(2T + 1) \langle j_\alpha j_\beta (JT) | V | j_\alpha j_\beta (JT) \rangle \tag{1.36}$$

The monopole Hamiltonian governs the evolution of effective spherical single-particle energies (ESPE) with the number of particles in the valence space, schematically:

$$\epsilon_j(\{n_i\}) = \epsilon_j(\{n_i = 0\}) + \sum_i a_{ij} n_i \tag{1.37}$$

Notice that the ESPEs not only evolve along isotopic and isotonic chains inside the valence space (shell evolution) but can change for different configurations in the same nucleus (configuration dependent or Type II shell evolution). It is very important to realize that even small defects in the centroids can produce large changes in the relative position of the different configurations due to the appearance of quadratic terms involving the number of particles in the different orbits.

The multipole Hamiltonian $\mathcal{H}_M$ can be written in two representations, particle-particle and particle-hole:

$$\mathcal{H}_M = \sum_{r \le s, t \le u, \Gamma} W^{\Gamma}_{rstu} Z^{+}_{rs\Gamma} \cdot Z_{tu\Gamma} \tag{1.38}$$

$$\mathcal{H}_M = \sum_{rstu\Gamma} [\gamma]^{1/2} \frac{(1+\delta_{rs})^{1/2}(1+\delta_{tu})^{1/2}}{4} \omega^{\gamma}_{rtsu} (S^{\gamma}_{rt} S^{\gamma}_{su})^0 \tag{1.39}$$

where S^{γ} is the product of one creation and one annihilation operator coupled to γ (i.e. $\lambda \tau$):

$$S^{\gamma}_{rs} = [a^{\dagger}_r a_s]^{\gamma} \tag{1.40}$$

The W and ω matrix elements are related by a Racah transformation:

$$\omega^{\gamma}_{rtsu} = \sum_{\Gamma} (-)^{s+t-\gamma-\Gamma} \begin{Bmatrix} r & s & \Gamma \\ u & t & \gamma \end{Bmatrix} W^{\Gamma}_{rstu}[\Gamma] \tag{1.41}$$

$$W^{\Gamma}_{rstu} = \sum_{\gamma} (-)^{s+t-\gamma-\Gamma} \begin{Bmatrix} r & s & \Gamma \\ u & t & \gamma \end{Bmatrix} \omega^{\gamma}_{rtsu}[\gamma] \tag{1.42}$$

The operators $S^{\gamma=0}_{rr}$ are just the number operators for the orbits r and the terms $S^{\gamma=0}_{rr'}$ produce the spherical Hartree–Fock particle-hole jumps. The latter must have null coefficients if the monopole Hamiltonian satisfies the Hartree–Fock self-consistency. The operator $Z^{+}_{rr\Gamma=0}$ creates a pair of particle coupled to $J = 0$. The terms $W^{\Gamma}_{rrss} Z^{+}_{rr\Gamma=0} \cdot Z_{ss\Gamma=0}$ represent different kinds of pairing Hamiltonians. The operators S^{γ}_{rs} are typical one-body operators of multipolarity γ. For instance, $\gamma = (J = 1, L = 0, T = 1)$ contains a $(\sigma \cdot \sigma)(\tau \cdot \tau)$ term which is nothing else but the Gamow-Teller component of the nuclear interaction. The terms $S^{\gamma}_{rs}\gamma = (J = 2, T = 0)$ are of quadrupole type $r^2 Y_2$. They are responsible for the existence of deformed nuclei, and they are specially large and attractive when $j_r - j_s = 2$ and $l_r - l_s = 2$.

A careful analysis of the available realistic effective nucleon-nucleon interactions obtained with different methods reveals that the multipole Hamiltonian is universal and dominated by BCS-like isovector and isoscalar pairing plus quadrupole-quadrupole and octupole-octupole terms of very simple nature $(r^{\lambda} Y_{\lambda} \cdot r^{\lambda} Y_{\lambda})$. As an example we list in Table 1.2 the strengths of the coherent multipole components of different interactions for the pf-shell.

Table 1.2 Strengths (in MeV) of the coherent multipole components of different interactions for the pf-shell

Interaction	Particle-particle		Particle-hole		
	$JT = 01$	$JT = 10$	$\lambda\tau = 20$	$\lambda\tau = 40$	$\lambda\tau = 11$
KB3	−4.75	−4.46	−2.79	−1.39	+2.46
FPD6	−5.06	−5.08	−3.11	−1.67	+3.17
GOGNY	−4.07	−5.74	−3.23	−1.77	+2.46
GXPF1	−4.18	−5.07	−2.92	−1.39	+2.47
BONNC	−4.20	−5.60	−3.33	−1.29	+2.70

1.4.2   Valence Spaces and Codes

An ideal valence space should incorporate the most relevant degrees of freedom for the nuclei under study and be computationally tractable. Classical $0\hbar\omega$ valence spaces are provided by the major oscillator shells p, sd and pf. As we move far from stability, other choices are compulsory; for instance, for the very neutron-rich nuclei around $N = 28$, a good choice is to take the sd-shell for protons and the pf-shell for neutrons, and for the very neutron-rich Cr, Fe, Ni and Zn, one should rather take $r_3 − (0g_{9/2}, 1d_{5/2})$ for the neutrons and pf for protons (in a major harmonic oscillator shell of principal quantum number p, the orbit $j = p + 1/2$ is called *intruder*, and the remaining ones are denoted by r_p; for instance, in the pf-shell, the intruder orbit is the $0f_{7/2}$, and r_3 includes the orbits $1p_{3/2}$, $1p_{1/2}$ and $0f_{5/2}$). To describe the intruders around N and/or $Z = 20$, a good valence space is $r_2 − pf$. For the nuclei above ^{100}Sn, the valence space $r_4 − h_{11/2}$ has been also widely used.

The solution of the secular problem of the SM-CI is computationally very demanding. Direct diagonalization is of very limited utility, and other algorithms like the Lanczos method, Monte Carlo shell model, quantum Monte Carlo diagonalization, density matrix renormalization group, etc., are employed. There are also a number of different extrapolation ansatzs. The Strasbourg-Madrid codes (Antoine, Nathan) [9] can deal with problems involving the basis of 10^{11} Slater determinants, using relatively modest computational resources. Other competitive codes which have been released publicly are Oxbash [13], Nushell [14] and Kshell [15].

1.5   Collectivity in Nuclei

For a given interaction, a many-body system would or would not display coherent features at low energy depending on the structure of the mean field around the Fermi level. So, when the spherical mean field around the Fermi surface favours the pairing interaction, as in the case of having only neutrons and protons on top of a doubly magic core, the nucleus tends to become superfluid. However, if the quadrupole-quadrupole interaction is dominant, for instance, if both protons and neutrons are available in open orbits, the nucleus acquires permanent deformation. In the extreme (unrealistic) limit in which the monopole Hamiltonian is negligible, the multipole interaction would maximize the deformation and together with the pairing interaction produce a kind of superfluid nuclear needles. Magic nuclei resist the strong multipole interaction, because the large gaps in the nuclear mean field at the Fermi surface block the correlations.

Let's consider a simple model consisting of two states that have diagonal energies that differ by Δ and an off-diagonal matrix element δ. The eigenvalues and eigenvectors of this problem are obtained diagonalizing the matrix:

$$\begin{pmatrix} 0 & \delta \\ \delta & \Delta \end{pmatrix} \tag{1.43}$$

In the limit $\delta \ll \Delta$, we can use perturbation theory, and no special coherence is found. On the contrary in the degenerate case, $\Delta \to 0$, the eigenvalues of the problem are $\pm \delta$, and the eigenstates are the 50% mixing of the unperturbed ones with different signs. They are the germ of the maximally correlated (or anticorrelated) states.

We can generalize this example by considering a degenerate case with N Slater determinants with equal (and attractive) diagonal matrix elements $(-\Delta)$ and off-diagonal ones of the same magnitude. The problem now is that of diagonalizing the matrix:

$$-\Delta \begin{pmatrix} 1 & 1 & 1 & \dots \\ 1 & 1 & 1 & \dots \\ 1 & 1 & 1 & \dots \\ \cdot & \cdot & \cdot & \cdot \cdot \cdot \\ \cdot & \cdot & \cdot & \cdot \cdot \cdot \\ \cdot & \cdot & \cdot & \cdot \cdot \cdot \end{pmatrix} \tag{1.44}$$

which has range 1 and whose eigenvalues are all zero except one which has the value $-N\Delta$. This is the coherent state. Its corresponding eigenvector is a mixing of the N unperturbed states with amplitudes $\frac{1}{\sqrt{N}}$.

1.5.1   Nuclear Superfluidity: Pairing Collectivity

The pairing Hamiltonian for one shell expressed in the m-scheme basis of two particles has a very similar matrix representation:

$$
-G
\begin{pmatrix}
1 & -1 & 1 & \cdots \\
-1 & 1 & -1 & \cdots \\
1 & -1 & 1 & \cdots \\
\cdot & \cdot & & \cdots \\
\cdot & \cdot & & \cdots \\
\end{pmatrix}
\tag{1.45}
$$

and its coherent solution is just the state of the two particles coupled to zero which gains an energy $-G\Omega$ ($\Omega = j+1/2$ is the degeneracy of the shell). It can be written as

$$
Z_j^\dagger |0\rangle = \frac{1}{\sqrt{\Omega}} \sum_{m>0} (-1)^{j+m} a_{jm}^\dagger a_{j-m}^\dagger |0\rangle
\tag{1.46}
$$

Using the commutation relations,

$$
\left[Z_j, Z_j^\dagger \right] = 1 - \frac{\hat{n}}{\Omega}; \quad \text{and} \quad \left[H, Z_j^\dagger \right] = -G(\Omega - \hat{n} + 2) Z_j^\dagger
\tag{1.47}
$$

where $\hat{n}$ is the operator number of particle, it is possible to construct the eigenstates of H for n particles consisting of $n/2$ pairs coupled to $J = 0$. These states are labelled as seniority zero states. The quantum number v (seniority) counts the number of particles not coupled to angular momentum zero:

$$
|n, v = 0\rangle = (Z_j^\dagger)^{\frac{n}{2}} |0\rangle \quad \text{and} \quad E(n, v = 0) = -\frac{G}{4} n(2\Omega - n + 2)
\tag{1.48}
$$

We can construct also eigenstates with higher seniority using the operators $B_J^\dagger$ which create a pair of particles coupled to $J \neq 0$. These operators satisfy the relation:

$$
\left[H, B_J^\dagger \right] |0\rangle = 0
\tag{1.49}
$$

States which contain m $B_J^\dagger$ operators have seniority $v = 2m$. Their eigen-energies relative to the seniority zero state are

$$
E(n, v) - E(n, v = 0) = \frac{G}{4} v(2\Omega - v + 2)
\tag{1.50}
$$

Notice that the gap is independent of the number of particles. The generalization to the odd number of particles is trivial.

For n protons and neutrons in the same shell of degeneracy Ω coupled to total isospin T, the eigenvalues of the $J = 0$ $(L = 0)$ $T = 1$ pairing Hamiltonian can be written as

$$E(\Omega, n, v, t, T) = -G((n - v)(4\Omega + 6 - n - v)/8 + t(t + 1)/2 - T(T + 1)/2) \tag{1.51}$$

where v is the sum of the seniorities of protons and neutrons and t the reduced isospin, one half of their difference.

The case of two particles in several shells is also tractable and has a great heuristic value. The problem in a matrix form reads:

$$\begin{pmatrix} 2\epsilon_1 - G\Omega_1 & -G\sqrt{\Omega_1\Omega_2} & -G\sqrt{\Omega_1\Omega_3} \dots \\ -G\sqrt{\Omega_2\Omega_1} & 2\epsilon_2 - G\Omega_2 & -G\sqrt{\Omega_2\Omega_3} \dots \\ -G\sqrt{\Omega_3\Omega_1} & -G\sqrt{\Omega_3\Omega_2} & 2\epsilon_3 - G\Omega_3 \dots \\ & & \dots \\ & & \dots \\ & & \dots \end{pmatrix} \tag{1.52}$$

There is a limit in which maximum coherence is achieved, when the orbits have the same Ω and they are degenerate. Then the coherent pair is evenly distributed among the shells, and its energy is $E = -G \sum_i \Omega$. All the other solutions remain at their unperturbed energies.

A textbook case of nuclear superfluidity is provided by the tin isotopes from $N = 52$ to $N = 80$. The five orbits comprised between the magic closures 50 and 82 are closely packed, and one should expect pairing dominance in several shells. The pairing gap is measured by the excitation energy of the first 2^+ state and should be independent of the neutron number. Indeed that is what the experiments tell us and what the SM-CI calculations reproduce nicely as can be seen in Fig. 1.1.

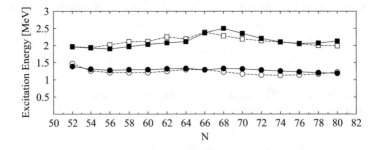

Fig. 1.1 Low energy excited states of the tin isotopes; experiment compared with SM-CI calculations in the r_4-$h_{11/2}$. space with the gcn50:82 interaction [16]; 2^+ circles, 4^+ squares, th. (filled) exp. (empty)

The problem can be turned into a dispersion relation as well. Let us write the most general solution as

$$|\alpha\rangle = \sum_j X_j^\alpha Z_j^\dagger |0\rangle \tag{1.53}$$

Plugging it in the Schrödinger equation, $H|\alpha\rangle = E_\alpha |\alpha\rangle$, we get

$$(2\epsilon_k - E_\alpha) X_k^\alpha = G \sum_j \sqrt{\Omega_j \Omega_k} X_j^\alpha \tag{1.54}$$

Multiplying by $\sqrt{\Omega_k}$ both sides and summing over k, we obtain the dispersion relation:

$$\frac{1}{G} = \sum_k \frac{\Omega_k}{2\epsilon_k - E_\alpha} \tag{1.55}$$

The dispersion relation can be solved graphically or iteratively. As we have seen before, we expect one coherent solution (the collective pair) to gain a lot of energy and the rest of the solutions to be very close to the unperturbed ones. If we assume that the single-particle energies are degenerate and take $\epsilon_k = <\epsilon>$, we obtain

$$E_\alpha = 2 <\epsilon> -G \sum_k \Omega_k \tag{1.56}$$

In this limit the energy gain is equivalent to the one in a single shell of degeneracy $\sum_k \Omega_k$.

For the case of many particles in non-degenerate orbits, the problem is usually solved in the BCS or Hartree–Fock–Bogoliubov approximations. Other approaches, which do not break the particle number conservation, are either the SM-CI or others based on it; these include the interacting boson model and its variants and different group theoretical approximations.

1.5.2   Vibrational Spectra: Quadrupole and Octupole Collectivity

In the semi-classical approach, vibrational spectra are described as the quantized harmonic modes of vibration of the surface of a liquid drop. The restoring force comes from the competition of the surface tension and the Coulomb repulsion. This is hardly germane to reality and to the microscopic description that we will develop in a simplified way. Let's just remind which are the characteristic features of a nuclear vibrator; first, a harmonic spectrum as the one shown in Fig. 1.2 and second, enhanced Eλ transitions between the states differing in one vibrational phonon.

Fig. 1.2 Schematic
depiction of a perfect
vibrational spectrum

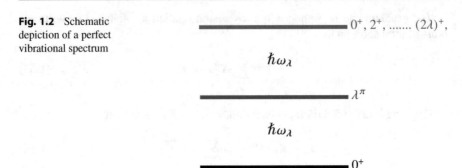

Imagine that for a given even-even nucleus, the orbits around the Fermi level are such as depicted in Fig. 1.3. Its ground state has $J^\pi = 0^+$, and, in the IPM, the lowest excited states correspond to promoting one particle from the occupied orbits to the empty ones. They are many, quasi-degenerate, and should appear at excitation energies Δ.

Let's take now into account the multipole Hamiltonian, which for simplicity will be of separable form, and choose as valence space just the particle-hole states, $|mi\rangle$, that correspond to making a hole in the orbit i and adding a particle in the orbit m, and then

$$\langle nj|V|mi\rangle = \beta_\lambda Q^\lambda_{nj} Q^\lambda_{mi} \tag{1.57}$$

the wave function can be developed in the p-h basis as

$$\Psi = \sum C_{mi}|mi\rangle \tag{1.58}$$

the Schödinger equation $H\Psi = E\Psi$ can thus be written as

$$C_{nj}(E - \epsilon_{nj}) = \sum_{mi} \beta_\lambda C_{mi} Q^\lambda_{nj} Q^\lambda_{mi} \tag{1.59}$$

Fig. 1.3 A valence space for
the description of the nuclear
vibrations (see text)

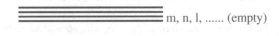

and

$$C_{nj} = \frac{\beta_\lambda Q_{nj}^\lambda}{E - \epsilon_{nj}} \sum_{mi} C_{mi} Q_{mi}^\lambda \tag{1.60}$$

trivially

$$1 = \beta_\lambda \sum_{nj} \frac{(Q_{nj}^\lambda)^2}{E - \epsilon_{nj}} \tag{1.61}$$

A graphical analysis of this equation shows that all its solutions except one are very close to the unperturbed values ϵ_{nj}. The remaining one is the lowest, and it is well separated from the others, very much as in the pairing case discussed before. Taking for all the ϵ_{nj} the average value $\overline{\epsilon}_{nj} = \Delta$, we obtain

$$E = \Delta + \beta_\lambda \sum_{nj} (Q_{nj}^\lambda)^2 \tag{1.62}$$

If the interaction is attractive $\beta_\lambda < 0$, the lowest state gains an energy which is proportional to β_λ, the strength of the multipole interaction, and to the coherent sum of the squared one-body matrix elements of the one-body multipole operators between the particle and hole orbits in the space. This mechanism of coherence explains the appearance of vibrational states in the nucleus and represents the basic microscopic description of the nuclear "phonons". Because the couplings β_λ are constant except for a global scaling, the onset of collectivity requires the presence of several quasi-degenerate orbits above and below the Fermi level. In addition, these orbits must have large matrix elements with the multipole operator of interest.

The wave function of the coherent (collective) state (phonon, vibration) has the following form:

$$\Psi_c(J = \lambda) = \frac{\sum_{nj} Q_{nj}^\lambda |nj\rangle}{\sum_{nj} (Q_{nj}^\lambda)^2} \tag{1.63}$$

The collective state is coherent with the transition operator Q^λ because the probability of its electromagnetic $E\lambda$ decay to the 0^+ ground state is very much enhanced:

$$B(E\lambda) \sim |\langle 0^+ | Q^\lambda | \Psi_c(J = \lambda)\rangle|^2 = \sum_{nj} (Q_{nj}^\lambda)^2 \tag{1.64}$$

which should be much larger than the single-particle limit (many Weisskopft units (WU)). Clearly, a large value of the $B(E\lambda)$ does not imply necessarily the existence of permanent deformation in the ground state. Notice also that nothing prevents that

$$|\beta_\lambda \sum_{nj} (Q_{nj}^\lambda)^2| > \Delta \qquad (1.65)$$

In this case the vibrational phonon is more bound than the ground state, and the model is no longer valid. What happens is that a phase transition from the vibrational to the rotational regime takes place as the nucleus acquires permanent deformation of multipolarity λ. The separation between filled and empty orbits does not hold anymore, and both have to be treated at the same footing.

1.6 Deformed Nuclei: Intrinsic vs. Laboratory Frame Approaches

The route to the description of permanently deformed nuclear rotors bifurcates now into laboratory frame and intrinsic descriptions. The latter include the deformed shell model (Nilsson) and the deformed Hartree–Fock approximation, plus the beyond mean field approaches such as angular momentum projection and configuration mixing with the generator coordinate method. The former, the SM-CI and in cases of full dominance of the quadrupole-quadrupole interaction group theoretical treatments like Elliott's SU(3) and its variants [17–19].

A case where the two approaches could be confronted is ^{48}Cr (four protons and four neutrons on top of ^{40}Ca) where an SM-CI description in the full pf-shell was for the first time possible more than two decades ago [20]. The mean field intrinsic description was a cranked Hartree–Fock–Bogoliubov (CHFB) approximation using the Gogny force. The results are presented in Fig. 1.4. Both calculations reproduce

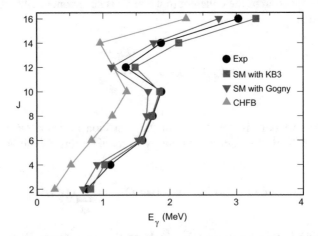

Fig. 1.4 The yrast band of ^{48}Cr; experimental data compared with the SM-CI with the KB3G interaction and the two-body matrix element of the Gogny functional and the CHFB calculations with the Gogny functional

Table 1.3 Quadrupole properties of the yrast band of ^{48}Cr, in e^2fm^4 and efm^2

J	$B(E2)_{exp}$	$B(E2)_{th}$	$Q_0(t)$
2	321(41)	228	107
4	330(100)	312	105
6	300(80)	311	100
8	220(60)	285	93
10	185(40)	201	77
12	170(25)	146	65
14	100(16)	115	55
16	37(6)	60	40

the rotor like behaviour at low and medium spin and the existence of a backbending at $J = 12$. However, the CHFB description misses the size of the moment of inertia due to the absence of neutron-proton pairing correlations in its wave functions. The Gogny force does contain the right proton-neutron $T = 0$ and $T = 1$ pairing as shown by the results of the SM-CI calculation with its two-body matrix elements. Hence the blame is on the CHFB method and not on the Gogny functional.

The laboratory frame wave functions are indeed collective as can be seen in Table 1.3 where we have listed the $B(E2)$s and compared with the experiment. From the calculated values, we can extract the intrinsic quadrupole moments $Q_0(t)$ using Eq. (1.85). They are roughly independent of J below the backbending as in a well-behaved Bohr–Mottelson rotor. From the intrinsic quadrupole moment, a deformation parameter can be calculated using

$$\beta = \frac{\sqrt{5\pi}}{3} \frac{Q_0(t)}{ZR^2} \tag{1.66}$$

The resulting value, $\beta = 0.28$, is in very good agreement with the CHFB result (Table 1.3).

1.6.1 The Nilsson Model

The Nilsson model is an approximation to the solution of the IPM plus a quadrupole-quadrupole interaction:

$$H = \sum_i h(\mathbf{r}_i) + \hbar\omega\kappa \sum_{i<j} Q_i \cdot Q_j \tag{1.67}$$

$$h(r) = -V_0 + \mathcal{T} + \frac{1}{2}m\omega^2 r^2 - V_{so}\mathbf{l} \cdot \mathbf{s} - V_B\mathbf{l}^2 \tag{1.68}$$

which amounts to linearizing the quadrupole-quadrupole interaction, replacing one of the operators by the expectation value of the quadrupole moment (or by the deformation parameter). Thus, the resulting physical problem is that of the IPM subject to a quadrupole field, which obviously breaks rotational symmetry:

$$H_{Nilsson} = \sum_i h(\mathbf{r}_i) - \frac{1}{3}\hbar\omega\delta Q_0(i) \qquad (1.69)$$

The problem is equivalent to the diagonalization of the quadrupole operator in the basis of the IPM eigenstates. The resulting (Nilsson) levels are characterized by their magnetic projection on the symmetry axis m, also denoted K and the parity.
The formulae below make it possible to build the relevant matrices:

$$\langle pl|r^2|pl\rangle = p + 3/2 \quad : \quad \langle pl|r^2|pl+2\rangle = -[(p-l)(p+l+3)]^{1/2} \qquad (1.70)$$

$$Q_0 = 2r^2C_2 = 2r^2\sqrt{4\pi/(2l+1)}Y^{20} \quad : \quad \langle jm|C_2|jm\rangle = \frac{j(j+1) - 3m^2}{2j(2j+2)} \qquad (1.71)$$

$$\langle jm|C_2|j+2m\rangle = \frac{3}{2}\left\{\frac{[(j+2)^2 - m^2][(j+1)^2 - m^2]}{(2j+2)^2(2j+4)^2}\right\}^{1/2} \qquad (1.72)$$

$$\langle jm|C_2|j+1m\rangle = -\frac{3m[(j+1)^2 - m^2]^{1/2}}{j(2j+4)(2j+2)} \qquad (1.73)$$

The intrinsic wave functions provided by the Nilsson model correspond to the Slater determinants built putting the neutrons and the protons in the lowest Nilsson levels (each one has degeneracy two, $\pm m$). Therefore, for even-even nuclei $K = 0$, for odd nuclei $K = m$ of the last half occupied orbit, and for odd-odd, there are different empirical rules, not always very reliable. The Nilsson diagrams for the sd-shell are plotted in Fig. 1.5. The laboratory frame wave functions are obtained rotating the intrinsic frame with the Wigner matrices, i.e. correspond to the solutions of the quantum rotor problem. In the even-even case, this leads trivially to the energy formula:

$$E(J) = \sum_i (\epsilon_i)_{Nilsson} + \frac{\hbar^2}{2I}J(J+1)$$

1.6.2 The SU3 Symmetry of the HO and Elliott's Model

The mechanism that produces permanent deformation and rotational spectra in nuclei is much better understood in the framework of the SU(3) symmetry of the isotropic harmonic oscillator and its implementation in Elliott's model [17]. The

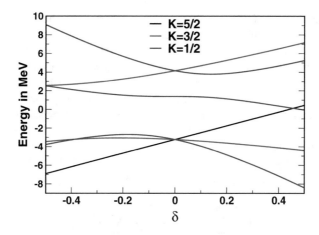

Fig. 1.5 Nilsson diagrams for the sd-shell: $K = 5/2$ (black), $K = 3/2$ (red) and $K = 1/2$ (blue). Energies of the states in MeV as a function of the deformation parameter

basic simplification of the model is threefold: (i) the valence space is limited to one major harmonic oscillator (HO) shell, (ii) the monopole Hamiltonian makes the orbits of this shell degenerate and (iii) the multipole Hamiltonian only contains the quadrupole-quadrupole interaction. This implies (mainly) that the spin-orbit splitting and the pairing interaction are put to zero. Let's then start with the isotropic HO which in units $m = 1\omega = 1$ can be written as

$$H_0 = \frac{1}{2}\left(p^2 + r^2\right) = \frac{1}{2}(\mathbf{p} + i\mathbf{r})(\mathbf{p} - i\mathbf{r}) + \frac{3}{2}\hbar = \hbar\left(\mathbf{A}^\dagger\mathbf{A} + \frac{3}{2}\right) \qquad (1.74)$$

with

$$\mathbf{A}^\dagger = \frac{1}{\sqrt{2\hbar}}(\mathbf{p} + i\mathbf{r}) \quad \mathbf{A} = \frac{1}{\sqrt{2\hbar}}(\mathbf{p} - i\mathbf{r}) \qquad (1.75)$$

which have bosonic commutation relations. H_0 is invariant under all the transformations which leave invariant the scalar product $\mathbf{A}^\dagger\mathbf{A}$. As the vectors are three-dimensional and complex, the symmetry group is U(3). We can build the generators of U(3) as bi-linear operators in the As. The antisymmetric combinations produce the three components of the orbital angular momentum L_x, L_y and L_z, which are in turn the generators of the rotation group O(3). From the six symmetric bi-linears, we can remove the trace that is a constant: the mean field energy. Taking it out we move into the group SU(3). The five remaining generators are the five components of the quadrupole operator:

$$q_\mu^{(2)} = \frac{\sqrt{6}}{2\hbar}(\mathbf{r} \wedge \mathbf{r})_\mu^{(2)} + \frac{\sqrt{6}}{2\hbar}(\mathbf{p} \wedge \mathbf{p})_\mu^{(2)} \qquad (1.76)$$

The generators of SU(3) transform single-nucleon wave functions of a given p (principal quantum number) into themselves. In a single-nucleon state, there are p oscillator quanta which behave as $l = 1$ bosons. When we have several particles, we need to construct the *irreps* of SU(3) which are characterized by the Young tableaux (n_1, n_2, n_3) with $n_1 \geq n_2 \geq n_3$ and $n_1 + n_2 + n_3 = Np(N$ being the number of particles in the open shell). The states of one particle in the p shell correspond to the representation $(p, 0, 0)$. Given the constancy of Np, the *irreps* can be labelled with only two numbers. Elliott's choice was $\lambda = n_1 - n_3$ and $\mu = n_2 - n_3$. In the Cartesian basis, we have $n_x = a + \mu$, $n_y = a$ and $n_z = a + \lambda + \mu$, with $3a + \lambda + 2\mu = Np$.

The quadratic Casimir operator of SU(3) is built from the generators:

$$\mathbf{L} = \sum_{i=1}^{N} \mathbf{l}(i) \qquad Q_\alpha^{(2)} = \sum_{i=1}^{N} q_\alpha^{(2)}(i) \tag{1.77}$$

as

$$C_{SU(3)} = \frac{3}{4}(\mathbf{L} \cdot \mathbf{L}) + \frac{1}{4}(Q^{(2)} \cdot Q^{(2)}) \tag{1.78}$$

and commutes with them. With the usual techniques of group theory, it can be shown that the eigenvalues of the Casimir operator in a given representation (λ, μ) are

$$C(\lambda, \mu) = \lambda^2 + \lambda\mu + \mu^2 + 3(\lambda + \mu) \tag{1.79}$$

Once these tools are ready, we come back to the physics problem as posed by Elliott's Hamiltonian:

$$H = H_0 + \chi(Q^{(2)} \cdot Q^{(2)}) \tag{1.80}$$

which can be rewritten as

$$H = H_0 + 4\chi C_{SU(3)} - 3\chi(\mathbf{L} \cdot \mathbf{L}) \tag{1.81}$$

The eigenvectors of this problem are thus characterized by the quantum numbers λ, μ and L. We can choose to label our states with these quantum numbers because O(3) is a subgroup of SU(3) and therefore the problem has an analytical solution:

$$E(\lambda, \mu, L) = N\hbar\omega(p + \frac{3}{2}) + 4\chi(\lambda^2 + \lambda\mu + \mu^2 + 3(\lambda + \mu)) - 3\chi L(L+1) \tag{1.82}$$

This important result can be interpreted as follows: For an attractive quadrupole-quadrupole interaction, $\chi < 0$, the ground state of the problem pertains to the representation which maximizes the value of the Casimir operator, and this corresponds to maximizing λ or μ (the choice is arbitrary). If we solve the problem in the Cartesian basis, this state is the one which has the maximum number of oscillator quanta in the z-direction, thus breaking the rotational symmetry at the intrinsic level. We can then speak of a deformed solution even if its wave function conserves the good quantum numbers of the rotation group, i.e. L and L_z. For that one (and every) (λ, μ) representation, there are different values of L which are permitted, for instance, for the representation $(\lambda, 0)$ $L = 0, 2, 4 \ldots \lambda$. And their energies satisfy the $L(L + 1)$ law, thus giving the spectrum of a rigid rotor. The problem of the description of the deformed nuclear rotors in the laboratory frame is thus formally solved.

We can describe the intrinsic states and their relationship with the physical ones using another chain of subgroups of SU(3). The chain we have used until now is SU(3)⊃O(3)⊃U(1) which corresponds to labelling the states as $\Psi([\tilde{f}](\lambda\mu)LM)$.

$[\tilde{f}]$ is the representation of U(Ω) ($\Omega = 1/2\,(p + 1)\,(p + 2)$) conjugate of the U(4) spin-isospin representation which guarantees the antisymmetry of the total wave function. For instance, in the case of ^{20}Ne, the fundamental representation (8,0) (four particles in $p = 2$) is fully symmetric, $[\tilde{f}] = [4]$, and its conjugate representation in the U(4) of Wigner [1, 1, 1, 1], fully antisymmetric. The other chain of subgroups, SU(3)⊃SU(2)⊃U(1), does not contain O(3), and therefore the total orbital angular momentum is not a good quantum number anymore. Instead we can label the wave functions as $\Phi([\tilde{f}](\lambda\mu)q_0\Lambda K)$, where q_0 is the intrinsic quadrupole moment whose maximum value is $q_0 = 2\lambda + \mu$. K is the projection of the angular momentum on the Z-axis, and Λ is an angular momentum without physical meaning. Both representations provide a complete basis; therefore it is possible to write the physical states in the basis of the intrinsic ones. Actually, the physical states can be projected out of the intrinsic states with maximum quadrupole moment as

$$\Psi([\tilde{f}](\lambda\mu)LM) = \frac{2L + 1}{a(\lambda\mu KL)} \int D_{MK}^{L}(\omega)\Phi_\omega([\tilde{f}](\lambda\mu)(q_0)_{max}\Lambda K)d\omega \quad (1.83)$$

Remarkably, this is the same kind of expression used in the unified model of Bohr and Mottelson, the Wigner functions D being the eigenfunctions of the rigid rotor and the intrinsic functions the solutions of the Nilsson model.

Elliott's model was initially applied to nuclei belonging to the sd-shell that show rotational features like ^{20}Ne and ^{24}Mg. The fundamental representation for ^{20}Ne is (8,0), and its intrinsic quadrupole moment, $Q_0 = (2\lambda + \mu + 3)\,b^2 = 19\,b^2 \approx 60\,efm^2$ (b is the length parameter of the HO). For ^{24}Mg we have (8,4) and $23\,b^2 \approx 70\,efm^2$. To compare these figures with the experimental values, we need to know the transformation rules from intrinsic to laboratory frame quantities and vice versa. In the Bohr Mottelson model, these are

$$Q_0(s) = \frac{(J+1)(2J+3)}{3K^2 - J(J+1)} Q_{spec}(J), \quad K \neq 1 \tag{1.84}$$

$$B(E2, J \rightarrow J-2) = \frac{5}{16\pi} e^2 |\langle J K 20 | J-2, K \rangle|^2 Q_0(t)^2 \quad K \neq 1/2, 1 \tag{1.85}$$

$$Q_{spec}(J) = < J J | z^2 - r^2 | J J > \tag{1.86}$$

The expression for the quadrupole moments is also valid in Elliott's model. However the one for the B(E2)s is only approximately valid for very low spins. Using them it can be easily verified that the SU(3) predictions compare nicely with the experimental results $Q_{spec}(2^+) = -23(3)\,\text{efm}^2$ and $B(E2)(2^+ \rightarrow 0^+) = 66(3)\,e^2\,\text{fm}^4$ for ^{20}Ne and $Q_{spec}(2^+) = -17(1)\,\text{efm}^2$ and $B(E2)(2^+ \rightarrow 0^+) = 70(3)\,e^2\,\text{fm}^4$ for ^{24}Mg.

Besides Elliott's SU(3) there are other approximate symmetries related to the quadrupole-quadrupole interaction which are of great interest. Pseudo-SU(3) applies when the valence space consists of a quasi-degenerate harmonic oscillator shell except for the orbit with maximum j; this space has been denoted by r_p before. Its quadrupole properties are the SU(3) ones of the shell with $(p-1)$ [18]. Quasi-SU(3) [19] applies in a regime of large spin-orbit splitting, when the valence space contains the intruder orbit and the $\Delta j = 2$, $\Delta l = 2$; $\Delta j = 4$, $\Delta l = 4$; etc., orbits which are obtained from it. Its quadrupole properties are described in Ref. [9]. These symmetries turn out to be at the root of the appearance of islands of inversion far from stability. They are more prominent at the neutron-rich side and occur when the configurations which correspond to the neutron shell closures at $N = 8$, 20, 28 and 40 are less bound than the intruder ones (more often deformed) built by promoting neutrons across the Fermi level gap. The reason of the inversion is that the intruder configurations maximize the quadrupole correlations and thus their energy gains. This is only possible when the orbits around the Fermi level can develop the symmetries of the quadrupole interaction. For instance, at $N = 20$ the intruder states in ^{32}Mg have four sd protons in quasi-SU(3), two sd neutron holes in pseudo-SU(3) and two pf neutrons in quasi-SU(3). This leads to a huge gain of correlation energy (typically 12 MeV) which suffices to turn the intruders into ground states.

1.7 Nuclear Deformation in the Laboratory Frame: SM-CI Approaches

As stated above, large-scale SM-CI calculations, when doable, are the spectroscopic tool of choice in theoretical nuclear structure. When they are interpreted adequately, they may provide us with the link between the experimental data and the "ab initio" approaches. Indeed, the monopole anomaly of the realistic NN interactions may turn out to be the fingerprint of residual three-body effects [21, 22]. A non-negligible fraction of the Segré chart is nowadays amenable to SM-CI descriptions. As explained in detail in [9], the choice of a valence space which can encompass the physics dictated by the effective interaction is the crucial one in SM work. Magic numbers provide the natural borders of the SM valence spaces, because they are supposed to correspond to large gaps in the spherical mean field. Nevertheless, sometimes, if the gaps are weakened, this may not hold anymore.

The pf-shell provides a valence space which can cope successfully with the physics of several $N = Z$ nuclei. From the point of view of the quadrupole coupling schemes which will be dealt with in the next section, ^{48}Cr and ^{52}Fe are good quasi-SU(3)-deformed nuclei. ^{56}Ni is doubly magic albeit the closed shell amounts just to 60–70% depending on the calculations. ^{60}Zn is transitional, and ^{64}Ge looks very much like a pseudo-SU(3) mildly deformed nucleus. Beyond, we need to change the valence space [25].

The next valence space, r$_3$g, has a core of ^{56}Ni and comprises the orbits $1p_{3/2}$, $0f_{5/2}$, $1p_{1/2}$ and $0g_{9/2}$. ^{68}Se is a natural inhabitant of this space. However, its region of applicability at or close to $N = Z$ does not extend very far. Already at ^{72}Kr, oblate-prolate coexistence sets in. This requires the inclusion of the quadrupole partners of the $0g_{9/2}$ orbit, the $1d_{5/2}$ and $2s_{1/2}$ ones, in the valence space [25]. This leads to the r$_3$gds space. The next (and last) doubly magic $N = Z$ nucleus, ^{100}Sn, can be taken as the core of another valence space, r$_4$h, which spans between the magic numbers 50 and 82.

1.7.1 The Quadrupole Interaction: Intrinsic States and Coherence

In order to gauge the quadrupole coherence of a given nucleus, it is customary to compare its quadrupole properties with perfect Bohr–Mottelson rotors, i.e. to verify that the E2 transition rates and the spectroscopic quadrupole moments of the states of the yrast band can be derived from a single intrinsic quadrupole moment using Eqs. (1.84), (1.85) and (1.86).

If $Q_0(s) \approx Q_0(t)$ and constant, we can speak of good rotors. The excitation energies of the yrast band should follow approximately the $J(J + 1)$ law as well. However, in light and medium mass nuclei, pairing might distort the lower part of the spectrum, giving extra binding to the 0^+ ground states. It is therefore advisable to verify the $J(J + 1)$ law excluding the ground state. Another caveat here has to do with the (bad) habit of blindly extracting deformation parameters β from the $B(E2)$ values, even in cases in which the existence of an intrinsic deformed state is not guaranteed at all.

The only rigorous (but far more demanding) way of characterizing the quadrupole deformation in the laboratory frame is through the quadrupole invariants introduced by Kumar [23]. Very recently, its scope has been enlarged to make it possible to extract not only the mean values of β and γ but also their variances. Indeed, when these are very large, the notion of intrinsic state makes no sense [24].

Quadrupole coherence may be associated with a single shell (Eq. (1.71)), with a full r_p space (pseudo-SU(3)), and in this case the intrinsic single-particle states are obtained diagonalizing the quadrupole operator using Eqs. (1.71), (1.72) and (1.73) or with a full HO major shell; thus the Nilsson-like orbits of Elliott's SU3 [17] are obtained. If this is done in the $\Delta j = 2$ HO sequences (quasi SU(3)), one can diagonalize the quadrupole operator using Eqs. (1.71) and (1.72) or the corresponding expressions in LS coupling as adopted in [19]. We shall work with the latter choice unless explicitly stated otherwise. The radial integrals are given in Eq. (1.70). We use adimensional r^2 and $Q_0 = 2r^2C_2$ and quadrupole effective charges equal to 0.5 and 1.5 for neutrons and protons, respectively. To recover dimensions, $r^2 \rightarrow r^2b^2$.

1.7.2 The Quadrupole Interaction in a Single Orbit

The intrinsic quadrupole moment for n particles in an orbit j with principal HO quantum number p is given by the formula:

$$\frac{Q_0}{b^2} = \sum_m 2r^2 \langle jm|C_2|jm \rangle = \sum_m (p + 3/2) \frac{j(j+1) - 3m^2}{j(j+1)} \tag{1.87}$$

If we fill orderly the magnetic sub-states with increasing $|m|$, we obtain prolate intrinsic states. If we do it the other way around, we obtain oblate intrinsic states. In Table 1.4, we list the Q_0 values for the $0g_{9/2}$ orbit and $N = Z$. For $n < (2j + 1)$ the oblate solutions have the larger Q_0 (and therefore the larger binding if the quadrupole interaction is dominant). For $n > (2j + 1)$ the prolate solutions lead. For $n = (2j + 1)$ both are degenerate.

1.7.3 SU(3) and Pseudo-SU(3)

For a full HO shell, $q_0 = 2n_z - n_x - n_y$, with $n_x + n_y + n_z = p$, where n_i are the numbers of oscillator quanta in each spatial direction. For $p = 2$, the intrinsic states are $[n_z n_x n_y] = [200], [110], [101], [020], [011], [002]$.

The quadrupole moments of the intrinsic states of the $p = 4$ shell are plotted in Fig. 1.6 (for $p = 3$ remove the upper row and change the values in the y-axis to

Table 1.4 Intrinsic quadrupole moments for n particles in the $0g_{9/2}$ orbit and $N = Z$ (in units of b^2)

n	2	4	6	8	10	12	14	16	18
Prolate	5.33	10.66	14.66	18.66	20	21.33	18.66	16	8
Oblate	−8	−16	−18.66	−21.33	−20	−18.66	−14.66	−10.66	−5.33

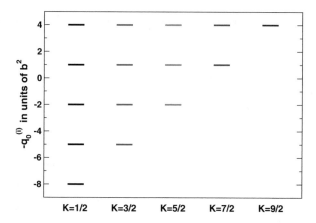

Fig. 1.6   Quadrupole moments of the intrinsic states of SU(3) for the $p = 4$ HO shell

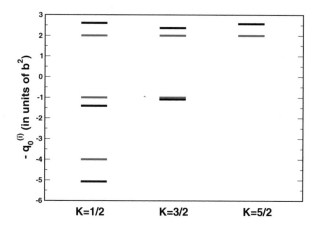

Fig. 1.7   Quadrupole moments of the intrinsic states of pseudo-SU(3) for the r_3 valence space (black) compared with those of SU(3) in the $p = 2$ HO shell (red)

$-6, -3, 0, +3$), and (for $p = 2$ remove the two upper rows and change the values in the y-axis to $-4, -1, +2$). Orderly fillings in the figure produce eigenstates of $h = qq - \lambda L(L + 1)$ and *a fortiori* rotational bands.

The values of the quadrupole moments in a pseudo-SU(3) space [18] in shell p would be the ones of SU(3) in shell $p-1$, except for the fact that the radial integrals are not the same. This introduces factors in the range $(p + 3/2)/(p + 1/2)$ and $\sqrt{(p+4)/p+2)}$. In Fig. 1.7 we have plotted the quadrupole moments in pseudo-SU(3) for the r_3 valence space compared with those of SU(3) in the $p = 2$ HO shell. Notice that the lowest prolate states have quadrupole moments that are about 20% larger than the SU(3) ones for $p = 2$.

Table 1.5 Intrinsic quadrupole moments for n holes in r_3 in the pseudo-SU(3) limit (in units of b^2)

n	2	4	6	8	10	12
Oblate	−9.74	−19.48	−22.06	−24.65	−27.20	−29.77
n	22	20	18	16	14	
Oblate	−5.14	−10.28	−15.43	−20.57	−25.17	

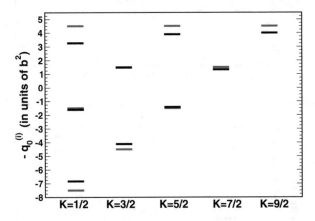

Fig. 1.8 Quadrupole moments of the intrinsic states of quasi-SU(3) for the p=4 HO shell

Table 1.6 Intrinsic quadrupole moments for n particles in the quasi-SU(3) sector of the $p = 4$ HO shell, in units of b^2, assuming $N = Z$

n	2	4	6	8	10	12	14	16
Prolate	15	30	39	48	51	54	57	60

In Table 1.5 we list the intrinsic quadrupole moments for n holes in r_3 for later use. Only the oblate solutions are explicitly included. To get the prolate ones, take for n holes the oblate value corresponding to 24-n, and change sign.

1.7.4   Quasi-SU3

In the case of a $\Delta j = 2$ HO sequence, the resulting scheme is very much like that of SU(3) except that some degeneracies are not present and the quadrupole collectivity is a bit smaller as shown in Fig. 1.8. The schematic quasi-SU(3) results are in red and the exact ones in black.

In Table 1.6 we have listed the intrinsic quadrupole moments for n particles in the quasi-SU(3) sector of the $p = 4$ HO shell. Only the prolate cases are considered. Remember that in all the SU3-like cases the total intrinsic quadrupole moment obtained from the eigenvalues of q_0 has to be increased by 3 b^2 as explained in Ref. [26].

1.8 Coexistence: Single-Particle, Deformed and Superdeformed States in ^{40}Ca

Let's describe the structure of ^{40}Ca with a SM-CI calculation in the valence space of two major shells and interpret the results in the framework of SU(3) and its variants. The orbits of the valence space are sketched in Fig. 1.9.

The relevant configurations are $[sd]^{24}$ 0p-0h, spherical; $[sd]^{20}$ $[pf]^4$ 4p-4h, deformed (ND); and $[sd]^{16}$ $[pf]^8$ 8p-8h, superdeformed (SD).

The results presented in Fig. 1.10 show clearly the importance of the correlations. In the 8p-8h configuration, the correlations amount to 18.5 MeV. From these, 5.5 MeV is due to pairing, and the remaining 13 MeV is most likely of quadrupole

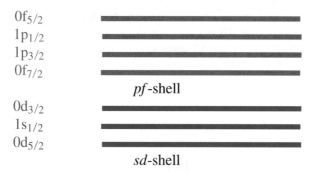

Fig. 1.9 The *sd-pf* valence space

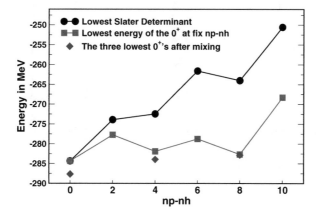

Fig. 1.10 Energies of the np-nh configurations: uncorrelated energies (blue), full mixing at fixed np-nh (black) and full mixing (red)

origin. In the 4p-4h configuration, the pairing contribution is similar, but the quadrupole one is smaller at about 4.5 MeV. The closed shell gains 5 MeV of pairing energy by mixing (30%) with the 2p-2h states, the ND band head 2 MeV and the SD band head essentially nothing.

Concerning the character of the solutions, we can see that for the 4p-4h intrinsic state of ^{40}Ca, the two neutrons and the two protons in the pf-shell can be placed in the lowest $K = 1/2$ quasi-SU3 level of the $p = 3$ shell. This gives a contribution to the intrinsic quadrupole moment of $q_0 = 25$ b^2. In the pseudo-sd-shell, $p = 2$, we are left with eight particles that contribute with $q_0 = 7$ b^2. In the 8p-8h configuration, the values are $q_0 = 35$ b^2 and $q_0 = 11$ b^2. Using the proper value of the oscillator length, it obtains:

- ^{40}Ca 4p-4h band Q_0=125 e fm^2
- ^{40}Ca 8p-8h band Q_0=180 e fm^2

These results are in very good accord with the data ($Q_0 = 120$ efm^2 and $Q_0 = 180$ efm^2). Assuming full SU(3) symmetry in both shells, we should get $Q_0 = 148$ efm^2 and $Q_0 = 226$ efm^2, respectively. The SM-CI results almost saturate the quasi-SU(3) bounds. The SU(3) values are a 25% larger.

Finally we compare the SM-CI-level scheme with the experiment in Fig. 1.11. The agreement is extremely satisfactory.

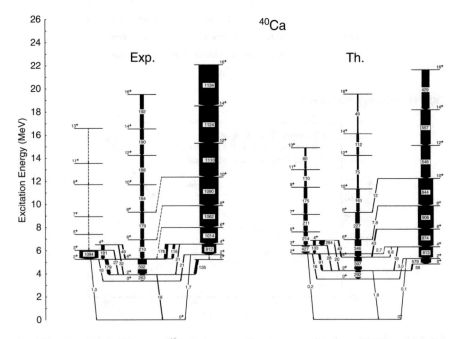

Fig. 1.11 Partial-level scheme of ^{40}Ca: Experiment (left) compared with the SM-CI results from [27] (right). The numbers in the arrows give the $B(E2)$s in e^2fm^4

1.8.1 ^{56}Ni: The Lightest Spin-Orbit Doubly Magic Nucleus

A crucial difference between the HO and the SO magic closures is that in the former the orbits below and above the Fermi level have different parities, whereas in the latter, they have the same. This implies that the lowest excited states are of positive parity and of 1p-1h nature. For instance, in ^{56}Ni the first excited state is a 2^+ at 2.7 MeV whose configuration is $(0f_{7/2})^{15} (1p_{3/2})^1$. Therefore, the E2 transition to the ground state is allowed and even enhanced due to the favoured quadrupole connexion between the two orbits at play. Indeed, the experimental value amounts to 9 WU, a rather large value for a doubly magic nucleus, much larger than the corresponding one in ^{40}Ca. However, the rest of their low energy spectra resemble to each other with the appearance of low-lying 0^+ states with multiparticle-hole configuration that can be the band heads of rotational structures. This is the case of the band built on the 0^+ at 5 MeV, which, according to theory, is a 4p-4h state. As in the ^{40}Ca case, a lot of correlation energy is needed to compensate for the cost of promoting four particles across a $N = Z = 28$ gap of about 5 MeV. But we can look at it the other way round; should the gap have been slightly smaller, the deformed 4p-4h configuration would have become the ground state and ^{56}Ni expelled of the doubly magic club. In fact if one uses the original Kuo-Brown interaction, that is what occurs. In reality, due to the presence of a substantial gap, it is the quasi-degeneracy of the orbits $1p_{3/2}$, $1p_{1/2}$ and $0f_{5/2}$, which enhances the quadrupole collectivity through the action of the pseudo-SU(3) coupling and favours the polarization of the $0f_{7/2}$ orbit, as discussed in previous sections. This is another example of the relevance of the shell evolution, because these orbits, which are split apart in ^{48}Ca, become much closer in ^{56}Ni when protons fill the $0f_{7/2}$ orbit, due to the proton-neutron monopole interaction. Higher-lying bands involve excitations into the $0g_{9/2}$ orbit [28]. All in all, this is another example of the coexistence of single particle and collective degrees of freedom already at the very doubly magic nuclei.

1.9   Islands of Inversion at the Neutron-Rich Shores

Among the nuclear structure topics at the forefront of present-day experimental and theoretical research, the study of very neutron-rich nuclei plays a central role, and future facilities will make it even more prominent. We refer to three very recent review articles that give a global view of the status of the field [29–31]. A unifying theme in this research is the so-called islands of inversion (IoI). We use this term for regions of nuclei, close to a magic neutron closure, that, instead of having the expected semi-magic nature, are deformed in their ground states. The name was coined in Ref. [32] for the region surrounding ^{31}Na ($N = 20$). Reference [31] contains a detailed account of the history of the IoIs. The physics of the IoI is a prime example of the competition between the spherical mean field (a.k.a. shell evolution) and the correlations which involve excitations across the Fermi level, the same type of configurations that are responsible for the phenomenon of shape coexistence, ubiquitous in the nuclear chart. The first IoI occurs at N=8 but was overlooked for many years, among other things, because the dominant physics there was related to the appearance of nuclear halos. We refer to [31] for the second ($N = 20$) and third ($N = 28$) ones and their merger and will dwell with the heavier ones in what follows.

1.9.1 ^{68}Ni and the $N = 40$ Island of Inversion

Twelve neutrons away from the $N = Z$ line, we reach the HO closure $N = 40$, which is the most debated one, as it only behaves like such for the protons in combination with the $N = 50$ SO closure for the neutrons in ^{90}Zr. The experimental spectrum shows a rather unclear situation: Three 0^+ states below 2.5 MeV of excitation energy, with a first excited 0^+ at 1.6 MeV and 2^+ at 2.0 MeV.

The natural valence space to capture the dynamics of this nucleus should contain at least the full pf-shell for the protons and the orbits below and above the $N = 40$ HO closure for the neutrons, i.e. $1p_{3/2}$, $1p_{1/2}$, $0f_{5/2}$, $0g_{9/2}$ and $1d_{5/2}$ (ideally the orbit $2s_{1/2}$ should be added as well). From the point of view of the development of quadrupole collectivity, this space is quite complete, because the neutron holes may live in a pseudo-SU3 regime as the proton particles do, whereas the neutron particles may take advantage of the quasi-SU(3) coupling scheme.

A monopole-adapted realistic interaction, dubbed LNPS-U, proposed for this valence space in reference [33], does provide a very satisfactory description of the level scheme of ^{68}Ni as can be seen in Fig. 1.12. According to this and other subsequent calculations, the ground state 0^+ is dominated at 65% by the doubly magic configuration $N = 40$ $Z = 28$. The second 0^+ can be said to be moderately deformed and oblate with the 2^+ and 4^+ states belonging to the same band and dominant 2p-2h (neutron) and 1p-1h (proton) configuration. The more interesting physics comes with the third 0^+ and the second 2^+ which are the germ of a highly

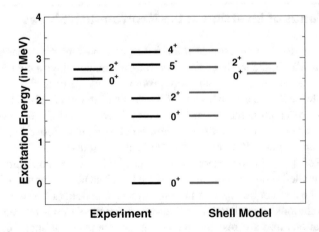

Fig. 1.12 ^{68}Ni, experiment *vs* the results of the SM-CI calculation with the interaction LNPS-U

deformed prolate band of many particle-hole structures as expected in the pseudo-plus quasi-SU(3) limit. In fact, this shape coexisting structure is the portal to the fourth IoI.

The shell evolution encoded in the LNPS-U interaction submits that at $Z = 20$, in ^{60}Ca the $N = 40$ gap is much smaller than at $Z = 28$, and the ordering of the orbits $0g_{9/2}$, $1d_{5/2}$ is inverted, a behaviour very close to that occurring in the $N = 20$ IoI between $Z = 8$, ^{28}O and doubly magic ^{34}Si. In the latter case, the orbits $0f_{7/2}$ and $1p_{3/2}$ are the ones which get inverted. It is the joint effect of the gap reduction and the quadrupole energy gains of the multiparticle-hole intruder states that produces the inversion of configurations characteristic of the IoI. The predictions of the LNPS-U calculation are displayed in Fig. 1.13 and compared with

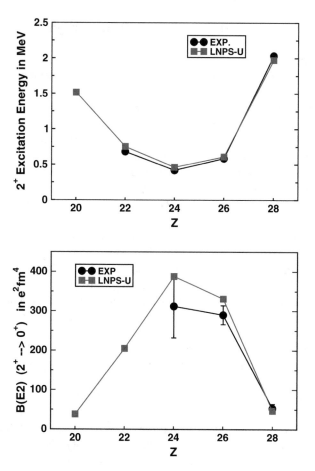

Fig. 1.13 $N = 40$ isotones, experiment *vs* the SM-CI results with the interaction LNPS-U. Excitation energy of the 2^+ (top) and $B(E2)$ (bottom)

the experimental data. The most recent measurement [34] has provided the first spectroscopic data on ^{62}Ti in full accord with the LNPS-U predictions.

Figure 1.13 shows the abrupt reduction of the excitation energy of the 2^+ and the simultaneous increase of the $B(E2)$s for ^{66}Fe, ^{64}Cr and ^{62}Ti, a signature of their location inside the IoI. As expected, their structure is very similar to that of the prolate excited band in ^{68}Ni. According to the calculations, ^{60}Ca should not be a doubly magic nucleus; instead it behaves as a pairing frustrated rotor fuelled by the neutron-neutron quadrupole interaction. A more detailed discussion can be found in ref. [31]. Therefore, neither the combination of the HO magic numbers $N = 40$, $Z = 20$ in ^{60}Ca nor $N = 40$, $Z = 40$ in ^{80}Zr survive to the competition with the quadrupole correlations.

1.9.2 ^{78}Ni: The $N = 50$ and the $N = 40$ IoI's Merge

The next milestone is ^{78}Ni, 22 neutrons away from $N = Z$. This nucleus has been a golden goal for experimentalists for decades, because it is a waiting-point nucleus candidate, whose structure, in particular whether it is doubly magic or not, has a large impact in the evolution of the r-process of nucleosynthesis. Now two HO closures are at play, $N = 50$ and $Z = 28$. From the SM-CI point of view, it has an appealing property that the valence space better suited for its description consists in two major oscillator shells, pf for the protons and sdg for the neutrons, a rather interesting circumstance. One can imagine a situation in which the $N = 50$ and $Z = 28$ gaps collapse. This would bring us close to the SU(3) limit. A back-of-an-envelope calculation using the formulas of Sect. 1.7 gives an intrinsic mass quadrupole moment $Q_0^m = 600\,\text{fm}^2$ or $\beta^m = 0.45$, a huge deformation indeed, larger than the record in the region at ^{76}Sr. In reality if the gaps are large enough to preserve the doubly magic nature of ^{78}Ni, the dominant quadrupole coupling scheme is pseudo-SU(3) both for the neutrons above $N = 50$ and for the protons above $Z = 28$, whereas the holes in the $0f_{7/2}$ proton orbit and in the $0g_{9/2}$ neutron orbit add prolate coherence to the intruder configurations. Nonetheless, the maximum quadrupole coherence attainable in the realistic scheme is far from the SU(3) limit.

For the pf-proton sdg-neutron valence space, an extension of the LNPS interaction was devised, guaranteeing a smooth transition between $N = 40$ and $N = 50$. The results of this study were published in [35], and we proceed to mention its more salient aspects, keeping an eye on Fig. 1.14. First of all, the calculation supports the doubly magic character of the ground state of ^{78}Ni, with about 70% of closed $N = 50$ and $Z = 28$. The states of 1p-1h structure, depicted to the left of Fig. 1.14, are very much correlated, but without losing their identity. The lowest one is a 2^+ at about 3 MeV excitation energy. However the first excited state according to the calculation is a 0^+ that, as can be gathered visually on the right part of Fig. 1.14, is the head of a well-deformed rotational band, which is the counterpart of the one found in ^{68}Ni on top of its third 0^+, discussed in the preceding section. Here again,

Fig. 1.14 ^{78}Ni, predictions of the SM-CI calculation with the interaction PFSDG

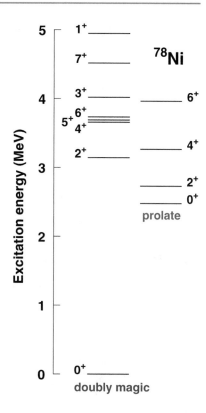

the coexistence of a deformed phase with the doubly magic ground state invites to surmise that another IoI might occur in the more neutron-rich isotones. It is indeed what the calculations predict, as can be seen in Fig. 1.15.

The figure shows in a very pictorial way how the deformed intruder configuration, which is excited in ^{78}Ni, becomes the ground state band as we move towards ^{70}Ca. At odds with the situation in $N = 40$, the semi-magic 0^+ state seems to survive at low excitation energy in ^{76}Fe and ^{70}Ca, causing a certain distortion of the energetics of the bands. On the contrary, the yrast band of ^{74}Cr is strictly identical to the excited band of ^{78}Ni and that of ^{72}Ti nearly so, albeit with a slightly smaller moment of inertia.

The first spectroscopy of ^{78}Ni has been obtained very recently in a ground-breaking experiment at RIKEN [36]. Two γ rays are reported which are consistent with the decays of two 2^+ states to the ground state, as predicted by the SM-CI calculations. Although the evidence is still scarce, the image of coexistence put forward by the calculation seems to be valid.

How are the fourth and fifth IoIs connected? We answer this question in Fig. 1.16. We have plotted the excitation energy of the first 2^+ as a proxy for magicity/deformation as a function of the neutron number. We compare theory and experiment where possible. The former results are obtained in the two valence

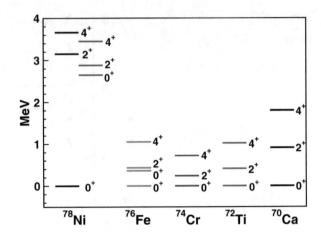

Fig. 1.15 The fifth IoI as predicted by the SM-CI calculation with the interaction PFSDG

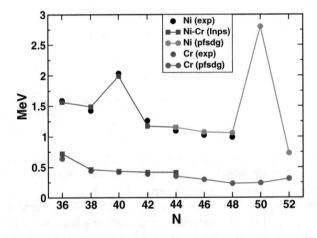

Fig. 1.16 The merging of the IoIs at $N = 40$ and $N = 50$ in the chromium isotopes, compared with the nickel chain. The excitation energy of the first 2^+ is used as control parameter

spaces discussed in this section and show clearly that the transition is smooth, with the crossover taking place around $N = 45$. Overall, the agreement with experiment is outstanding. In the nickel chain, we observe the persistence of the magic numbers $N = 40$ and $N = 50$. On the contrary, both closures are washed out in the chromium chain (in the iron and titanium as well, but we have not plotted them not to make the figure too busy), and the 2^+ excitation energy remains constant all along. Therefore we can say that the two IoIs merge. What turns out to be the same behaviour is found in the $N = 20$ and $N = 28$ IoIs for the neon, sodium and magnesium isotopes: a kind of universal behaviour indeed.

1.10   Epilogue

These lectures include a very basic introduction to the microscopic approach to the structure of the nucleus in the laboratory frame. Large-scale SM-CI calculations are a safe way to approach the exact solution of the many-body problem. However the explosion of the dimensions of the basis to be used prevents the brute force treatment. Hence, one must resort to the use of effective theories to design valence spaces that contain the more relevant degrees of freedom, renormalizing consistently the Hamiltonian and the transition operators. But, another ingredient, particular to the nuclear system, is the uncertainty of our knowledge of the nucleon-nucleon interaction, and it's poorly known many-body terms. As we have discussed in detail, close-to-exact solutions of the nuclear many-body problem for few nucleon systems (VMC, NCSM) require the explicit use of three-body forces in order to explain the full body of experimental data. Nuclear interactions obtained in χ-EFT do produce such three-body interactions naturally.

Our endeavour, monopole-adapted interaction, plus multipole-guided valence spaces, offers a solid bridge between the purely phenomenological approach, based on the fit of all the two-body matrix elements in a given valence space to the available experimental data (pioneered by B. H. Wildenthal and his USD interaction [37]), and the "ab initio" campaign (see, for instance, [4]).

We submit, and we have explained it in these lectures that once the monopole behaviour is corrected, the pairing and the multipole-multipole contents of the realistic interactions obtained via renormalized G-matrices make it possible to reproduce a large amount of experimental data with great accuracy and in some cases to predict new physics, as, for instance, the existence of IoIs. It is important to recall that the monopole Hamiltonian is responsible for all the extensive properties of the nucleus, basically its saturation point in the Coester plot, one of the long-standing problems in nuclear physics which appears in the nuclear spectroscopy hidden in the shell evolution. No reason for the latter to be right if the former is wrong. In a nutshell, we rely in these three pillars: the multipole part of the realistic interactions, a monopole Hamiltonian which reproduces what is experimentally known of the spherical mean field evolution with N and Z and the invaluable help of the SU(3) heuristic to define the minimal valence spaces to be used in the SM-CI approach.

How to make explicit the connection with the "ab initio" program? Indeed we all will be more than happy if there were a single QCD complying NN+NNN+? interaction and a rigorous reduction of these bare interactions to tractable Fock space bases. This is not the case yet, because the number of χ-EFT interactions is still too large, and sometimes the better ones for the many-body problem are not the best for the two- and three-body cases (see the contribution of R. Roth in this volume for a detailed description of the present state of the field). At present, the final step of the "ab initio" approach, from a spectroscopic point of view, consists in the derivation of two-body interactions to be used in SM-CI diagonalizations (VS-IMSRG). Hence, they are subject to the same dimensionality limitations of the standard SM-CI calculations. Obviously, they also rely in a physically sound choice of the valence spaces. We are afraid that the sensitivity of the outcome of the SM-CI calculations to (not so big) modifications of the spherical mean field can make it difficult for the "pure ab initio" approach to reach an optimum level of spectroscopic quality. However, the mere fact of understanding the monopole crisis of the realistic interactions in terms of three-body forces is a huge step ahead in our mastering on the nuclear dynamics. Another important asset of the "ab initio" calculations are the studies of the basic physics involved in the quenching of the $\sigma\tau$ (and related) operators in the nucleus, most notably in the neutrinoless double beta decay process. But, in our opinion, a first meeting point, and a much simpler one, would be to produce a kind of "ab initio" monopole Hamiltonian including three-body terms.

Finally, let us mention that in order to overcome the dimensionality barriers of the SM-CI calculations, there is a new line of thought that proposes the use of beyond mean field techniques, like projected Hartree–Fock plus generator coordinator configuration mixing, in shell model valence spaces, which we find very promising [38, 39].

Disclaimer In these lectures we have drawn freely from many standard nuclear structure books. In particular:

- P. Ring and P. Schuck, The Nuclear Many-Body Problem (Springer 1980)
- K. Heyde, The Nuclear Shell Model (Springer 1994)
- A. Bohr and B. Mottelson, Nuclear Structure, vols. I y II, (World Scientific 1998)
- J. Suhonen, From Nucleons to Nucleus (Springer 2007)
- A. De Shalit and H. Feshbach, Theoretical Nuclear Physics, vol I, Nuclear Structure (Wiley 1974)
- P. Brussaard and P. Glaudemans, Shell Model Applications in Nuclear Spectroscopy (North Holland 1977)
- I. Talmi, Simple Models of Complex Nuclei (Harwood, 1993)
- G. Brown, Unified Theory of Nuclear Models and Forces (North Holland 1971)

For the less standard aspects of the presentation, we follow the work of the Strasbourg Madrid Shell Model collaboration, notably the review of Ref. [9].

Acknowledgments AP's work is supported in part by the Ministerio de Ciencia, Innovación y Universidades (Spain), Grant CEX2020-001007-S funded by MCIN/AEI/10.13039/501100011033 and Grant PGC-2018-94583.

References

1. S.C. Pieper, V.R. Pandharipande, R.B. Wiringa, J. Carlson, Phys. Rev. C **64**, 011401 (2001)
2. P. Navratil, W.E. Ormand, J.P. Vary, B.R. Barret, Phys. Rev. Lett. **87**, 172501 (2001)
3. E. Epelbaum, H. Krebs, D. Lee, Ulf-G. Meissner, Phys. Rev. Lett. **106**, 192501 (2011)
4. S.R. Stroberg, H. Hergert, S.K. Bogner, J.D. Holt, Ann. Rev. Nucl. Part. Sci. **69**, 307 (2019)

5. V.R. Pandharipande, I. Sick, P.K.A, deWitt Huberts, Rev. Mod. Phys. **69**, 981 (1997)
6. D.M. Brink, E. Boeker, Nucl. Phys. A **91**, 1 (1967)
7. J. Dechargé, D. Gogny, Phys. Rev. C **21**, 1569 (1980)
8. J.M. Cavedon et al., Phys. Rev. Lett. **49**, 978 (1982)
9. E. Caurier, G. Martínez-Pinedo, F. Nowacki, A. Poves, A.P. Zuker, Rev. Mod. Phys. **77**, 427 (2005)
10. T.A. Brody, M. Moshinsky, *Tables of Transformation Brackets for Nuclear Shell-Model Calculations* (Gordon and Breach Science Publishers, New York, 1967)
11. K. Heyde, *The Nuclear Shell Model* (Springer, Berlin,1994)
12. M. Dufour, A.P. Zuker, Phys. Rev. C **54**, 1641 (1996)
13. B. A. Brown et al., OXBASH code, MSU-NSCL technical report 524 (1985)
14. B.A. Brown, W.D.M. Rae, Nucl. Data Sheets **120**, 115 (2014)
15. N. Shimizu, T. Mizusaki, Y. Utsuno, Y. Tsunoda, Comput. Phys. Commun. **244**, 372 (2019)
16. E, Caurier, F. Nowacki, A. Poves, K. Sieja, Phys. Rev. C **82**, 064304 (2010)
17. J.P. Elliott, Proc. R. Soc. Lond. Ser. A **245**, 128 (1956)
18. A. Arima, M. Harvey, K. Shimizu, Phys. Lett. B **30**, 517 (1969); K. Hecht, A. Adler, Nucl. Phys. A **137**, 129 (1969)
19. A.P. Zuker, J. Retamosa, A. Poves, E. Caurier, Phys. Rev. C **52**, R1741 (1995)
20. E. Caurier, J.L. Egido, G. Martínez-Pinedo, A. Poves, J. Retamosa, L.M. Robledo, A.P. Zuker, Phys. Rev. Lett. **75**, 2466 (1995)
21. A.P. Zuker, Phys. Rev. Lett. **90**, 042502 (2003)
22. P. Navratil, W.E. Ormand, Phys. Rev. Lett. **88**, 152502 (2002)
23. K. Kumar, Phys. Rev. Lett. **28**, 249 (1972)
24. A. Poves, F. Nowacki, Y. Alhassid, Phys. Rev. C **101**, 054307 (2020)
25. A.P. Zuker, A. Poves, F. Nowacki, S.M. Lenzi, Phys. Rev. C **92** 024320 (2015)
26. J. Retamosa, J.M. Udias, A. Poves, E. Moya de Guerra, Nucl. Phys. A **511**, 211 (1990)
27. E. Caurier, J. Menéndez, F. Nowacki, A. Poves, Phys. Rev. C **75** 054317 (2007)
28. D. Rudolf et al., Phys. Rev. Lett. **82**, 3763 (1999)
29. O. Sorlin, M.-G. Porquet, Prog. Part. Nucl. Phys. **61**, 602 (2008)
30. T. Otsuka, A. Gade, O. Sorlin, T. Suzuki, Y. Utsuno, Rev. Mod. Phys. **92**, 015002 (2020)
31. F. Nowacki, A. Poves, A. Obertelli, Prog. Part. Nucl. Phys. **120** 103866 (2021)
32. E.K. Warburton, J.A. Becker, B.A. Brown, Phys. Rev. C **41**, 1147 (1990)
33. S.M. Lenzi, F. Nowacki, A. Poves, K. Sieja, Phys. Rev. C **82**, 054301 (2010)
34. M. L. Cortes et al., Phys. Lett. B **800**, 135071 (2020)
35. F. Nowacki, A. Poves, E. Caurier, B. Bounthong, Phys. Rev. Lett. **117**, 272501 (2016)
36. R. Taniuchi et al., Nature **569**, 53 (2019)
37. B.H. Wildenthal, Prog. Part. Nucl. Phys. **11**, 5 (1984)
38. A. Sánchez-Fernández, B. Bally, T.R. Rodríguez, Phys. Rev. C **104**, 054306 (2021)
39. D. Duy Duc, F. Nowacki, Phys. Rev. C **105**, 054314 (2022)

Low-Energy Coulomb Excitation and Nuclear Deformation

2

Magda Zielińska

Abstract

Coulomb excitation is one of the rare methods available to obtain information on static electromagnetic moments of short-lived excited nuclear states. In the scattering of two nuclei, the electromagnetic field that acts between them causes their excitation. The process selectively populates low-lying collective states and is therefore ideally suited to study nuclear collectivity. While these experiments used to be restricted to stable isotopes, the advent of new facilities providing intense beams of short-lived radioactive species has opened the possibility to apply this powerful technique to a much wider range of nuclei. In this chapter, we discuss observables that can be measured in a Coulomb-excitation experiment and their relation to nuclear-structure parameters and, in particular, the nuclear shape. Possible solutions for normalisation of the measured γ-ray intensities and requirements for particle-detection systems are also presented.

2.1 Introduction

In the Coulomb-excitation process, excited states in colliding nuclei are populated via the mutually generated, time-dependent electromagnetic field that acts between them. The contribution from the short-range nuclear interaction can be neglected if the distance between the collision partners is sufficiently large. This condition is fulfilled for all scattering angles if the total kinetic energy in the centre of mass is well below the Coulomb barrier, which usually translates into beam energies of a few MeV/A. For higher beam energies (tens of MeV/A or more), selection of very

M. Zielińska (✉)
IRFU/DPhN, CEA, Université Paris-Saclay, Gif-sur-Yvette, France
e-mail: magda.zielinska@cea.fr

forward projectile scattering angles is necessary to ensure a purely electromagnetic excitation process. These two experimental approaches are commonly referred to as low- and intermediate-energy Coulomb excitation, while the term "high-energy Coulomb excitation" is often applied to processes at ultrarelativistic energies (few hundreds of MeV/A or more), where straight-line trajectories of collision partners are a very good approximation. This chapter is focused on low-energy Coulomb-excitation studies; for an introduction to Coulomb excitation at higher beam energies, we refer the reader to, e.g. Ref. [1].

For nuclei with $A \approx 30$ and heavier, analysis of inelastic scattering and transfer reaction data collected at beam energies of few MeV/A led to the formulation of the Cline's safe-distance criterion for low-energy Coulomb excitation [2], which states that if the distance of closest approach between the surfaces of the collision partners exceeds 5 fm, the influence of the nuclear interaction on excitation cross sections is below 0.5%. If one of the collision partners is lighter than $A \approx 30$, sometimes a more restrictive approach has to be used. For example, nuclear effects have been observed at distances beyond 5 fm in several experiments using ^{12}C and ^{16}O beams (see, e.g. [3–8]), and under such conditions, a conservative approach of adopting 6.5 fm as the minimum separation distance has been suggested.

The Coulomb-excitation cross sections depend on electromagnetic matrix elements between the low-lying states in the nucleus of interest, including diagonal $E2$ matrix elements, which are related to spectroscopic quadrupole moments. The decay of Coulomb-excited states is governed by the same set of electromagnetic matrix elements, although the impact of specific matrix elements on the excitation and decay processes may be very different. In the excitation process, the dominant multipolarities are $E2$ and $E3$, and other multipolarities have usually a weak influence on the low-energy Coulomb-excitation cross sections, as discussed in Sect. 2.3.1. They may, however, impact the decay, where especially $M1$ and $E1$ multipolarities play an important role.

The quantities that can be determined from a low-energy Coulomb-excitation study are thus:

- Those among $E2$ and $E3$ matrix elements coupling the low-lying excited states, which have the largest influence on the observed excitation cross sections. Under certain experimental conditions, the first-order perturbation theory is sufficient to describe the excitation process and relate the measured cross section to populate an I_f state to the $\langle I_f \| EL \| I_{g.s.} \rangle$ matrix element ($L = 2, 3$), as described in Sect. 2.3.1. In general, a set of coupled differential equations needs to be solved in order to link Coulomb-excitation cross sections to electromagnetic matrix elements. This formalism is introduced in Sect. 2.2.1.
- Quadrupole moments of short-lived excited states, which affect their excitation cross sections via the reorientation effect discussed in Sect. 2.3.2.2.
- $M1$ matrix elements, which can be determined from particle-γ correlations and angular dependence of the excitation cross section; an example of such analysis is given in Sect. 2.5.2.

- Relative signs of electromagnetic matrix elements, influencing excitation cross sections via interference effects discussed in Sect. 2.3.2.1.

Moreover, the transitional and diagonal $E2$ matrix elements can be linked to deformation parameters using non-energy-weighted quadrupole sum rules [2, 9], as presented in Sect. 2.4.1.

Finally, one should mention that since the low-energy Coulomb-excitation process selectively populates states that are connected to the ground state by enhanced $E2$ and $E3$ transitions, for some exotic nuclei such studies led to the first observation of certain excited states. As examples one could cite the experiments at ISOLDE, which identified the 2_1^+ states in ^{78,80}Zn [10], rotational ground-state bands in ^{97,99}Rb [11] and multiple levels in ^{98}Y [12].

The quantities measured in Coulomb-excitation experiments are, most commonly, γ-ray yields in coincidence with at least one of the collision partners. It is, however, also possible to measure Coulomb-excitation cross sections by detecting only scattered particles or only γ rays. These possibilities will be reviewed in Sect. 2.5 together with a presentation of selected experimental setups.

While this chapter is illustrated by numerous examples of low-energy Coulomb-excitation experiments, we do not aim to provide a comprehensive review of recent results obtained using this experimental method, which can be found elsewhere (see, for instance, [13]).

2.2 Theoretical Description of Excitation and Decay Processes

In "safe" Coulomb-excitation studies, when the observed excitation is due only to the well-known electromagnetic interaction, the relevant cross sections can be calculated with high precision from a given set of electromagnetic matrix elements. While a full quantum-mechanical treatment is possible, a semi-classical approach is typically employed to overcome difficulties arising from the long range of the electromagnetic interaction. In this approximation [14], the quantum-mechanical treatment is limited to the excitation process, while the kinematics of the collision is described using classical equations of motion. In this way, the calculations are significantly simplified, while their accuracy remains better than typical experimental uncertainties. For the semi-classical approach to be valid, the wave packets of the collision partners must not overlap, i.e. de Broglie wavelength of the projectile must be small compared to the distance of closest approach d. This condition can be expressed using the Sommerfeld parameter η:

$$\eta \equiv \frac{d}{2\lambdabar} = \frac{Z_p Z_t e^2}{\hbar v} \gg 1 \,, \tag{2.1}$$

where v is the initial beam velocity. Typical η values in low-energy Coulomb excitation induced by heavy ions range from a few tens to a few hundreds, but when light nuclei are involved (i.e. protons, deuterons, α particles), η values drop below

10 and a full quantum-mechanical analysis is required. In general, the semi-classical treatment is expected to deviate from exact calculations by terms of the order of $1/\eta$ [14].

For excitation to be possible, the perturbation of the electromagnetic potential experienced by the nucleus of interest needs to be sudden, i.e. the collision time τ_{coll} (which is of the order of a/v, where a is the impact parameter) should not be longer than the fluctuation time of nuclear wave functions given by $\tau_{nucl} = \hbar/\Delta E$, where ΔE is the excitation energy difference between the initial and final states. This is usually quantified by introducing the adiabaticity parameter ξ:

$$\xi \equiv \frac{\tau_{coll}}{\tau_{nucl}} = \frac{a\,\Delta E}{\hbar v} \, . \tag{2.2}$$

If $\xi \gg 1$, the changes of the electromagnetic field are too gradual for excitations to occur. This would be the case of very low beam energies, but the adiabaticity condition also introduces a limitation of energy transfer achievable in low-energy Coulomb excitation. For commonly used beam energies of a few MeV/A, it leads to an energy transfer cutoff of about 1–2 MeV. This basically eliminates a potential issue related to the semi-classical approximation. The classical description of the kinematics makes it impossible to introduce corrections to the trajectories of the collision partners due to the energy transfer between them, as it is not known at which point of the trajectory such transfer occurred. Consequently, for the semi-classical approximation to be valid, the transferred energy should not modify the collision kinematics in a significant way, i.e. it should be small with respect to the total kinetic energy in the centre-of-mass frame. Due to the adiabaticity cutoff, this condition is satisfied in typical low-energy Coulomb-excitation experiments. Moreover, the effect of energy transfer on the kinematics is usually accounted for, in an approximative way, by averaging the parameters of hyperbolic trajectories for elastic scattering (no energy transfer) and those resulting from decreasing the incident energy by the energy transferred in the collision. With this treatment, Ref. [14] reports that total cross sections resulting from the semi-classical approximation deviate by less than 5% from those resulting from the full quantum-mechanical analysis for $\eta > 3$, while to obtain the same accuracy in the angular distribution $\eta > 10$ is needed.

2.2.1 Coulomb-Excitation Process

The electromagnetic potential describing the interaction between the target and projectile nuclei can be presented as a sum of three terms:

(i) The monopole-monopole term ($Z_1 Z_2 e^2/r^2(t)$) describing the elastic scattering and yielding the classical hyperbolic orbits, described by the vector $\mathbf{r}(t)$, of the collision partners

(ii) The monopole-multipole term, which describes inelastic scattering (excitation of one collision partner by the other one)
(iii) The multipole-multipole term of the order of $1/\eta^2$ [14], which is usually neglected

In this representation, the Schrödinger equation for the nucleus of interest will contain only the term (ii), and it will take the form:

$$i\hbar \frac{\partial}{\partial t}|\Psi(\mathbf{r}, t)\rangle = \left(\hat{H}^0 + \hat{V}(\mathbf{r}(t))\right)|\Psi(\mathbf{r}, t)\rangle, \tag{2.3}$$

where $\hat{H}^0$ is the free Hamiltonian of the nucleus of interest and $\hat{V}(\mathbf{r}(t))$ is the time-dependent potential generated by the collision partner.

The solution of (2.3) can be presented as a linear combination of eigenfunctions of the free Hamiltonian of the nucleus of interest (i.e. wave functions of individual nuclear states) with time-dependent coefficients $a_n(t)$, commonly referred to as excitation amplitudes:

$$|\Psi(\mathbf{r}, t)\rangle = \sum_n a_n(t) \exp\left(\frac{-i E_n t}{\hbar}\right)|\phi_n(\mathbf{r}, t)\rangle, \tag{2.4}$$

where E_n is the energy of the $|\phi_n(\mathbf{r}, t)\rangle$ state:

$$\hat{H}^0|\phi_n(\mathbf{r}, t)\rangle = E_n|\phi_n(\mathbf{r}, t)\rangle. \tag{2.5}$$

By substituting (2.4) into the Schrödinger equation (2.3), one obtains a set of coupled equations for the excitation amplitudes $a_n(t)$:

$$i\hbar \sum_n \frac{da_n}{dt} \exp\left(\frac{-i E_n t}{\hbar}\right)|\phi_n(\mathbf{r}, t)\rangle = \sum_n a_n(t) \exp\left(\frac{-i E_n t}{\hbar}\right) \hat{V}(\mathbf{r}(t))|\phi_n(\mathbf{r}, t)\rangle, \tag{2.6}$$

which due to the orthonormality of the $|\phi_n(\mathbf{r}, t)\rangle$ functions can be simplified after evaluating the expression following the application of the ket $\langle \phi_k(\mathbf{r}, t)|$ for a specific state k:

$$\frac{d}{dt}a_k(t) = -\frac{i}{\hbar} \sum_n a_n(t) \langle \phi_k(\mathbf{r}, t)|\hat{V}(\mathbf{r}(t))|\phi_n(\mathbf{r}, t)\rangle \exp\left(\frac{-i(E_n - E_k)t}{\hbar}\right). \tag{2.7}$$

As the nucleus of interest is in its ground state prior to the collision, the initial condition of (2.7) is $a_k(t \to -\infty) = \delta_{k0}$ (index 0 denotes the ground state).

Under the condition that the collision time is very much smaller than the lifetimes of the excited nuclear states, the excitation amplitudes after the collision can be determined from $a_k(t \to \infty)$ and are related to excitation probabilities P_k and cross sections to populate the $|\phi_k(\mathbf{r}, t)\rangle$ states:

$$P_k = |\lim_{t \to \infty} a_k(t)|^2 , \tag{2.8}$$

$$\left(\frac{d\sigma}{d\Omega}\right)_k = \left(\frac{d\sigma}{d\Omega}\right)_{Ruth} \times P_k , \tag{2.9}$$

where $(d\sigma/d\Omega)_{Ruth}$ is the Rutherford cross section:

$$\left(\frac{d\sigma}{d\Omega}\right)_{Ruth} = \left(\frac{d}{4}\right)^2 \frac{1}{\sin^4(\theta_{CM}/2)} , \tag{2.10}$$

and d denotes the distance of the closest approach in a head-on collision.

Equation (2.7) can be expressed via matrix elements of the electromagnetic multipole operator. In order to do that, the potential $\hat{V}(\mathbf{r}(t))$ needs to be expanded in a multipole series:

$$\hat{V}(\mathbf{r}(t)) = \sum_{\lambda=1}^{\infty} \sum_{\mu=-\lambda}^{\lambda} \frac{4\pi Z_2 e}{2\lambda + 1} (-1)^{\mu} S_{\lambda\mu}^{E,M}(\mathbf{r}, t) \times \hat{M}(^E_M\lambda, -\mu) , \tag{2.11}$$

where $\hat{M}(^E_M\lambda, \mu)$ is the electromagnetic multipole operator of the order λ (E stands for the electric part and M for magnetic), μ denotes the projection of λ and collision functions $S_{\lambda\mu}^{E,M}(\mathbf{r}, t)$ have the following form:

$$S_{\lambda\mu}^{E}(\mathbf{r}, t) = \frac{Y_{\lambda\mu}(\vartheta(t), \varphi(t))}{r(t)^{\lambda+1}} \tag{2.12}$$

for electric excitation, and:

$$S_{\lambda\mu}^{M}(\mathbf{r}, t) = \frac{1}{\lambda c} \times \frac{\frac{d\mathbf{r}(t)}{dt}(\mathbf{r} \times \nabla)}{r(t)^{\lambda+1}} Y_{\lambda\mu}(\vartheta(t), \varphi(t)) \tag{2.13}$$

for magnetic excitation. In the equations above, $Y_{\lambda\mu}(\vartheta, \varphi)$ are spherical harmonics, and Z_2 is the atomic number of the collision partner.

The Wigner–Eckhart theorem (2.14) can be used to relate matrix elements of the $\hat{M}(^E_M\lambda, \mu)$ operators to reduced matrix elements:

$$\langle I_k, m_k | \hat{M}(^E_M\lambda, \mu) | I_n, m_n \rangle = \frac{1}{\sqrt{2I_n + 1}} (I_n, m_n, \lambda, \mu | I_k, m_k) \langle I_k \| \hat{M}(^E_M\lambda) \| I_n \rangle , \tag{2.14}$$

where $(I_n, m_n, \lambda, \mu | I_k, m_k)$ are Clebsch–Gordan coefficients.

Applying this theorem and substituting the $\hat{V}(\mathbf{r}(t))$ potential (2.11) into (2.7) leads to a set of coupled differential equations relating the excitation amplitudes $a_k(t)$ with reduced matrix elements of the electromagnetic operator:

$$\frac{\mathrm{d}}{\mathrm{d}t}a_k(t) = \frac{i}{\hbar}\frac{4\pi Z_2 e}{\sqrt{2I_n+1}}\sum_n a_n(t)\exp\left(\frac{-i(E_n-E_k)t}{\hbar}\right) \tag{2.15}$$

$$\times\sum_{\lambda=1}^{\infty}\sum_{\mu=-\lambda}^{\lambda}\frac{(-1)^{\mu}}{2\lambda+1}(I_n,m_n,\lambda,\mu|I_k,m_k)S_{\lambda\mu}^{E,M}(\mathbf{r},t)\langle I_k\|\hat{M}(\substack{E\\M}\lambda)\|I_n\rangle .$$

Solving the set of coupled equations (2.15) for a given set of $\langle I_k\|\hat{M}(\substack{E\\M}\lambda)\|I_n\rangle$ matrix elements yields the populations of nuclear states following Coulomb excitation. One should note that in the above equation the index k runs not only over individual excited states but also their magnetic substates, which are treated as independent states in this description. Due to the properties of the $S_{\lambda\mu}^{E,M}(\mathbf{r},t)$ functions, the solutions of (2.15) are usually derived in the frame of reference with the origin in the centre of mass of the target nucleus and one axis defined along the symmetry axis of the hyperbolic trajectory, pointing towards the projectile.

2.2.2 Electromagnetic Decay of Coulomb-Excited States

An important simplification of the theoretical description comes from the fact that it is possible to completely separate in time the excitation process and the subsequent electromagnetic decay. Indeed, the collision time is of the order of 10^{-19}–10^{-20} s, while typical lifetimes of low-lying excited states are 10^{-14} s or longer.

The same set of matrix elements that describes the excitation process also governs the γ-ray decay of the excited states. Reduced matrix elements are related to reduced transition probabilities $B(\substack{E\\M}\lambda)$ via:

$$B(\substack{E\\M}\lambda; I_i \to I_f) = \frac{1}{2I_i+1}|\langle I_f\|\hat{M}(\substack{E\\M}\lambda)\|I_i\rangle|^2 . \tag{2.16}$$

Those, in turn, can be used to express the probability of a decay via γ-ray emission:

$$P(\substack{E\\M}\lambda; I_i \to I_f) = \frac{8\pi(\lambda+1)}{\lambda((2\lambda+1)!!)^2}\frac{1}{\hbar}\left(\frac{E_\gamma}{\hbar c}\right)^{2\lambda+1} \times B(\substack{E\\M}\lambda; I_i \to I_f) . \tag{2.17}$$

Using excitation amplitudes determined from (2.15), it is possible to calculate γ-ray intensities following Coulomb excitation. One should note here that alternative decay paths, such as internal conversion, should also be taken into account.

If the γ-ray emission occurs from a state with an uneven magnetic substate population, the angular distribution of the γ rays can be anisotropic. The polarisation of the Coulomb-excited nucleus is described by the statistical tensor:

$$\rho_{k\chi}(I) = \sqrt{2I+1} \sum_{m,m'} (-1)^{I-m'} \begin{pmatrix} I & k & I \\ -m' & \chi & m \end{pmatrix} a^*_{Im'} a_{Im} \, , \tag{2.18}$$

where a_{Im} are the excitation amplitudes for the magnetic substate m of a state of spin I. In (2.18) the Clebsch–Gordan coefficient is replaced by the Wigner's 3j symbol according to the relation:

$$(J_1, m_1, J_2, m_2 | J, m) = (-1)^{-J_1+J_2-m} \sqrt{2J+1} \begin{pmatrix} J_1 & J_2 & J \\ m_1 & m_2 & -m \end{pmatrix} . \tag{2.19}$$

The angular distribution of γ radiation can then be expressed as:

$$\frac{d^2\sigma}{d\Omega_p d\Omega_\gamma} = \sigma_{Ruth}(\theta_p) \frac{1}{2\gamma(I_i)\sqrt{\pi}} \sum_{\chi, even\ k} \rho^*_{k\chi}(I_i) \sum_{\lambda,\lambda'} \delta_\lambda \delta^*_{\lambda'} F_k(\lambda\lambda' I_i I_f) Y_{k\chi}(\vartheta_\gamma, \varphi_\gamma) \, , \tag{2.20}$$

where index p denotes the projectile, $\sigma_{Ruth}(\theta_p)$ is the Rutherford cross section, $F_k(\lambda\lambda' I_i I_f)$ are γ-γ correlation coefficients (defined, e.g. in Ref. [15]) and $\gamma(I_i)$ is the decay constant including all multipolarities and final states:

$$\gamma(I_i) = \sum_{\lambda, f} |\delta_\lambda(I_i \rightarrow I_f)|^2 \, , \tag{2.21}$$

where δ_λ are amplitudes of the $I_i \rightarrow I_f$ transition of multipolarity λ, defined as:

$$\delta_\lambda = i^{n(\lambda)} \sqrt{P(^E_M\lambda; I_i \rightarrow I_f)} \, , \tag{2.22}$$

and $n(\lambda)$ is equal to λ for electric transitions and to $\lambda + 1$ for magnetic transitions. The probabilities $P(^E_M\lambda; I_i \rightarrow I_f)$ are defined by (2.17).

The above expressions are valid in the frame of reference that was used to derive the excitation amplitudes, which was related to the collision kinematics. In contrast, the γ-ray angular distributions have to be derived in the laboratory frame, and thus an appropriate transformation of the statistical tensor, taking into account possible relativistic effects, must be applied.

The observed angular distribution may be influenced by the deorientation effect, i.e. the depolarisation of the nuclear-state alignment caused by the interaction of the nuclear magnetic moment with electrons cascading to the lowest atomic shells, observed when the nucleus of interest, which may be in a highly ionised atomic state, recoils into a vacuum. Typical time constants of these atomic transitions are of the order of 10^{-12} s, similar to lifetimes of nuclear excited states. Consequently, the deorientation effect may influence the polarisation of the nucleus before γ-ray emission, which attenuates the observed γ-ray angular distribution. The precise modelling of this process is not obvious, and thus attenuation coefficients introduced to account for deorientation usually rely on simplified empirical models (see, e.g. [16, 17]).

Equation (2.20) assumes that both the scattering angle and the γ-ray emission angle are precisely known. In reality, both of them are determined with a certain limited precision related to the sizes and properties of the particle and γ-ray detectors. As the excitation amplitudes, and hence the statistical tensor defined by (2.18), depend on the particle scattering angle, the size of a particle detector is usually accounted for by performing a numerical integration of the calculated γ-ray intensities over the angular range covered by it. In contrast, for the γ-ray detectors, it is more convenient to introduce attenuation factors defined as, e.g. in Ref. [18], leading to the following expression for the γ-ray angular distribution:

$$\frac{d^2\sigma}{d\Omega_p d\Omega_\gamma} = \sigma_{Ruth}(\theta_p)\frac{1}{2\gamma(I_i)\sqrt{\pi}} \tag{2.23}$$

$$\times \sum_{\chi,\text{even }k} \rho_{k\chi}^{\prime*}(I_i,\theta_p) \sum_{\lambda,\lambda'} \delta_\lambda \delta_{\lambda'}^* F_k(\lambda\lambda' I_i I_f) G_k Q_k Y_{k\chi}(\vartheta_\gamma,\varphi_\gamma) ,$$

where:

- $\rho_{k\chi}^{\prime*}(I_i,\theta_p)$ is the statistical tensor $\rho_{k\chi}^*(I_i)$ transformed to the laboratory frame. This transformation is parametrised by the particle scattering angle θ_p.
- G_k and Q_k are attenuation coefficients accounting for the deorientation effect and the finite size of γ-ray detectors, respectively, which can be derived as described in, e.g. Refs. [16, 17] (G_k) and [18] (Q_k).

2.2.3 Coulomb-Excitation Codes

Due to the complexity of the formalism presented in Sects. 2.2.1 and 2.2.2, dedicated codes have been developed for calculation of low-energy Coulomb-excitation probabilities and the subsequent γ-ray decay. The two most widely used are CLX/DCY [19] and GOSIA [20]. They are both based on the Winther and de Boer code [21] and rely on the semi-classical approximation discussed in the beginning of Sect. 2.2. As a main input, they require masses and atomic numbers of both collision partners, beam energy, the level scheme of the nucleus of interest and the relevant electromagnetic matrix elements coupling the declared states.

The CLX code calculates Coulomb-excitation cross sections as a function of the scattering angle θ_{CM} at the given beam energy corresponding to the middle of the target. Its output can be used as an input for the DCY code, which calculates the corresponding γ-ray yields taking into account the particle and γ-ray detection geometry and the effects influencing the γ-ray angular distribution discussed in

Sect. 2.2.2, such as possible deexcitation by internal conversion, deorientation effect and attenuation due to the finite size of the γ-ray detectors.

The GOSIA code [20] provides similar simulation capabilities as CLX/DCY but, moreover, offers a possibility to perform a multi-dimensional χ^2 fit of electromagnetic matrix elements to the measured γ-ray yields. Known matrix elements and complementary spectroscopic data such as lifetimes, branching ratios and $E2/M1$ mixing ratios can also be included as independent data points in the fitting procedure. Data sets corresponding to various projectile-target combinations and different beam energies or scattering angles can be fitted simultaneously. In order to speed up the calculations, the fitting procedure uses γ-ray yields calculated for a specific scattering angle and beam energy, rather than integrating them at each minimisation step over the range covered by particle detectors and the range of incident energies resulting from slowing down of beam ions in the target material. To account for these effects, correction factors CF are applied to the measured γ-ray yields. Before the minimisation procedure, for each possible γ-ray transition, GOSIA calculates from a given set of matrix elements the "point" yield (Y_{point}), corresponding to a specific scattering angle and beam energy, and the yield integrated over the possible scattering angles and incident energies (Y_{intg}). Their ratio gives the correction factor CF for each transition:

$$CF = \frac{Y_{point}}{Y_{intg}} . \tag{2.24}$$

These correction factors are then applied to the measured γ-ray yields Y_{exp}:

$$Y_{exp}^{corr} = Y_{exp} \times CF . \tag{2.25}$$

As this procedure depends, albeit weakly, on the actual values of matrix elements, it should be repeated periodically over the course of the χ^2 minimisation procedure until a self-consistent solution is found.

For ease of use, the GOSIA code requires as input the measured γ-ray yields rather than absolute excitation cross sections, and their normalisation is performed internally by the code. To relate experimentally measured and calculated γ-ray intensities, normalisation constants C_m are introduced in the χ^2 function used by GOSIA:

$$\chi^2 = \sum_m \sum_i (C_m Y_c^{i,m} - (Y_{exp}^{corr})^{i,m})^2 / \sigma_{i,m}^2 + \sum_n \left(D_c^n - D_{exp}^n \right)^2 / \sigma_n^2 , \tag{2.26}$$

where:

$(Y_{exp}^{corr})^{i,m}$: experimental γ-ray intensity for the ith transition in the mth experiment, multiplied by the correction factor defined by (2.24)

$Y_c^{i,m}$: γ-ray intensity calculated using the current set of matrix elements for the ith transition in the mth experiment

C_m: normalisation constant for the mth experiment

D_c^n and D_{exp}^n: spectroscopic data calculated using the current set of matrix elements and their experimental values, respectively

$\sigma_{i,m}$, σ_n: experimental uncertainties of γ-ray intensities (including the correction factor) and of the spectroscopic data, respectively

The normalisation constants C_m include the time-integrated beam current, the absolute efficiency of particle and γ-ray detection and the particle solid angle factor. The experimental γ-ray yields are compared with the products $C_m Y_c^{i,m}$, which depend both on the electromagnetic matrix elements and the adopted normalisation. Therefore, additional constraints on the normalisation constants have to be provided; otherwise a GOSIA fit may converge but produce unreliable results. Possible normalisation techniques are discussed in Sect. 2.5.1.

Additionally, a term related with the user-defined "observation limit", χ_{UPL}^2, may be added to the χ^2 function defined by (2.26). If the calculated intensity of any unobserved γ-ray transition, divided by the intensity of the normalising transition specified by the user, Y_{cN}^m, exceeds this upper limit $u(m)$, it will affect the least-squares fit via:

$$\chi_{UPL}^2 = \sum_m \sum_i \left(\frac{Y_c^{i,m}}{Y_{cN}^m} - u(m) \right)^2 \times \frac{1}{u^2(m)} . \qquad (2.27)$$

As in (2.26), index m runs over experiments, and i over the possible γ-ray transitions. Introduction of proper upper limits should prevent finding unphysical fit solutions involving γ-ray transitions which have not been experimentally observed.

Since the influence of individual matrix elements on the excitation process varies strongly, some of them have a very limited, or even null, effect on the calculated χ^2. Consequently, the commonly applied methods of error estimation based on the inversion of the covariance matrix cannot be applied. Instead, for each matrix element, GOSIA defines in the vicinity of the χ^2 minimum a "maximum correlation path", i.e. a curve in the matrix element space $\bar{x}$, for which the effect of varying the matrix element in question, x_i, is to the greatest extent balanced by changes of other matrix elements. The uncertainty is then found by requesting that the integral of the normalised probability distribution contained within error bars equals to the confidence limit of 68.3%:

$$\frac{\int_l \exp\left(-\frac{1}{2}\chi^2(\bar{x})\right) d\bar{x}}{\int \exp\left(-\frac{1}{2}\chi^2(\bar{x})\right) d\bar{x}} = 68.3\% , \qquad (2.28)$$

where the integration in the numerator is performed along the maximum correlation path l and in the denominator over all possible values of $\bar{x}$.

2.3   First- and Higher-Order Effects in Coulomb Excitation

In typical low-energy Coulomb-excitation studies, the sets of $\langle I_f \| \hat{M}(^E_M\lambda) \| I_i \rangle$ matrix elements that influence the excitation and decay process are large. In order to determine them unambiguously, the number of experimental data (γ-ray intensities following Coulomb excitation and other spectroscopic data related to electromagnetic matrix elements, such as lifetimes, branching ratios, mixing ratios) must be at least equally large. The potential experienced by the nucleus of interest (2.11) depends on the atomic number of the collision partner Z_2 and its trajectory $\mathbf{r}(t)$, given by the centre-of-mass scattering angle θ_{CM}. Consequently, independent experimental information can be obtained by measuring γ-ray intensities corresponding to various ranges of θ_{CM} scattering angles and, if possible, using several beam-target combinations involving the nucleus of interest.

It is also possible to simplify the description of the excitation process by performing the experiments under conditions where multi-step excitation is strongly suppressed, and only states that can be reached in a one-step process from the ground state are populated. Most of early Coulomb-excitation studies in the 1950–1960s were performed in this way, but when heavy-ion beams with $A > 40$ and high-resolution Ge detectors became more widely available, this approach has mostly given way to more complex multi-step experiments that necessitate the full theoretical formalism described in Sect. 2.2.1. However, limitations in terms of available beam energies appeared again in early days of post-accelerated ISOL (Isotope Separator OnLine) beam facilities. Moreover, certain physics questions can be better addressed with a high-precision one-step Coulomb-excitation measurement than with a complex multi-step study that may suffer from, e.g. ambiguities related to relative signs of $E2$ matrix elements, as discussed in the following two sections. Relative simplicity of the first-order approximation also makes it easier to demonstrate particular features of the excitation process with respect to the scattering angle, excitation energy and transition multipolarity.

2.3.1   First-Order Effects

If the interaction between the colliding nuclei is weak, i.e. the excitation probability is much smaller than 1, Coulomb-excitation amplitudes can be calculated within the first-order perturbation theory. In this case, the solution of (2.7) can be written as:

$$a_f = \frac{1}{i\hbar} \int_{-\infty}^{\infty} \exp\left(\frac{i(E_f - E_i)t}{\hbar}\right) \langle \phi_f(\mathbf{r}, t) | \hat{V}(\mathbf{r}(t)) | \phi_i(\mathbf{r}, t) \rangle \, dt. \qquad (2.29)$$

Multipole expansion of $\hat{V}(\mathbf{r}(t))$ (2.11) yields the following expressions for the excitation cross sections, which linearly depend on the reduced transition probabilities $B(^E_M\lambda; I_i \rightarrow I_f)$:

$$\sigma_{E\lambda} = \left(\frac{Z_2 e}{\hbar v}\right)^2 \left(\frac{d}{2}\right)^{-2\lambda+2} B(E\lambda; I_i \rightarrow I_f) \int_{\theta_{min}}^{\theta_{max}} \frac{\mathrm{d}f_{E\lambda}(\xi, \theta)}{\mathrm{d}\Omega} \mathrm{d}\Omega \qquad (2.30)$$

for the electric excitation and

$$\sigma_{M\lambda} = \left(\frac{Z_2 e}{\hbar c}\right)^2 \left(\frac{d}{2}\right)^{-2\lambda+2} B(M\lambda; I_i \rightarrow I_f) \int_{\theta_{min}}^{\theta_{max}} \frac{\mathrm{d}f_{M\lambda}(\xi, \theta)}{\mathrm{d}\Omega} \mathrm{d}\Omega \qquad (2.31)$$

for the excitation caused by the magnetic field, where d is the distance of the closest approach in backscattering, Z_2 is the atomic number of the collision partner and ξ is the adiabaticity parameter defined by (2.2). The excitation functions $f_{E\lambda \atop M}(\xi, \theta)$ are integrated over the angular range corresponding to the scattered particle detection in the centre-of-mass frame. While we refer the reader to Ref. [14] for the exact formulae, we will briefly discuss here the properties of these functions. For $\xi > 0$ they display dome-like shapes as a function of θ, with a maximum shifting towards larger scattering angles with increasing ξ, which implies that for higher excitation energies, or lower beam energy, larger scattering angles are preferable. The integrals of the $f_{E\lambda \atop M}(\xi, \theta)$ functions over the full range of scattering angles, $f_{E\lambda \atop M}(\xi)$, decrease roughly exponentially with ξ. One should note here that while increasing of beam energy leads to lower ξ and, consequently, enhanced excitation cross section, at some point the distance of the closest approach for the highest centre-of-mass scattering angle will go below the value required for a "safe" Coulomb-excitation process. With a further increase of beam energy, the solid angle corresponding to "safe" Coulomb excitation will decrease, thus reducing the cross section for this process. This is illustrated in Fig. 2.1.

The excitation functions $f_{E\lambda \atop M}(\xi, \theta)$ also strongly depend on multipolarity λ; those for $E3$ transitions are about one order of magnitude smaller than those for $E2$, etc. While the excitation functions for the electric dipole transitions have the highest values, typical $B(E1)$ values are orders of magnitude smaller than transition probabilities for higher multipolarities, which makes excitation via $E1$ negligible. Moreover, one can immediately notice that while the denominator of (2.30) includes v, i.e. the initial projectile velocity, it is replaced by c in (2.31). This results in suppression of magnetic excitations by a factor of $(v/c)^2$ with respect to electric excitation, making them negligible in low-energy Coulomb-excitation studies for which (v/c) rarely exceeds 0.1.

The excitation process depends on the kinematics and the mass and atomic numbers of the target and projectile nuclei. The first-order approximation is usually sufficiently accurate to describe the population of states excited from the ground state in experiments employing a light beam or a light target or when small centre-of-mass scattering angles are used. For example, multiple recent Coulomb-excitation experiments aiming at high-precision measurements of the $B(E2; 2_1^+ \rightarrow 0_1^+)$ values in Sn isotopes were performed under conditions that minimised the role of multi-step excitations. A series of experiments at Oak Ridge National Laboratory [22, 23] were performed in strongly inverse kinematics, with a ^{12}C target bombarded by ^{126,128,130,134}Sn beams of 3 MeV/A energy. The impact of the diagonal matrix elements $\langle 2_1^+ || E2 || 2_1^+ \rangle$ on the obtained $\langle 2_1^+ || E2 || 0_1^+ \rangle$ values was estimated to be below 0.001 eb, i.e. over one order of magnitude smaller than the experimental uncertainty dominated by the statistics. A follow-up high-precision study was carried out in very similar experimental conditions for stable even-even $^{112-124}$Sn nuclei [24] in order to map the evolution of the $B(E2; 2_1^+ \rightarrow 0_1^+)$ values in Sn isotopes over a broad range of A. In a complementary experimental campaign performed at IUAC, New Delhi [25, 26], a reaction partner with a much higher Z was used: a ^{58}Ni beam of 3 MeV/A impinged on ^{112,116,118,120,122,124}Sn targets. In this case, however, only events with the Ni beam particles scattered at forward angles were selected, which again resulted in observation of the 2_1^+ state only. Another example of a recent one-step Coulomb-excitation experiment aiming at a precise $B(E2; 2_1^+ \rightarrow 0_1^+)$ measurement is the study of ^{80}Zn [10], performed at ISOLDE with a 2.8-MeV/A ^{80}Zn beam scattered on a ^{108}Pd target. Even though in this case a much heavier reaction partner was used, and the angular range covered by the particle detector extended over a broad range up to $\theta_{CM} \approx 120°$, the high excitation energy of the 2_1^+ state, 1.492 MeV, strongly limited the multi-step excitation probability. Indeed, the influence of the unobserved 4_1^+ excitation on the extracted $B(E2; 2_1^+ \rightarrow 0_1^+)$ value was estimated to be below 0.5% [10]. For the same reason, the first-order approximation is also suitable to describe the population of 2_1^+ and 3_1^- states in ^{132}Sn, located at 4.04-MeV and 4.35-MeV excitation energy, respectively, in the experiment that used a 5.5 MeV/A ^{132}Sn beam from HIE-ISOLDE and a ^{206}Pb target [27]. In this case, the probability of multi-step processes is further limited by the beam energy, which is significantly higher than in other studies mentioned in this section, resulting in a shorter collision time for events corresponding to the "safe" Coulomb-excitation process.

2.3.2 Higher-Order Effects

If the electromagnetic field acting between the collision partners is strong enough and the collision lasts sufficiently long, multi-step excitation becomes possible, and higher-order contributions to the excitation cross section need to be taken into account. These contributions lead to the experimental sensitivity to the spectroscopic quadrupole moments of excited states via the reorientation effect, as well as to the relative signs of transitional matrix elements.

The cross section for a two-step excitation proceeding via an intermediate state I_n is approximately given by:

$$\sigma\left((E2; I_i \rightarrow I_n) \times (E2; I_n \rightarrow I_f)\right) \approx \sigma(E2; I_i \rightarrow I_n) \times \sigma(E2; I_n \rightarrow I_f)/\sigma_{Ruth} .$$
(2.32)

Consequently, transition probabilities between the states that can only be reached in multi-step excitation are related to ratios of intensities of γ rays deexciting the final and intermediate states. For example, under the assumption of quadrupole moments equal to zero, the $4_1^+ \rightarrow 2_1^+/2_1^+ \rightarrow 0_1^+$ intensity ratio observed in Coulomb excitation of a weakly deformed even-even nucleus depends exclusively on the $\langle 4_1^+ \| E2 \| 2_1^+ \rangle$ matrix element; changing the $\langle 2_1^+ \| E2 \| 0_1^+ \rangle$ matrix element would influence the total number of counts in both transitions, but not the ratio.

As discussed earlier, the importance of multi-step excitation increases with atomic numbers of collision partners, centre-of-mass scattering angle and beam energy. The latter dependence is illustrated by Fig. 2.1, presenting calculated cross sections for Coulomb excitation of the 2_1^+ and 4_1^+ states in ^{74}Zn scattered on a ^{208}Pb target. At beam energies below 4 MeV/A, the excitation cross section for the 4_1^+ state increases more rapidly as a function of beam energy than that for the 2_1^+ state. For example, when going from 2 MeV/A to 4 MeV/A, a gain of one order of magnitude is observed for the 2_1^+ state, while it is almost three orders of magnitude for the 4_1^+ state. The gradual decrease of cross sections observed for beam energies above 4 MeV/A is due to the calculation being limited to angles corresponding to (i) "safe" Coulomb-excitation process and (ii) detection of one of the collision partners between $20° < \theta_{LAB} < 55°$, i.e. angular coverage of the experimental setup at ISOLDE (see Sect. 2.5.2). One should note, however, that the total multi-step excitation cross sections decrease more rapidly with beam energy than those for the one-step process. This effect, due to shortening of the collision time, leads to the restriction of the intermediate-energy Coulomb-excitation process to single-step excitations.

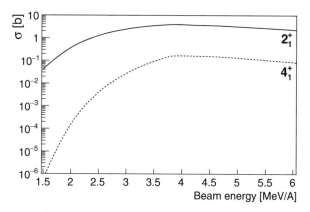

Fig. 2.1 Cross sections for Coulomb excitation of the 2_1^+ and 4_1^+ states in ^{74}Zn scattered on ^{208}Pb, calculated as a function of the beam energy assuming $E2$ matrix elements from Ref. [28]. The cross sections were integrated over the scattering angles corresponding to $20° < \theta_{LAB} < 55°$, assuming detection of any of the reaction partners but excluding angles where the 5-fm "safe-distance" criterion for Coulomb excitation is not valid

2.3.2.1 Relative Signs of Transitional Matrix Elements

In the excitation of a state that can be populated in a one-step process directly from the ground state or in a two-step process via an intermediate state, interference effects will appear, resulting from competition of these two excitation modes. As an example, we can consider a 2_2^+ state in an even-even nucleus. For each of its two possible excitation paths, the contribution to the total excitation amplitude is proportional to the relevant matrix elements: $\langle 2_2^+ || E2 || 0_{g.s.}^+ \rangle$ for the direct excitation and the product of $\langle 2_2^+ || E2 || 2_1^+ \rangle$ and $\langle 2_1^+ || E2 || 0_{g.s.}^+ \rangle$ for the two-step process. As the excitation probability $P(2_2^+)$ is given by the square of the sum of the excitation amplitudes, it will include not only quadratic terms ($\langle 2_2^+ || E2 || 0_{g.s.}^+ \rangle^2$ and $\langle 2_2^+ || E2 || 2_1^+ \rangle^2 \langle 2_1^+ || E2 || 0_{g.s.}^+ \rangle^2$) but also a term resulting from interference between these two excitation paths, namely, $\langle 2_2^+ || E2 || 0_{g.s.}^+ \rangle \langle 2_2^+ || E2 || 2_1^+ \rangle \langle 2_1^+ || E2 || 0_{g.s.}^+ \rangle$. As this term includes non-squared matrix elements, depending on their relative phases, its contribution can result in an increase of the population of the state in question (constructive interference, i.e. positive sign of the interference term) or in its decrease (destructive interference). This effect leads to experimental sensitivity of Coulomb-excitation data to relative signs of electromagnetic matrix elements.

In typical multi-step Coulomb-excitation experiments, excited states are populated through several excitation patterns involving multiple intermediate states, and thus the measured cross sections are influenced by more complex interference terms. Therefore it is highly recommended to adopt a transparent sign convention. A common choice is to assume that signs of all in-band transitional $E2$ matrix elements are positive. Moreover, for each band head, one needs to impose a sign for one transition linking it with a state in a different band. The signs of all remaining matrix elements can be determined relative to those.

To illustrate how important the influence of the interference terms on the excitation cross sections may be, we briefly discuss an example of a Coulomb-excitation study performed with a ^{42}Ca beam impinging on ^{208}Pb and ^{197}Au targets [29, 30]. In this measurement, performed with the AGATA γ-ray tracking array [31] at LNL, Legnaro, the signs of two matrix elements coupling low-spin states in ^{42}Ca, $\langle 2_2^+ || E2 || 0_1^+ \rangle$ and $\langle 2_2^+ || E2 || 2_1^+ \rangle$, were found to strongly influence the observed excitation cross sections. Figure 2.2 presents the effect of the sign of the $\langle 2_2^+ || E2 || 0_1^+ \rangle$ matrix element on the intensities of γ-ray transitions following Coulomb excitation of ^{42}Ca on ^{208}Pb (the influence of the sign of the $\langle 2_2^+ || E2 || 2_1^+ \rangle$ matrix element on the same observables is discussed in Ref. [32]). The calculations were performed using magnitudes of matrix elements obtained in Refs. [29, 30]. Those of transitional $E2$ matrix elements were strongly constrained by the fact that the lifetimes of all excited states in ^{42}Ca included in the analysis were known with precision ranging from 2% for the 0_2^+ state to about 20% for the 4_2^+ state. As shown in Fig. 2.2, in the range of scattering angles covered by the particle detector (105–142°), the influence of the interference term on Coulomb-excitation cross sections leads to changes in the population of excited states of a factor of two or more. Varying the magnitudes of the transitional matrix elements within the experimental uncertainties resulting from lifetime measurements would only

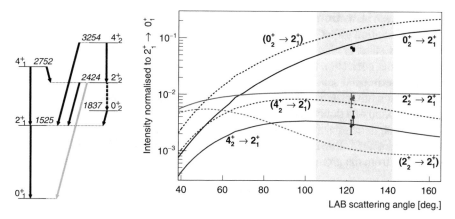

Fig. 2.2 *Left panel:* Low-lying levels of ^{42}Ca, with their energies given in keV. Transition marked with the dashed arrow is too weak to be observed experimentally, but the corresponding matrix element impacts excitation cross sections of observed states, as discussed in Sect. 2.3.2.3. *Right panel:* Transition intensities in ^{42}Ca normalised to that of $2_1^+ \to 0_1^+$, following Coulomb excitation of ^{42}Ca on ^{208}Pb in the experimental conditions of the measurement described in Refs. [29,30]. The solid lines denote the calculations using the final set of matrix elements obtained in the analysis, while for the dashed lines, an opposite sign of the $\langle 2_2^+ \| E2 \| 0_1^+ \rangle$ matrix element (corresponding to the transition marked in grey in the left panel) has been imposed. The labels in brackets are used for the latter curves. The experimental intensity ratios are also shown, both those measured with the ^{208}Pb target (*circles*) and in a shorter run with a ^{197}Au target (*squares*). The latter are rescaled to take into account the slight difference in cross sections and, for clarity, offset on the X axis. The shaded area reflects the angular coverage of the particle detection system

produce cross-section changes at the level of the precision of the measured lifetimes, i.e. about 20%. Under these experimental conditions, the sensitivity of the data to the relative signs of matrix elements is undeniable. Figure 2.2 also shows that the evolution of the γ-ray intensities with the scattering angle is non-trivial; while the effect of the interference term on the population of the 2_2^+ state increases monotonically with the scattering angle, those of the 0_2^+ and 4_2^+ states evolve in a different way, with a slight enhancement observed for forward scattering angles.

The sensitivity of low-energy Coulomb excitation to the relative signs of electromagnetic matrix elements is usually discussed in the context of electric quadrupole transitions. However, an attempt has also been made to measure the relative sign of the electric dipole and octupole matrix elements in ^{226}Ra [33]. Although the measurement was not fully conclusive due to the limited statistics and insufficient precision of the $B(E1)$ values, the fit of matrix elements to the experimental transition intensities and other spectroscopic data was shown to be better for the negative relative phase of the $E3$ and $E1$ matrix elements (normalised χ^2 of 0.61) than for a positive one (normalised χ^2 of 1.76) [33]. One should note here that ^{226}Ra represented a very favourable case for such study due to enhanced $B(E3)$ and $B(E1)$ values and low excitation energies of the relevant states, and that in typical low-energy Coulomb-excitation experiments the effect of $E1$ matrix elements on excitation cross sections is negligible, as discussed in Sect. 2.3.1.

2.3.2.2 Reorientation Effect

The reorientation effect [34], i.e. a change in the population of magnetic substates of a given state I_f due to the interaction with the collision partner, produces a second-order correction to its Coulomb-excitation cross section. This process leads to the experimental sensitivity to diagonal $E2$ matrix elements of excited nuclear states, which are related to their quadrupole moments Q_s. It can be understood as a particular kind of a double-step excitation, when the intermediate state and the final state are magnetic substates of the same excited state. The second-order perturbation theory yields the following dependence of the excitation probability of the I_f state populated from the ground state on the diagonal matrix element of the former:

$$P(I_f) \approx \alpha \langle I_f \| E2 \| I_{gs} \rangle^2 (1 + \beta \langle I_f \| E2 \| I_f \rangle) . \tag{2.33}$$

The factors α and β depend on the atomic numbers of the collision partners, the reaction kinematics (beam energy, scattering angles) and nuclear spins involved. For the exact formula, we refer the reader to Ref. [14]. As an example, the left panel of Fig. 2.3 presents the dependence of the excitation cross section calculated for the 2_1^+ state in ^{76}Zn, Coulomb-excited on a ^{196}Pt target, on the laboratory scattering angle θ_{LAB} and the $\langle 2_1^+ \| E2 \| 2_1^+ \rangle$ matrix element. A positive sign of the diagonal matrix element, corresponding to an oblate shape, leads to an enhanced excitation

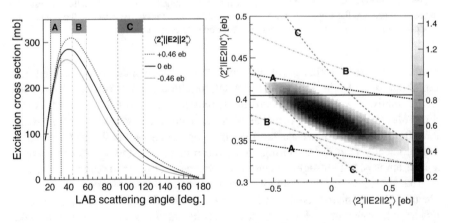

Fig. 2.3 Reorientation effect example: influence of the $\langle 2_1^+ \| E2 \| 2_1^+ \rangle$ matrix element on the excitation of the 2_1^+ state in ^{76}Zn, Coulomb excited on a ^{196}Pt target at 2.8 MeV/A beam energy. *Left panel:* Excitation cross section of the 2_1^+ state, integrated over 10° bins in θ and 360° in ϕ, for $\langle 2_1^+ \| E2 \| 2_1^+ \rangle = 0$ and the extreme values of $\langle 2_1^+ \| E2 \| 2_1^+ \rangle$ predicted by the rotational model (discussed in Sect. 2.4) assuming the literature value of $\langle 2_1^+ \| E2 \| 0_1^+ \rangle = 0.38\ eb$ [28]. *Right panel:* Correlation between the $\langle 2_1^+ \| E2 \| 0_1^+ \rangle$ and $\langle 2_1^+ \| E2 \| 2_1^+ \rangle$ matrix elements resulting from the experimental data from data [28], for three ranges of projectile scattering angles (A, B, C) marked in the left panel. A part of the χ^2 surface corresponding to the comparison of the experimentally determined cross sections with the values calculated using the GOSIA2 code [35] is presented with the condition that $\chi^2 < \chi_{min}^2 + 1$. Solid horizontal lines represent the $B(E2; 2_1^+ \rightarrow 0_1^+)$ value with its uncertainty determined from an independent measurement [10]

cross section, while its negative sign (i.e. prolate shape) results in a cross-section decrease with respect to the $\langle 2_1^+ ||E2||2_1^+ \rangle = 0$ case. As shown in Fig. 2.3, this effect strongly depends on the scattering angle. For low θ_{CM} angles, where the excitation process can be well described within the first-order approximation, it is negligible. The impact of the reorientation effect on the excitation cross section increases with the θ_{CM} scattering angle, i.e. longer collision time and shorter distance of closest approach. Consequently, by measuring the excitation cross section as a function of the scattering angle, one can disentangle the contribution from one- and two-step excitation processes and determine both the transitional $\langle I_f ||E2||I_i \rangle$ and diagonal $\langle I_f ||E2||I_f \rangle$ matrix elements. This is illustrated in the right panel of Fig. 2.3, which shows the constraints on the $\langle 2_1^+ ||E2||2_1^+ \rangle$ and $\langle 2_1^+ ||E2||0_1^+ \rangle$ matrix elements resulting from the 2_1^+ excitation cross sections measured for three distinct ranges of scattering angle defined in the left panel of Fig. 2.3. For the angular range A, the reorientation effect has a weak influence on the measured cross section, and therefore the borders of the band in the $\langle 2_1^+ ||E2||2_1^+ \rangle$, $\langle 2_1^+ ||E2||0_1^+ \rangle$ plane resulting from this data set are almost horizontal, i.e. provide strong constraints on $\langle 2_1^+ ||E2||0_1^+ \rangle$, but rather weak ones on $\langle 2_1^+ ||E2||2_1^+ \rangle$. The slopes of bands B and C are considerably more steep, in accordance with a marked difference visible for these ranges between the cross sections plotted for negative and positive values of the diagonal $E2$ matrix element (left panel of Fig. 2.3). The widths of the bands corresponding to each range reflect uncertainties of the measured excitation cross sections. The bands for ranges A, B and C have a common intersection, which demonstrates internal consistency of the experimental data.

In the course of the data analysis for ^{76}Zn, for each combination of the $\langle 2_1^+ ||E2||0_1^+ \rangle$ and $\langle 2_1^+ ||E2||2_1^+ \rangle$ matrix elements, the expected 2_1^+ excitation cross sections for ranges A, B and C were calculated and compared with the experimental results. This gives rise to a two-dimensional χ^2 distribution, whose minimum defines the $\langle 2_1^+ ||E2||0_1^+ \rangle$ and $\langle 2_1^+ ||E2||2_1^+ \rangle$ matrix elements that optimally reproduce the experimental data. A 1σ-uncertainty contour surrounds the region of the χ^2 surface for which $\chi^2 < \chi^2_{min} + 1$ [35]. The final uncertainties are obtained by projecting the 1σ uncertainty contour on the respective axes.

The 1σ uncertainty contour presented in the right panel of Fig. 2.3 extends over a broad range of possible $\langle 2_1^+ ||E2||2_1^+ \rangle$ values, even exceeding the values deduced using the rotational model (2.42) from the literature value of $\langle 2_1^+ ||E2||0_1^+ \rangle$, which should normally be interpreted as the maximum values of $\langle 2_1^+ ||E2||2_1^+ \rangle$. This means that the experimental sensitivity of this particular measurement to the $\langle 2_1^+ ||E2||2_1^+ \rangle$ matrix element was insufficient. This is due to the low total statistics of only about 300 counts in the $2_1^+ \rightarrow 0_1^+$ transition. Its increase by about a factor of 4 would narrow down the bands in the $\langle 2_1^+ ||E2||2_1^+ \rangle$, $\langle 2_1^+ ||E2||0_1^+ \rangle$ plane by about one half, leading to the precision on the $\langle 2_1^+ ||E2||2_1^+ \rangle$ of about 0.3 eb, which could give the first idea of the underlying deformation. From this estimation one can see that for this type of analysis usually at least a thousand counts in the peak corresponding to the deexcitation of the state of interest are needed. Moreover, the particle detector should cover an angular range in which the impact of the quadrupole moment on the excitation cross section changes in a meaningful way. For example, while this impact is maximised for large scattering angles, it remains rather constant throughout, e.g. the $140° < \theta_{CM} < 180°$ range. Consequently, if such data were subdivided into a few ranges of scattering angle, the slopes of resulting bands in the $\langle I_f||E2||I_f \rangle$, $\langle I_f||E2||I_i \rangle$ plane would be similar. In such case, even for a high-statistics measurement, the precision of the $E2$ matrix elements determined from the analysis would be limited.

Another possibility is to combine Coulomb-excitation data with an independently measured $\langle I_f||E2||I_i \rangle$ matrix element, resulting from, e.g. a lifetime measurement. Such information, which would be represented as a horizontal band in the $\langle I_f||E2||I_f \rangle$, $\langle I_f||E2||I_i \rangle$ plane, would provide an additional contribution to the χ^2 function and lead to a reduction of the uncertainty of the $\langle I_f||E2||I_f \rangle$ matrix element. In the case of a differential cross-section measurement, one could easily verify the consistency between Coulomb-excitation cross sections and the measured lifetime: the horizontal band representing the lifetime with its uncertainty should overlap with the 1-σ contour of the χ^2 surface resulting from the Coulomb-excitation analysis, as it is the case in Fig. 2.3[1]. However, this approach may also be applied to measurements of the integral excitation cross section, which without additional lifetime data provide only a correlation between the $\langle I_f||E2||I_i \rangle$ and $\langle I_f||E2||I_f \rangle$ matrix elements. In this case there is no internal consistency check, and the conclusions of the analysis strongly rely on the quality of the lifetime information. This has been demonstrated by the study of ^{70}Se performed at REX-ISOLDE [36]. The level of statistics for the $2_1^+ \rightarrow 0_1^+$ decay obtained in this experiment, namely 139(13) counts, was clearly insufficient to perform a subdivision of the data according to the scattering angle. Hence, in order to deduce the $\langle 2_1^+||E2||2_1^+ \rangle$ matrix element, the integral Coulomb-excitation cross section was combined with the known 2_1^+ lifetime of 1.5(3) ps [37], leading to a conclusion that the sign of the $\langle 2_1^+||E2||2_1^+ \rangle$ matrix element was negative, and its magnitude exceeded the rotational estimate, calculated from the $\langle 2_1^+||E2||0_1^+ \rangle$ value deduced from the lifetime as discussed in Sect. 2.4. This result was at odds with multiple model calculations predicting an oblate shape, i.e. a positive $\langle 2_1^+||E2||2_1^+ \rangle$ value in ^{70}Se (see, e.g. [38] for an overview of those). However, a later measurement of lifetimes in ^{70}Se [38], which made use of γ-γ coincidences and thus provided a better control of the side feeding than that of Ref. [37] relying on singles γ-ray spectra, yielded a substantially longer 2_1^+ lifetime of 3.2(2) ps. The intersection of the areas of possible $\langle 2_1^+||E2||2_1^+ \rangle$, $\langle 2_1^+||E2||0_1^+ \rangle$ values resulting from the Coulomb-excitation data [36] and the revised 2_1^+ lifetime [38] seems to favour a positive value of the $\langle 2_1^+||E2||2_1^+ \rangle$ matrix element, i.e. an oblate shape of the 2_1^+ state.

---

[1] While typically information from lifetime measurements is used in this type of analysis, no such data exist for ^{76}Zn, and consequently a $B(E2; 2_1^+ \rightarrow 0_1^+)$ value determined in an earlier Coulomb-excitation measurement [10] has been plotted in Fig. 2.3 for illustration purposes.

One should stress that if a measurement aiming at the determination of transition probabilities, in particular of the $B(E2; 2_1^+ \rightarrow 0_1^+)$ value, is not performed under conditions where the first-order approximation is valid, the influence of the unknown quadrupole moment of the state in question needs to be taken into account. As discussed in Sect. 2.5.2, detection of scattered projectiles or target recoils corresponding to low scattering angles, where the reorientation effect could be neglected, is not always feasible due to high flux of elastically scattered beam particles. In the example of Fig. 2.3, one could attempt to extract the $B(E2; 2_1^+ \rightarrow 0_1^+)$ value from range-A data by imposing limits on possible $\langle 2_1^+||E2||2_1^+\rangle$ values using a model assumption. In the simplest case of rotational limits, this would lead to a higher uncertainty of the $\langle 2_1^+||E2||0_1^+\rangle$ value than that resulting from the analysis involving also the data from ranges B and C. Also, in this particular case, no improvement can be expected from narrowing down range A in order to limit it to the angles where the influence of the $\langle 2_1^+||E2||2_1^+\rangle$ matrix element on the excitation probability is negligible. While gating on the innermost part of the particle detector (e.g. $\theta_{LAB} < 25°$; see the differential cross section in the left panel of Fig. 2.3) would probably produce the intended result, i.e. a practically horizontal band in the $\langle 2_1^+||E2||2_1^+\rangle$, $\langle 2_1^+||E2||0_1^+\rangle$ plane, due to the low statistics the width of such band would be larger than the limits on the $\langle 2_1^+||E2||0_1^+\rangle$ value resulting from the analysis involving all three ranges A, B and C.

Sometimes the assumption is made that the influence of the $\langle 2_1^+||E2||2_1^+\rangle$ matrix element on the 2_1^+ excitation cross section is negligible, and the $\langle 2_1^+||E2||0_1^+\rangle$ value with its uncertainty is determined assuming a certain single value of $\langle 2_1^+||E2||2_1^+\rangle$, usually $\langle 2_1^+||E2||2_1^+\rangle = 0$. This would be equivalent to a projection of the 1-σ contour of the χ^2 surface on the Y axis for $\langle 2_1^+||E2||2_1^+\rangle = 0$ and would result in a significant underestimation of the $\langle 2_1^+||E2||0_1^+\rangle$ uncertainty due to the correlations between the matrix elements being ignored. In the example presented in Fig. 2.3, this underestimation would be as large as about a factor of 2.

The impact of the reorientation effect on Coulomb-excitation cross sections can be comparable to that of multi-step excitations, and, consequently, the latter should be carefully evaluated. Excitation of higher-lying states via the state in question, or feeding from above involving such states, may impact the measured γ-ray intensities, even though the feeding transitions are not directly observed; this will be addressed in Sect. 2.3.2.3. Moreover, appearance of interference terms introduced in Sect. 2.3.2.1 can lead to similar modifications of the differential excitation cross section as those related to the influence of the quadrupole moment. For this reason, two values of the spectroscopic quadrupole moment were often reported in early low-energy Coulomb-excitation measurements: one corresponding to a positive sign of the $\langle 0_1^+||E2||2_1^+\rangle\langle 2_1^+||E2||2_2^+\rangle\langle 2_2^+||E2||0_1^+\rangle$ interference term and the other one for a negative sign. Including other spectroscopic data in Coulomb-excitation data analysis, especially lifetimes, helps tremendously to disentangle competing contributions to the cross sections, especially if this information is complemented by high-statistics differential excitation cross-section measurements. The use of different beam-target combinations in the same experiment is also recommended. Finally, the graphical method presented in this section can be quite useful when trying to decide on the best way of subdividing the existing data according to the scattering angle.

2.3.2.3 Effect of Unobserved States and Transitions on Excitation Cross Sections

When calculating excitation probabilities of states forming a collective band, it is important not to truncate the calculations to only those states that have been observed in a Coulomb-excitation study, as it will result in erroneous calculations for the uppermost states. For very collective bands, several states can be affected by this, and in the data analysis this effect often leads to unreasonably large diagonal matrix elements that need to be introduced for highest-energy members of a band in order to reproduce the experimentally measured excitation cross sections. Therefore, in the analysis of Coulomb-excitation data, "buffer states" are usually added on the top of all bands. If the matrix elements coupling them to the observed states are not known from experimental studies, which is often the case for nuclei far from stability, model assumptions are necessary. For rotational structures relations provided by the axial rigid-rotor approximation, introduced in Sect. 2.4, may be used, and in other cases, estimates can be made on the basis of predictions of modern nuclear-structure theories. While the influence of the higher-lying unobserved states can be reasonably accounted for without knowing their exact excitation energies and matrix elements involved, it is a good practice to investigate the effect of assumed values on the calculated cross sections and, if it is not negligible, to include it in systematic uncertainties of the matrix elements resulting from the data analysis.

Due to the strong dependence of the decay probability on the transition energy (2.17) large energy differences are preferred in the decay process. Consequently, the decay branches leading to states lying close in energy to the state in question are hindered. Such $E2$ decay branches, even if they are too weak to be observed experimentally, may strongly affect the population of states in Coulomb excitation, since, as discussed in Sect. 2.3.1, the excitation process is enhanced for low ξ, i.e. small differences in excitation energy. Consequently, the knowledge of precise branching ratios is crucial in Coulomb-excitation data analysis.

Finally, in high-precision Coulomb-excitation studies, in principle it is possible to measure directly the effect on excitation cross sections of matrix elements corresponding to unobserved transitions. As an example, Fig. 2.4 illustrates the influence of the $\langle 2_2^+ \| E2 \| 0_2^+ \rangle$ matrix element (left panel) and the spectroscopic quadrupole moment of the 2_2^+ state (right panel) on the measured γ-ray intensities in the $^{42}\text{Ca} + {}^{208}\text{Pb}$ experiment introduced in Sect. 2.3.2.1. Clearly, even though the $2_2^+ \rightarrow 0_2^+$ transition is too weak to be observed and prior to the study of Ref. [29] only an upper limit for the branching ratio has been known, the corresponding matrix element strongly affects excitation cross sections of the observed states, and hence it could be determined from the intensities of other transitions measured in the Coulomb-excitation experiment. As seen from Fig. 2.4, the effects of the $\langle 2_2^+ \| E2 \| 0_2^+ \rangle$ and $\langle 2_2^+ \| E2 \| 2_2^+ \rangle$ matrix elements on the population of both the 0_2^+ and 2_2^+ states are opposite, which implies that an increase of $\langle 2_2^+ \| E2 \| 0_2^+ \rangle$ can be compensated by a larger value of $\langle 2_2^+ \| E2 \| 2_2^+ \rangle$, leading to an equally good

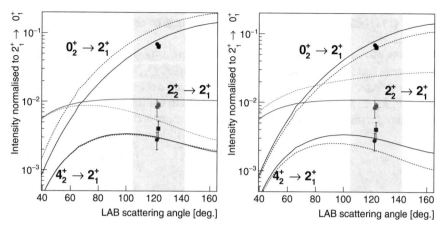

Fig. 2.4 Same as Fig. 2.2, but the dotted lines correspond to the following: (*left panel*) $\langle 2_2^+ \| E2 \| 0_2^+ \rangle$ equal to zero and (*right panel*) quadrupole moment of the 2_2^+ state equal to zero

reproduction of the $2_2^+ \to 2_1^+$ and $0_2^+ \to 2_1^+$ transition intensities. The population of the 4_2^+ state, however, is sensitive only to the latter, which makes it possible to extract both matrix elements independently. The existing correlation between them results in relatively large uncertainties of the two matrix elements ($\langle 2_2^+ \| E2 \| 0_2^+ \rangle = 26 \, {}^{+5}_{-3} \, efm^2$, $\langle 2_2^+ \| E2 \| 2_2^+ \rangle = -55 \, {}^{+15}_{-15} \, efm^2$ [29]). The value of the $\langle 2_2^+ \| E2 \| 0_2^+ \rangle$ matrix element corresponds only to a 0.3% branch in the decay of the 2_2^+ state, and yet its effect on the population of the states in the side band of ^{42}Ca is clearly visible in Fig. 2.4. As mentioned in Sect. 2.3.2.1, the precision of the 4_2^+ lifetime, crucial for distinguishing between the effects of the $\langle 2_2^+ \| E2 \| 0_2^+ \rangle$ and $\langle 2_2^+ \| E2 \| 2_2^+ \rangle$ matrix elements, is about 20%, lower than those of most lifetimes of low-lying states in ^{42}Ca. Improving it would help constrain the magnitudes of all matrix elements governing the deexcitation of this state and in turn increase the sensitivity to the $\langle 2_2^+ \| E2 \| 0_2^+ \rangle$ and $\langle 2_2^+ \| E2 \| 2_2^+ \rangle$ matrix elements.

2.4   Quadrupole Moments and Nuclear Deformation

The spectroscopic electric quadrupole moment Q_s of a state of spin I is defined as the diagonal matrix element of the $\mu = 0$ component of the $\hat{M}(E2, \mu)$ quadrupole operator, with states of maximum m value ($m = I$):

$$eQ_s = \sqrt{\frac{16\pi}{5}} \langle I, m = I | \hat{M}(E2, \mu = 0) | I, m = I \rangle =$$

$$= \sqrt{\frac{16\pi}{5}} \frac{1}{\sqrt{2I+1}} (I, I, 2, 0 | I, I) \langle I \| \hat{M}(E2) \| I \rangle , \qquad (2.34)$$

where $\langle I \| \hat{M}(E2) \| I \rangle$ is the reduced diagonal matrix element of the $\hat{M}(E2, \mu)$ operator.

In order to relate the intrinsic quadrupole moment Q_0, defined in the principal-axis frame, and the spectroscopic quadrupole moment Q_s, a transformation between these two frames must be applied. This transformation relies on model assumptions, and, e.g. for an axially symmetric rotor, it results in the following relation between the reduced matrix elements of the $\hat{M}(E2, \mu)$ operator and Q_0:

$$\langle I_f K \| \hat{M}(E2) \| I_i K \rangle = \sqrt{2I_i + 1}(I_i, K, 2, 0 | I_f, K) \sqrt{\frac{5}{16\pi}} e Q_0 , \qquad (2.35)$$

where K is the projection of the nuclear spin I on the symmetry axis of the nucleus. Combining (2.34) and (2.35) makes it possible to relate Q_s to Q_0:

$$Q_s = (I, K, 2, 0 | I, K)(I, I, 2, 0 | I, I) Q_0 = \frac{3K^2 - I(I + 1)}{(I + 1)(2I + 3)} Q_0 . \qquad (2.36)$$

For $I = K$, which is in particular true for the ground states, the spectroscopic quadrupole moment vanishes for spins of 0 and 1/2, even for deformed states, i.e. when Q_0 is different from zero.

The electric quadrupole operator can be expressed as:

$$\hat{M}(E2, \mu) = \int \hat{\rho}(\mathbf{r}) r^2 Y_{2\mu}(\theta, \phi) \, d^3r . \qquad (2.37)$$

Assuming a uniform charge density leads to the following expression for its $\mu = 0$ component:

$$\hat{M}(E2, \mu = 0) = \frac{3Z}{4\pi R_0^3} \int \int_0^R r^4 \, Y_{20}(\theta, \phi) \, dr \, \sin\theta \, d\theta \, d\phi \qquad (2.38)$$

$$= \frac{3Z}{4\pi R_0^3} \int \frac{1}{5} R^5 Y_{20}(\theta, \phi) \, \sin\theta \, d\theta \, d\phi .$$

The radius of an axially deformed nucleus can be expressed as:

$$R(\theta, \phi) \approx R_0 \left(1 + \beta_2 Y_{20}(\theta, \phi) \right) , \qquad (2.39)$$

which leads to the commonly used relation between Q_0 and β_2:

$$Q_0 = \frac{3}{\sqrt{5\pi}} Z R_0^2 \beta_2 \left(1 + \frac{2}{7} \sqrt{\frac{5}{\pi}} \beta_2 + \ldots \right)$$

$$\approx \frac{3}{\sqrt{5\pi}} Z R_0^2 \beta_2 \left(1 + 0.36 \beta_2 \right) . \qquad (2.40)$$

The axial rigid-rotor model is often used to estimate the limits of the spectroscopic quadrupole moments, or equivalently the diagonal $E2$ matrix elements, from known $B(E2)$ values. Equation (2.35) leads to the following relation:

$$\langle I_f K \| \hat{M}(E2) \| I_f K \rangle = \frac{(I_f, K, 2, 0|I_f, K)}{(I_i, K, 2, 0|I_f, K)} \sqrt{B(E2; I_i \rightarrow I_f)(2I_f + 1)} \,,$$

$$(2.41)$$

which for a 2_1^+ state in an even-even nucleus simplifies to:

$$\langle 2_1^+ \| \hat{M}(E2) \| 2_1^+ \rangle = 5 \sqrt{B(E2; 2_1^+ \rightarrow 0_1^+)} \times \frac{(2, 0, 2, 0|2, 0)}{(0, 0, 2, 0|2, 0)} \approx 1.2 \times \langle 2_1^+ \| \hat{M}(E2) \| 0_1^+ \rangle \,.$$

$$(2.42)$$

In (2.41) and (2.42) a prolate shape of the nucleus is assumed, i.e. a negative value of the diagonal $E2$ matrix for $K = 0$ states. For an oblate shape, the signs of diagonal matrix elements resulting from (2.41) and (2.42) should be inverted.

2.4.1 Quadrupole Sum Rules

As discussed above, the nuclear shape can be inferred indirectly from transition probabilities or spectroscopic quadrupole moments relying on model assumptions. An alternative model-independent approach, proposed by Kumar and Cline [2, 9], takes advantage of specific properties of electromagnetic multipole operators. Since these operators are spherical tensors, their products coupled to zero angular momentum are rotationally invariant. The expectation values of these products are observables, and they can be, on the one hand, related to the parameters describing the shape of the charge distribution and, on the other hand, to reduced electromagnetic matrix elements, defined in the laboratory system.

In the principal-axis system, the coefficients of the expansion into spherical harmonics of the quadrupole electric operator, $\hat{M}(E2, \mu)$, defined via (2.37), can be represented using two variables Q and δ. Their expectation values are equivalent to the standard β_2 and γ parameters describing the quadrupole shape, but instead of the mass distribution, they describe the charge distribution:

$$\hat{M}(E2, 0) = Q \cos \delta \,,$$

$$\hat{M}(E2, 1) = \hat{M}(E2, -1) = 0 \,,$$

$$(2.43)$$

$$\hat{M}(E2, 2) = \hat{M}(E2, -2) = \frac{1}{\sqrt{2}} Q \sin \delta \,.$$

The simplest two invariants read:

$$\left[\hat{M}(E2) \times \hat{M}(E2)\right]^0 = \frac{1}{\sqrt{5}} Q^2, \tag{2.44}$$

$$\left[\left[\hat{M}(E2) \times \hat{M}(E2)\right]^2 \times \hat{M}(E2)\right]^0 = -\sqrt{\frac{2}{35}} Q^3 \cos 3\delta \tag{2.45}$$

where the superscript on the square bracket is the angular momentum J to which the operators inside the bracket are coupled. Their expectation values for a state I_n can be expressed through reduced $E2$ matrix elements as follows:

$$\langle I_n | Q^2 | I_n \rangle = \frac{\sqrt{5} \, (-1)^{2I_n}}{\sqrt{2I_n + 1}} \sum_m M_{nm} M_{mn} \begin{Bmatrix} 2 & 2 & 0 \\ I_n & I_n & I_m \end{Bmatrix}, \tag{2.46}$$

$$\langle I_n | Q^3 \cos 3\delta | I_n \rangle = -\sqrt{\frac{35}{2}} \frac{(-1)^{2I_n}}{2I_n + 1} \sum_{ml} M_{nl} M_{lm} M_{mn} \begin{Bmatrix} 2 & 2 & 2 \\ I_n & I_m & I_l \end{Bmatrix}, \tag{2.47}$$

where $M_{ab} \equiv \langle I_a || E2 || I_b \rangle$ and the expression in curly brackets is a $6j$ coefficient. For states with an axial prolate shape $\langle \cos 3\delta \rangle = 1$; for axial oblate states $\langle \cos 3\delta \rangle = -1$; and for maximally triaxial or spherical states $\langle \cos 3\delta \rangle = 0$.

A finite value of $\langle Q^2 \rangle$ may result from both a static deformation, β_{stat}, as observed, for example, for well-deformed rotational nuclei, and a dynamic deformation, β_{dyn}, resulting from vibrational or non-collective motion. These can be distinguished based on the dispersion of $\langle Q^2 \rangle$:

$$\sigma(Q^2) = \sqrt{\langle Q^4 \rangle - (\langle Q^2 \rangle)^2}. \tag{2.48}$$

In order to obtain the $\langle Q^4 \rangle$ invariant, products of four matrix elements need to be considered, which requires a level of detail that is difficult to achieve experimentally. Similarly, fluctuations in $\langle Q^3 \cos 3\delta \rangle$ can be evaluated, requiring even more extensive experimental data.

While the simplest two invariants $[\hat{M}(E2) \times \hat{M}(E2)]^0$ and $\left[[\hat{M}(E2) \times \hat{M}(E2)]^2 \times \hat{M}(E2)\right]^0$ can be constructed in only one way, it is no longer true for higher-order invariants. For example, the $\langle Q^4 \rangle$ invariant can be obtained by coupling two $\hat{M}(E2)$ operators to spin 2, $\left[\hat{M}(E2) \times \hat{M}(E2)\right]^2$, and then coupling such rank-2 tensor with another one constructed in the same way: $\left[[\hat{M}(E2) \times \hat{M}(E2)]^2 \times [\hat{M}(E2) \times \hat{M}(E2)]^2\right]^0$. However, coupling of each pair of operators to intermediate spins 0 and 4 is also possible. These three ways to construct the $\langle Q^4 \rangle$ invariant will correspond to different sets of reduced $E2$ matrix elements, and in principle comparing the resulting $\langle Q^4 \rangle$ values could be used to check for internal consistency of the obtained set of matrix elements. Unfortunately, it is very rare to obtain this level of detail in experimental studies.

According to (2.46), in order to obtain information on the axial deformation of a given state from the $\langle Q^2 \rangle$ invariant, one needs to know all reduced $E2$ matrix elements linking it to all other states that can be reached from the state in question via a single $E2$ transition. In general, if the spin of the state in question is different than 0 or 1/2, it means that also its diagonal $E2$ matrix element has to be known. For the lowest-order invariant, $\langle Q^2 \rangle$, all matrix elements enter the sum in squares, and thus for its determination, knowledge of reduced transition probabilities ($B(E2)$ values) is sufficient. However, for higher-order invariants, relative signs of $E2$ matrix elements play a role. For example, the $\langle Q^3 \cos 3\delta \rangle$ invariant is constructed from triple products of $E2$ matrix elements, $\langle I_n || E2 || I_l \rangle \langle I_l || E2 || I_m \rangle \langle I_m || E2 || I_n \rangle$, where $|I_n\rangle$ is the state in question and $|I_l\rangle$ and $|I_m\rangle$ are the intermediate states. The influence of such triple products of transitional matrix elements on excitation cross sections was discussed in Sect. 2.3.2.1. Moreover, if $|I_l\rangle = |I_m\rangle$, the triple product involves a diagonal matrix element.

In order to obtain the $\langle \cos 3\delta \rangle$ value from the $\langle Q^3 \cos 3\delta \rangle$ invariant, the following approximation is usually made:

$$\langle Q^3 \cos 3\delta \rangle \approx \langle Q^2 \rangle^{3/2} \langle \cos 3\delta \rangle . \tag{2.49}$$

Assuming that the charge and mass distributions are identical, one can relate the Q^2 and δ parameters to the β_2 and γ parameters of Bohr's model by [39]:

$$\langle Q^2 \rangle = q_0^2 \langle \beta_2^2 \rangle \tag{2.50}$$

and

$$\langle Q^3 \cos 3\delta \rangle = q_0^3 \langle \beta_2^3 \cos 3\gamma \rangle , \tag{2.51}$$

where $q_0 = \frac{3}{4\pi} Z e R_0^2$ and $R_0 = 1.2 A^{1/3}$ fm. The expressions depend on the definition of the collective variables; for example, if β and γ are the Nilsson model ellipsoidal deformation parameters, more complex formulae given in the appendix of Ref. [40] emerge.

While the sums in (2.46) and (2.47) formally run over all intermediate states that can be reached from the state in question via a single $E2$ transition, usually only a few key states contribute to the invariant in a meaningful way. In particular, for the ground state of an even-even nucleus, the contributions to $\langle Q^2 \rangle$ are dominated by the coupling to the 2_1^+ state, which typically amounts to well over 90% of the total. Thus the following approximation can be made:

$$\langle Q_{0_1^+}^2 \rangle \approx |\langle 2_1^+ || E2 || 0_1^+ \rangle|^2 \tag{2.52}$$

leading to the well-known expression linking β_2 and $B(E2; 0_1^+ \to 2_1^+)$:

$$\sqrt{\langle \beta_2^2 \rangle} \approx \frac{4\pi}{3ZR_0^2} \sqrt{B(E2; 0_1^+ \to 2_1^+)/e^2} \, . \tag{2.53}$$

Similarly, the largest contributions to $\langle Q^3 \cos 3\delta \rangle$ for the ground state come from the $\langle 0_1^+ ||E2||2_1^+ \rangle \langle 2_1^+ ||E2||2_1^+ \rangle \langle 2_1^+ ||E2||0_1^+ \rangle$ and $\langle 0_1^+ ||E2||2_1^+ \rangle \langle 2_1^+ ||E2||2_2^+ \rangle$ $\langle 2_2^+ ||E2||0_1^+ \rangle$ products [41]. The situation becomes much more complicated for excited states, and the number of intermediate states that need to be included in the sum rules varies from one case to another. While theoretical approaches can, in principle, provide complete sets of electromagnetic matrix elements, this is not always true for experiments. Systematic studies addressed this convergence issue on the basis of beyond-mean-field [39] and shell-model calculations [42–44]. For example, the $\langle Q^2 \rangle$ and $\langle Q^3 \cos 3\delta \rangle$ values were obtained for 0_1^+ and 0_2^+ states in ^{100}Mo from Coulomb-excitation data, pointing to a coexistence of a moderately deformed triaxial ground state with a more deformed prolate excited structure [39]. Three intermediate 2^+ states were included in the calculation of invariants. The analysis of contributions of individual products of matrix elements to the invariants obtained for the 0_2^+ state showed that the couplings to each of the three intermediate states were almost equally important. As general Bohr Hamiltonian calculations for ^{100}Mo [39] reproduced well the experimental transition probabilities in this nucleus, the same theoretical approach has been used to evaluate the influence of possible couplings to higher-lying states on the invariants obtained for the 0_1^+ and 0_2^+ states. To this end, the expectation values of the $\langle Q^2 \rangle$ and $\langle Q^3 \cos 3\delta \rangle$ invariants were calculated either directly, by acting with the electromagnetic operator on the wave functions of the 0_1^+ and 0_2^+ states, or from the matrix elements resulting from the theoretical calculations, limiting the sum to three intermediate states as it has been done in the case of experimental values. In all cases, differences of at most 3% were observed, which shows that three intermediate states are sufficient to converge in this case.

A similar analysis was performed for ^{66}Zn using the shell model [42]. The $\langle Q^2 \rangle$ invariant obtained for the ground state from $E2$ matrix elements calculated within the shell model was strongly dominated by the coupling to the 2_1^+ state, and two intermediate 2^+ states were sufficient to obtain $\langle \gamma \rangle$ within $1°$ from the value corresponding to the full convergence. The situation was shown to be very different for the 0_2^+ state, where the convergence was very slow, and even for the $\langle Q^2 \rangle$ invariant six intermediate states were necessary. For the $\langle Q^3 \cos 3\delta \rangle$ invariant, the convergence was observed starting from eight intermediate states. Moreover, unphysical $\langle \cos 3\delta \rangle > 1$ values resulted from calculations where only one or two 2^+ states were considered. Consequently, since only two matrix elements coupling the 0_2^+ state to 2^+ states were determined experimentally, i.e. $\langle 0_2^+ ||E2||2_1^+ \rangle$ and $\langle 0_2^+ ||E2||2_2^+ \rangle$, the $\langle Q^2 \rangle$ value deduced from those should be considered to be a lower limit.

From the theory point of view, there are no limitations to the number of $E2$ operators entering the invariants, and thus the quadrupole sum rules are widely used to relate $E2$ matrix elements resulting from theoretical calculations to the underlying shapes and deformation softness. Recently, for example, such an analysis making use of shell model predictions and invariants up to the sixth rank ($\langle Q^6 \rangle$) has pointed to considerable γ softness for nuclei in a wide range of masses [45]. The practical limitations arise from the number and precision of experimentally available $E2$ matrix elements: in general, with an increasing number of $E2$ operators, the number of required $E2$ matrix elements also increases, and the uncertainty propagation leads to larger relative uncertainties of the resulting invariant quantities. In the following, we briefly introduce some experimental studies that yielded particularly rich sets of electromagnetic matrix elements, making it possible to discuss dispersions $\sigma(Q^2)$ and even $\sigma(Q^3 \cos 3\delta)$.

For many years, the Coulomb-excitation study of ^{186,188,190,192}Os and ^{194}Pt [46] has been unmatched in terms of richness of experimental information on nuclear shapes obtained using the quadrupole sum rule approach. To this end, a series of experiments with ^{40}Ar, ^{58}Ni, ^{136}Xe and ^{208}Pb beams were performed at various experimental facilities (University of Rochester, Australian National University, Brookhaven National Laboratory, Lawrence Berkeley National Laboratory). In each of the nuclei under study, $\langle Q^2 \rangle$ and $\langle Q^3 \cos 3\delta \rangle$ invariants were obtained for multiple states in the ground-state band and the γ band, as well as for the $K = 4$ band head, and were found to be remarkably constant with spin, consistent with a rotational behaviour. The $\langle Q^2 \rangle$ values for the ground states were found to decrease with atomic mass, which was accompanied by a gradual transition from prolate-triaxial shapes observed for ^{186,188}Os, via maximally triaxial shapes in ^{190,192}Os to a more oblate shape in ^{194}Pt. Moreover, dispersions $\sigma(Q^2)$ and $\sigma(Q^3 \cos 3\delta)$ were obtained for the ground states and point to them being rather rigid in β but with some γ softness increasing with mass. However, important inconsistencies are visible between some $\sigma(Q^2)$ and $\sigma(Q^3 \cos 3\delta)$ values obtained assuming different intermediate spins in the construction of the $\langle Q^4 \rangle$ and $\langle Q^6 \cos^2 3\delta \rangle$ invariants. This is especially true for ^{194}Pt, which suggests that the set of matrix elements obtained for this nucleus, although very rich, is insufficient for convergence of these invariant quantities.

Recent studies using the GRETINA array [47] in combination with the 4π CHICO2 particle-detection setup [48] reached a similar level of detail. Quadrupole invariants $\langle Q^3 \cos 3\delta \rangle$ were obtained for multiple states in both ground-state and γ bands in ^{72}Ge [49] and ^{76}Ge [50]. The corresponding $\langle \cos 3\delta \rangle$ values appear to be rather constant with spin and correspond to almost maximum triaxiality. The fluctuations of $\langle Q^2 \rangle$ obtained for states in the ground-state band and the γ bands in ^{76}Ge are small, pointing to a considerable rigidity in β. Moreover, for the 0_1^+, 2_1^+ and 2_2^+ states in ^{76}Ge, also the $\sigma(Q^3 \cos 3\delta)$ dispersions were extracted. Also in this case, certain differences between the $\sigma(Q^2)$ and $\sigma(Q^3 \cos 3\delta)$ values obtained using different subsets of matrix elements were observed, although they are in general less pronounced than in Ref. [46]. Nonetheless, the values determined for the 0_1^+ state seem fully consistent both in terms of magnitude and uncertainty and suggest its rigid triaxial deformation, supporting the conclusions from level staggering observed in the γ band in this nucleus [51].

In the study of ^{42}Ca [29, 30] introduced in Sect. 2.3.2.1, a non-zero value of the $\langle Q^2 \rangle$ invariant obtained for the ground state could be viewed as a surprise, as this nucleus, lying only two neutrons above the closed $N = 20$ shell, is widely considered to be spherical in its ground state. However, the magnitude of the $\sigma(Q^2)$ dispersion was shown to be comparable with $\langle Q^2 \rangle$, leading to the conclusion that the wave function of the 0_1^+ state exhibits large fluctuations around the spherical shape. This is consistent with the maximum triaxiality obtained for this state, $\langle \cos 3\delta \rangle = 0.06(10)$, which results from averaging over all possible quadrupole-deformed shapes ranging from prolate to oblate. In contrast, the $\sigma(Q^2)$ value obtained for the band head of the highly deformed structure in ^{42}Ca is considerably lower than the $\langle Q^2 \rangle$ value obtained for this state, which suggests a large static deformation. Additionally, the triaxiality parameter $\langle \cos 3\delta \rangle$ for the 0_2^+ state, corresponding to $\gamma = 13(^{+5}_{-6})°$, provided experimental evidence for its slightly non-axial character. In this analysis, the $\langle Q^4 \rangle$ invariants could only be obtained assuming coupling of $E2$ operators to intermediate spin 0 due to the incompleteness of the obtained set of matrix elements. Notably, a γ band is suggested to be built on the 2_3^+ state in ^{42}Ca, but its 4^+ member, which according to model calculations [30] would strongly contribute to the $\langle Q^4 \rangle$ invariant with $E2$ operators coupled to intermediate spin 4, has not been identified experimentally.

All the examples discussed above come from stable beam studies. The currently available radioactive beam intensities make it difficult to obtain this level of precision and detail for exotic nuclei, and hence only in a very few cases the simplest invariants have been determined. A notable example is the low-energy Coulomb-excitation study of ^{74,76}Kr [52], which was performed in the early 2000s at GANIL, Caen, and can be viewed as a highly successful proof-of-principle experiment for determination of quadrupole moments of excited states in short-lived exotic nuclei. From the complete set of $E2$ matrix elements coupling the 0_1^+, 0_2^+, 2_1^+, 2_2^+ and 2_3^+ states in ^{74,76}Kr, including their relative signs, the $\langle Q^2 \rangle$ and $\langle Q^3 \cos 3\delta \rangle$ were obtained for the ground states and the 0_2^+ states. The two shape-coexisting configurations in ^{74}Kr strongly mix due to their proximity in energy, which is reflected by, e.g. a large $\rho^2(E0; 0_2^+ \rightarrow 0_1^+)$ value of 0.085(19) and significant perturbations of the energies of low-spin members of bands built on the $0_{1,2}^+$ states. Consistently with strong mixing, the $\langle Q^2 \rangle$ and $\langle Q^3 \cos 3\delta \rangle$ values obtained for the 0_1^+ and 0_2^+ states in ^{74}Kr are rather similar, pointing to a moderate triaxial deformation. The excitation energy of the 0_2^+ state in ^{76}Kr is higher than in ^{74}Kr, which leads to a reduction of the mixing of the 0_1^+ and 0_2^+ states, resulting in a clearer distinction in their shapes; the former has a moderate prolate deformation with $\langle \cos 3\delta \rangle \approx 1$, i.e. close to axial symmetry, while the latter is much more deformed and mostly oblate with some triaxiality. This provides a definite experimental proof of the prolate-oblate shape coexistence scenario predicted by multiple theoretical approaches, e.g. [53–55].

Another recent example illustrating how quadrupole sum rules can be used to probe shapes of exotic nuclei is the study of ^{96,98}Sr [12, 56]. The $\langle Q^2 \rangle$ invariants obtained for the 0_1^+ and 0_2^+ states in ^{98}Sr provide a firm proof of shape coexistence in this nucleus and confirm a dramatic change in deformation between the well-deformed ground state and the weakly deformed 0_2^+ state, consistent with the behaviour of in-band $E2$ matrix elements and spectroscopic quadrupole moments of the $2_{1,2}^+$ states. While the experimental information on ^{96}Sr is too limited to perform a proper sum rule analysis, a good approximation of the $\langle Q^2 \rangle$ value for the ground state can be obtained from the $B(E2; 2_1^+ \rightarrow 0_1^+)$ value via (2.52). The similarity of thus obtained $\langle Q^2(0_1^+) \rangle$ value in ^{96}Sr with the $\langle Q^2(0_2^+) \rangle$ value in ^{98}Sr strongly supports the conclusion of Refs. [12, 56] about these two configurations interchanging in energy at $N = 60$. More-over, the $\langle Q^3 \cos 3\delta \rangle$ value for the ground state, obtained using the approx-imation of Ref. [41], i.e. from the $\langle 0_1^+||E2||2_1^+ \rangle \langle 2_1^+||E2||2_1^+ \rangle \langle 2_1^+||E2||0_1^+ \rangle$ and $\langle 0_1^+||E2||2_1^+ \rangle \langle 2_1^+||E2||2_2^+ \rangle \langle 2_2^+||E2||0_1^+ \rangle$ products, corresponds to $\gamma = 21(5)°$. The contribution from the latter product of $E2$ matrix elements is at the level of only a few percent of the total, and the obtained γ value is linked to the significant reduction of the $\langle 2_1^+||E2||2_1^+ \rangle$ matrix element in ^{98}Sr with respect to the rotational estimate (2.42). Interestingly, no such reduction of diagonal $E2$ matrix elements has been observed for higher-spin members of the ground-state band in ^{98}Sr.

2.5 Experimental Considerations

Thanks to advances in detector and accelerator technologies, goals and modalities of Coulomb-excitation experiments evolved with time. Early studies of the 1950s–1970s used light ion beams to populate excited states in target nuclei, and the excitation process could usually be well described using the first- and second-order perturbation theory. Normalisation of the measured excitation cross section to that of the Rutherford scattering was common. These measurements provided $B(E2; 2_1^+ \rightarrow 0_1^+)$ values in many stable nuclei, as well as the spectroscopic quadrupole moments of the first excited state. With availability of heavier ion beams and development of high-resolution γ-ray detection arrays, complex multi-step excitation studies became possible, which often relied on known lifetimes in the nuclei of interest to normalise the measured cross sections. These experiments were performed with a variety of particle detectors, often combining large angular coverage and high granularity. In order to extract electromagnetic matrix elements from extensive sets of Coulomb-excitation data, coupled-channel codes were developed, including the CLX/DCY and GOSIA codes discussed in Sect. 2.2.3. Finally, the advent of radioactive beam facilities opened the possibility to apply low-energy Coulomb excitation to short-lived exotic nuclei. This led to new challenges related to intensity and purity of available radioactive beams, as well as the need to deal with high background of radioactive decay of beam ions, which provided constraints for the particle detectors used in such studies. As complementary experimental data, such as excited-state lifetimes, were often lacking, normalisation of the measured cross sections to the excitation of target nuclei became common.

2.5.1 Normalisation of Excitation Cross Sections

In order to extract electromagnetic matrix elements from Coulomb-excitation data, the measured γ-ray intensities have to be converted to absolute excitation cross sections. While this procedure is performed internally by analysis codes such as GOSIA, it is important to understand its principle and implications. Possible normalisation techniques are discussed below.

2.5.1.1 Normalisation to Elastic Scattering

Normalisation to the measured Rutherford scattering cross section is the most direct option, although not widely used in typical experiments using particle-γ coincidences, as it requires precise knowledge of the angular range covered by the particle-detection system and its efficiency, as well as of the acquisition dead time. Examples of such studies [22–24] were already mentioned in Sect. 2.3.1. One should note that in these cases the uncertainties related to geometry were limited due to the detection of the recoiling target nuclei corresponding to high centre-of-mass scattering angles, where the Rutherford cross section changes more slowly.

A particular class of Coulomb-excitation measurements, where normalisation to the elastic scattering is always applied, are those that determine populations of Coulomb-excited states by means of particle spectroscopy. This technique was extensively used in early Coulomb-excitation studies with light beams (protons, α particles, ^{16}O, etc.) as such experiments offered certain advantages over those making use of γ-ray detection that at this time were usually performed with NaI(Tl) scintillators of limited energy resolution. Namely, the determination of the cross sections in this type of measurements does not require a knowledge of internal conversion coefficients and branching ratios and is simply obtained by comparing the integrals of the peaks corresponding to individual excited states to that of the elastic-scattering peak. Moreover, contrary to measurements with standard γ-ray detectors, in such studies beam intensities of a few hundred pnA can be used, which can compensate the strong reduction of excitation cross sections observed for states at high excitation energy. An example of an early Coulomb-excitation study performed with a magnetic spectrograph can be found in Ref. [57]. Interestingly, one of the very first low-energy Coulomb-excitation experiments involving short-lived unstable nuclei was also performed in this mode. In this measurement [58], ^{8}Li ($T_{1/2} = 0.8$ s) was produced in a ^{9}Be(^{7}Li,^{8}Li)^{8}Be transfer reaction and separated in a superconducting solenoid magnet. The resulting beam of 14.6-MeV ^{8}Li ions impinged on a ^{nat}Ni target. In the energy spectra of the scattered ^{8}Li projectiles, measured at several angles corresponding to the "safe" Coulomb-excitation process, in addition to an elastic-scattering peak a peak corresponding to the first excited state in ^{8}Li (1^{+}, E=0.908 MeV) was observed, and from the measured differential cross section, a $B(E2; 2^{+}_{gs} \rightarrow 1^{+})$ value was determined [58].

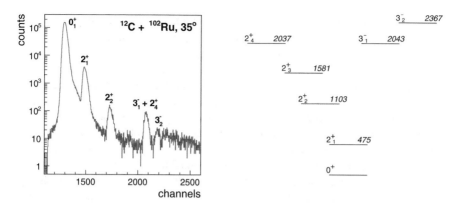

Fig. 2.5 *Left panel:* Spectrum of ^{12}C ions scattered at laboratory angle $\theta_{LAB} = 35°$ from a ^{102}Ru target. The peaks are labelled with the spins and parities of the corresponding states in ^{102}Ru. *Right panel:* Low-spin part of the ^{102}Ru level scheme presenting the states observed in the Coulomb-excitation experiment. Level energies are given in keV

To illustrate quality of the experimental data that can be obtained using this technique, Fig. 2.5 presents a particle spectrum from a recent measurement of ^{102}Ru, Coulomb-excited with a 53-MeV ^{12}C beam [59]. The ^{12}C ions scattered from the target were momentum analysed with a Q3D magnetic spectrograph, and the obtained energy resolution was close to 100 keV full width at half maximum (FWHM). The excitation probability of the 3_2^- state measured for ^{12}C ions scattered at 35° degrees is at the level of 0.1%, which is sufficient to properly fit the corresponding peak in the particle spectrum resulting from about 3.5 hours of data collection at $\approx$ 5 pnA intensity. The limitation of beam current in this particular measurement was due to ion-source problems, and under normal conditions it would increase by at least a factor of 20, which would yield similar quality data in about 10 minutes. For comparison, twice as large population of the 4_2^+ state in ^{42}Ca corresponded to about 600 counts per day in the γ-ray transition depopulating this state in the experiment of Refs. [29, 30].

Although largely superseded by measurements with γ-ray detectors, this technique still represents a very attractive option, especially to populate higher-lying low-spin states. On the other hand, because of a 20-fold (or more) decrease of energy resolution with respect to γ-ray spectroscopy, its applicability is limited to nuclei with low density of excited states. Moreover, target purity and quality are crucial, as elastic scattering on impurities may easily result in peaks in the particle spectra that are more intense than those related to inelastic scattering on the nucleus of interest. It should also be noted that in order to obtain good energy resolution in such experiments, very thin targets need to be used. The ^{102}Ru target used for the measurement presented in Fig. 2.5 was approximately 40 μg/cm^2 thick, much thinner than the 1–2 mg/cm^2 thickness often used for γ-ray detection.

2.5.1.2 Normalisation to Lifetimes in the Nucleus of Interest

In Coulomb-excitation experiments that involve population of multiple states, the measured γ-ray intensities can easily be converted to absolute excitation cross sections if one or more $B(E2)$ values connecting the ground state with the observed excited states are known. In even-even nuclei, an independent measurement of the 2_1^+ state lifetime, $\tau(2_1^+)$, is usually used for this purpose. In this case, from a comparison of experimental and calculated $2_1^+ \rightarrow 0_1^+$ transition intensities, one can obtain the normalisation factor, accounting for the integrated beam current, detection efficiency and solid angle covered by the particle detector. This factor can then be applied to the remaining γ-ray intensities. In the GOSIA code, this is achieved by fitting the normalisation constants C_m in (2.26) during the χ^2 minimisation procedure, and requires no additional calculations by the user. Precise information both on the lifetime of the state in question and its population in the Coulomb-excitation process are required. The latter depends on the level of statistics, precision of the efficiency calibration and branching ratio for the transition of interest, if a state different than the first excited one is used for normalisation. If in a GOSIA analysis multiple lifetimes of states with observed $E2$ decay to the ground state are declared, they will all influence the calculation of the C_m normalisation constants. The most important contribution to the χ^2 function will involve the state with the highest combined precision of γ-ray intensity and lifetime, i.e. typically the 2_1^+ state for even-even nuclei. In such case, even though the $\langle 2_1^+ \| E2 \| 0_1^+ \rangle$ matrix element formally enters the fit, it is bound to identically reproduce the lifetime value, and therefore no independent $B(E2; 2_1^+ \rightarrow 0_1^+)$ value is obtained from the analysis. Finally, one should note that in odd-mass and odd-odd nuclei the strongest observed γ rays may correspond to mixed $E2/M1$ transitions. If the corresponding mixing ratio is not precisely known, it may be better to use a less intense pure $E2$ transition for normalisation purposes, as demonstrated, e.g. for ^{97}Rb [11, 35].

2.5.1.3 Normalisation to Target Excitation

For many short-lived nuclei, especially on the neutron-rich side of the valley of stability, lifetimes of excited states are not known, and therefore a different solution for the normalisation of the measured Coulomb-excitation cross sections has to be adopted. One possibility is to use the number of elastically scattered beam particles, as described in Sect. 2.5.1.1. Alternatively, one can make use of the fact that the observed excitation of target nuclei can usually be described with high precision using literature values of relevant matrix elements, and the normalisation obtained in this way can be applied to γ-ray intensities measured for beam nuclei.

The observed number of γ rays emitted from an excited state in the target nucleus reads:

$$N_t = L \times \frac{\rho d N_A}{A_t} \times b_t \epsilon_\gamma (E_t) \epsilon_{\text{part}} \sigma_t , \qquad (2.54)$$

where σ_t is the cross section for exciting the state of interest, integrated over the angular range covered by the particle detector and the range of incident energies resulting from beam slowing down in the target, b_t is the total γ-ray branching ratio for a given transition, $\epsilon_\gamma(E_t)$ is the absolute efficiency for detecting a γ ray of energy E_t, ϵ_{part} is the intrinsic detection efficiency of the particle detector, ρd is the thickness of the target in mg/cm^2, N_A is Avogadro's number, A_t is the mass of the target in mg per mol and L is the beam current integrated over the measurement time.

A similar equation can be written for the number of γ rays in a transition observed for the projectile, assuming the same angular range for particle detection:

$$N_p = L \times \frac{\rho d N_A}{A_t} \times b_p \epsilon_\gamma(E_p) \epsilon_{part} \sigma_p . \tag{2.55}$$

If a ratio of (2.55) and (2.54) is taken, the dependence on particle-detection efficiency and integrated beam current cancels out:

$$\frac{N_p}{N_t} = \frac{b_p \epsilon_\gamma(E_p) \sigma_p}{b_t \epsilon_\gamma(E_t) \sigma_t} . \tag{2.56}$$

One should note that if beam or target is known to have contaminants, it is necessary to account for them when evaluating numbers of counts in the nuclei of interest. This is discussed in detail in, e.g. [35].

A version of the GOSIA code, GOSIA2, was developed to handle the simultaneous analysis of both target and projectile excitation. In this approach, the χ^2 functions for the target and projectile nuclei, defined by (2.26), are minimised in turns, with the normalisation factors C_m shared as parameters across both functions. Those for the target nucleus are obtained in the fit using literature values of relevant matrix elements and the γ-ray intensities of the target deexcitation, as explained in Sect. 2.5.1.2. The C_m factors obtained in this way are subsequently imposed in the fit performed for the projectile nuclei. The solution of the analysis is given by the global minimum of the total χ^2 function defined as the sum of χ^2 functions for both reaction partners. In common situations where only two matrix elements are used to describe the excitation of the nucleus under study, i.e. $\langle 2_1^+ \| E2 \| 0_1^+ \rangle$ and $\langle 2_1^+ \| E2 \| 2_1^+ \rangle$, a two-dimensional plot of the χ^2 surface may be used to determine their uncertainties, as described in Sect. 2.3.2.2.

2.5.2 Particle Detectors for Stable and Radioactive Beam Experiments

Particle detectors are used in almost all low-energy Coulomb-excitation studies, as they enable a clear selection of the scattering kinematics, which is necessary to properly describe the excitation process leading to the observed γ-ray decay. In radioactive beam experiments, the use of particle detection is mandatory to select Coulomb-excitation events from the background of γ rays emitted in radioactive decay of the beam ions and their daughters.

2.5.2.1 Experiments with No Particle Detection

In principle, stable beam Coulomb-excitation experiments do not require particle detectors, and experiments relying solely on γ-ray detection were relatively common in the 1970s–1990s. They are traditionally referred to as "thick-target" studies, as in order to avoid Doppler shift of γ-ray energies, which would significantly broaden the observed γ-ray peaks, they are performed with targets sufficiently thick to stop the recoiling nuclei. One should note here that the lifetimes of excited states under study should be long enough to ensure that their decay occurs when the recoil is at rest, which usually corresponds to lifetimes of at least a few picoseconds. The beam energy must be below "safe" energy for backscattering; otherwise the measured excitation cross sections may be affected by nuclear interaction. In order to extract electromagnetic matrix elements from the measured γ-ray yields, Coulomb-excitation cross sections need to be integrated over all possible scattering angles and velocities, which strongly limits experimental sensitivity to higher-order effects.

Measurements without particle detectors can also be performed in a strongly inverse kinematics (i.e. with the mass of the projectile exceeding by a factor of 5 or more that of the target nucleus), which results in beam particles being scattered in a narrow range of forward laboratory angles. Excellent examples of such measurements are studies of candidates for "mixed-symmetry" states[2] in the ^{128,130,132}Xe isotopes [63, 64]. In order to enable Doppler correction sufficient to separate numerous γ-ray transitions observed in these experiments, a target of ^{12}C was used. Under these conditions, the maximum laboratory scattering angle for ^{128}Xe nuclei was $\approx 5.3°$. The use of a light target favoured one-step excitation, making it the most suitable choice for the addressed physics case, i.e. search for enhanced $M1$ decays from the 2^+ states being a signature of their "mixed-symmetry" character. As an example, eight 2^+ states at energies ranging from 443 keV to 2718 keV have been observed in the experiment on ^{128}Xe [63].

While experiments without particle detection represent a valid option, they are definitely in the minority, as there are significant gains from the use of particle-γ coincidences. However, γ-ray singles data can be taken in parallel to coincidence data, providing a measurement under significantly different experimental conditions. If one complements a "thick-target" measurement performed in normal kinematics with a particle detector placed at backward angles, imposing particle-γ coincidence will lead to a selection of events both in terms of scattering angle and incident energy (even if the detector is not sensitive to energy, when the scattering occurs deep enough in the target, the backscattered beam ions will be absorbed in the target material before they reach the detector). Consequently, a combination of data

---

[2] "Mixed-symmetry" states are a special category of collective states predicted in models that treat the proton and neutron fluids separately. First predicted by Faessler [60], they were extensively studied within the framework of the proton-neutron interacting boson model (IBM-2) [61]. The terminology "mixed symmetry" arises from the properties of the wave functions, which contain at least one pair of proton and neutron bosons that is antisymmetric under the exchange of the proton and neutron labels. A detailed discussion of properties of such states can be found in Ref. [62].

collected in both trigger modes can lead to a better final sensitivity to second-order effects, such as quadrupole moments or relative signs of $E2$ matrix elements.

2.5.2.2 Requirements for Particle Detectors and Selected Examples

A variety of particle detectors are used for low-energy Coulomb-excitation studies. Important differences are usually observed between setups intended for stable and radioactive beam studies. As discussed in Sect. 2.3.2, multi-step excitation probability increases with the θ_{CM} angle, which leads to enhanced sensitivity to higher-order effects. As the total excitation cross section is given by the product of the excitation probability and the Rutherford cross section (2.9), due to the strong decrease of the latter with the scattering angle, measurements with weak exotic beams tend to be performed at relatively low θ_{CM} angles; otherwise the counting rates would be insufficient. In contrast, in stable beam studies, where typical beam intensities are of the order of particle nanoampers (i.e. 10^9 pps or more), detection of beam particles scattered in the backward hemisphere is usually a better option, especially if properties of higher-lying excited states are to be investigated. Moreover, particle detectors placed at forward scattering angles in typical stable beam measurements would be exposed to a high flux of scattered particles. This eliminates from this application detector types that are not sufficiently radiation hard under such conditions (e.g. silicon detectors), leaving as possible options those that are usually not energy-sensitive, such as gas-filled parallel-plate avalanche counters (PPAC) or solar cells.

As discussed in Sect. 2.3.2, measurements of differential cross sections can be used to disentangle various higher-order effects in Coulomb excitation. Therefore particle detectors with a broad angular coverage are preferred, with a granularity in θ that can be used to subdivide the collected data according to the scattering angle. Moreover, since γ rays from Coulomb-excited states are usually emitted from nuclei in flight, their energies need to be Doppler corrected. For this reason, segmentation in ϕ is also important. Since the energy resolution of Doppler-corrected spectra also depends on the geometry of the γ-ray detectors and energy loss in the target, a very fine segmentation in ϕ is rarely needed. For example, simulations of an annular particle detector in combination with the GALILEO array [65] showed that an increase of segmentation from $\Delta\phi = 51°$ to $\Delta\phi = 26°$ would improve the FWHM of Doppler-corrected 1332-keV γ rays emitted in flight by ^{60}Ni nuclei, scattered on a 1 mg/cm^2 ^{208}Pb target, from 15 keV to 11 keV, with 9 keV being the asymptotic limit [66]. However, segmentation in ϕ can also be beneficial for high-statistics measurements, as from the analysis of particle-γ correlations sensitivity to γ-ray multipolarity may be obtained, in particular to $E2/M1$ mixing ratios.

Since inelastic scattering is a two-body process, even when only one of the reaction partners is detected, the entire scattering kinematics can be reconstructed. For that, however, it is necessary to identify the detected nucleus, i.e. distinguish target recoils from scattered beam particles measured at the same θ_{LAB} angle. In normal kinematics ($A_{beam} < A_{target}$), only the latter are observed in the backward hemisphere, and therefore the particle detectors placed at these angles do not have to provide any additional information about the properties of the detected ion. This

Fig. 2.6 *Left panel:* SPIDER setup [66] installed in the GALILEO reaction chamber, with a LaBr$_3$ detector visible in the front (photo courtesy M. Rocchini). *Right panel:* "Munich chamber" [67] at HIL, Warsaw, equipped with 101 PIN diodes (photo courtesy K. Wrzosek-Lipska)

provides an important simplification of the experimental setup and data analysis, and one more reason why a vast majority of experiments using intense stable beams are performed in normal kinematics, with particle detectors in the backward hemisphere. Most commonly, segmented Si detectors are used in such studies. For example, the recently developed modular silicon array SPIDER [66] consists of seven trapezoidal Si detectors segmented into eight annular strips and arranged in a conical geometry (see the left panel of Fig. 2.6). For experiments with stable beams, SPIDER is installed in backward angles, covering θ_{LAB} angles between 124° and 161°. Granularity of the particle-detection system can also be obtained by using an array of small non-segmented detectors, such as PIN diodes or solar cells. An example of such array is the "Munich chamber" [67] at HIL, Warsaw, presented in the right panel of Fig. 2.6, which can accommodate up to 110 PIN diodes, each of 0.5 cm × 0.5 cm size. Its compact size (10 cm in diameter) makes it possible to place γ-ray detectors very close to the target, thus increasing the detection efficiency. Another interesting system for Coulomb excitation with stable beams was the LUNA setup [68] in JAERI, Tokai, which used 23 mm × 23 mm plastic and YAP:Ce scintillators combined with position-sensitive photomultipliers. The measured position resolution ranged from 0.5-mm FWHM at the centre of the detector to 1.2 mm at its edges, which translated into an excellent energy resolution: 5-keV FWHM was obtained after Doppler correction for the 596-keV line in ^{74}Ge, Coulomb-excited on an 1.7-mg/cm^2 ^{nat}Pb target [68]. A schematic plot of this setup, which was compact enough to fit in a reaction chamber of 11-cm diameter, is shown in Fig. 2.7. The YAP:Ce detectors placed at backward angles, covering θ_{LAB} angles between 106° and 153°, and the backward halves of the top and bottom plastic scintillators were used for the detection of scattered beam projectiles in normal-

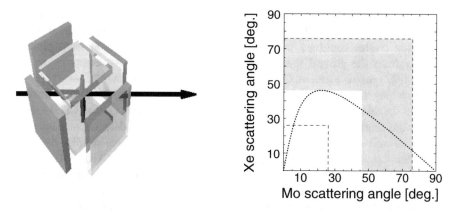

Fig. 2.7 *Left panel:* Schematic drawing of the LUNA setup [68] in JAERI, Tokai. Beam direction is marked with an arrow, and target position is indicated in the middle. The four detectors in the front, plotted as transparent, are plastic scintillators, and the remaining two are YAP:Ce. *Right panel:* Dotted line presents the correlation between scattering angles of ^{136}Xe and ^{98}Mo in an inverse-kinematics Coulomb-excitation experiment of ^{98}Mo [69] performed with the LUNA setup. The limits of the forward plastic detectors of LUNA are marked with a dashed line, and the shaded area denotes the part of the detector used for Coulomb-excitation analysis (see text for details)

kinematics studies. The forward part of the setup was used for inverse-kinematics experiments, as discussed later in this chapter.

In the forward hemisphere, both scattered beam projectiles and recoiling target nuclei can be observed. Moreover, in inverse kinematics ($A_{beam} > A_{target}$), two kinematic solutions are possible, and thus at a certain θ_{LAB} angle one may observe, in addition to recoiling target nuclei, two groups of scattered beam particles differing in energy, corresponding to $\theta_{CM} < 90°$ and $\theta_{CM} > 90°$, respectively. This is illustrated by the example of ^{136}Xe scattered on ^{98}Mo, presented in the right panel of Fig. 2.7: for each θ_{LAB} angle below $46°$, one can observe scattered beam nuclei corresponding to two different recoil scattering angles. Consequently, particle detectors placed in the forward hemisphere need to provide information on the energy of the ion or, alternatively, a precise timing signal that can be used to determine the difference in time of flight between the two collision partners, resulting in their unambiguous identification. Typical setups for radioactive beam studies involve a segmented Si detector at forward angles. Standard choices include annular Micron S2 (48 rings, 16 sectors) and S3 models (24 rings, 32 sectors) as well as CD composed of four independent quadrants (24 rings, 96 sectors in total) [70]. The latter is extensively used in measurements at ISOLDE, and in the standard configuration 25 mm downstream from the target, it covers $16° < \theta_{LAB} < 53°$, which for typical normal-kinematics Coulomb-excitation experiments ensures that the θ_{CM} ranges corresponding to scattered beam and recoil detection overlap and together cover angles up to over $120°$ in the centre-of-mass frame. In measurements with intense exotic beams, an extension of such systems to higher θ_{CM} angles may be beneficial. For example, in a Coulomb-excitation study of ^{72}Zn performed at

ISOLDE [73, 74], it was possible to observe an effect of the two-step excitation on the population of the 2_2^+ state for $\theta_{CM} > 140°$, while the excitation at lower scattering angles could be well reproduced assuming only a one-step process directly from the ground state. This resulted in experimental sensitivity to the $\langle 2_2^+ \| E2 \| 2_1^+ \rangle$ matrix element. Combining this result with the known $2_2^+ \to 2_1^+ / 2_2^+ \to 0_1^+$ branching ratio made it possible to determine also the $\langle 2_2^+ \| M1 \| 2_1^+ \rangle$ matrix element. In this experiment, the T-REX setup was used [75], i.e. the CD detector at forward angles was complemented by another identical detector at backward angles, as well as an array of eight rectangular $\Delta E - E$ telescopes installed between them. Similar solutions are used in the BAMBINO setup [71] at TRIUMF and the particle-detection part of the JANUS array [72] at the ReA3 reaccelerator facility, NSCL, which both consist of two Micron S3 detectors located 30 mm up- and downstream from the target position. One should note, however, that in order to obtain meaningful information from measurements performed at high θ_{CM} angles, intense beams need to be used in order to compensate for the reduction of the cross sections. For example, the study of ^{72}Zn was performed at $3\ 10^7$ pps beam intensity.

CHICO2 [48] and its predecessor CHICO [76] use a different principle to distinguish between scattered beam projectiles and recoiling target nuclei, which is based on precise timing information rather than energy measurement. The array consists of two hemispheres holding ten PPACs each, with active area covering 69% of 4π and an outstanding position resolution of $1.6°$ in θ and $2.5°$ in ϕ. A valid Coulomb-excitation event requires the detection of both collision partners in coincidence, and their identification is based both on kinematic considerations and the measured time-of-flight difference between them. A drawback of this solution is that the time resolution of CHICO PPACs (1.2 ns) implies a considerable base for the time-of-flight measurement. In the adopted geometry, the shortest flight path between the target and the detector is about 13 cm, which makes this device rather bulky (36-cm chamber diameter). Contrary to silicon detectors, CHICO2 can be operated at high counting rates (>500 kHz) without radiation damage and therefore can be used both with stable and radioactive beams. In the latter case, its mass resolution capabilities were crucial to reject events resulting from scattering of intense stable contaminants present in beams delivered by the CARIBU facility [77]. For example, in a study of ^{110}Ru Coulomb-excited on ^{208}Pb [78], events involving a ^{131}Xe beam contaminant were clearly identified.

More unusual solutions used for particle detection at forward angles involve, e.g. placing absorbers in front of detectors to shield them selectively from one of the collision partners. This has been successfully implemented in experiments performed with the JUROGAM array [79] at JYFL, Jyväskylä, where Coulomb-excitation cross sections for ^{128,130}Xe scattered on a ^{nat}Fe target were measured in parallel to a RDDS lifetime measurement [80, 81]. In these studies, an array of solar cells covering θ_{LAB} angles between $6°$ and $38°$ was used, shielded with a 20-mg/cm^2 Au foil which was thick enough to stop Xe ions while letting Fe recoils through. Coulomb-excitation experiments performed at IUAC, New Delhi, which use a position-sensitive PPAC covering an angular range of $15° < \theta_{LAB} < 45°$, are based on a different concept. The positions of γ-ray detectors are selected in such a way that even though it is not possible to distinguish scattered projectiles and target recoils using information from the PPAC, in the γ-ray spectra the peaks corresponding to the two types of particles are clearly separated, see e.g. [25]. Finally, one can make use of the fact that in inverse-kinematics Coulomb excitation scattered projectiles are forward focused, and there exists always a maximum θ_{LAB} angle, over which only target recoils are observed. This was, for example, used in the measurements with the LUNA system [68]. For example, in a study of ^{98}Mo Coulomb-excited by a ^{136}Xe beam [69], imposing a cut of $\theta_{LAB} > 46°$, as marked by the shaded area in the right panel of Fig. 2.7, ensured that only ^{98}Mo target recoils were detected.

2.5.3   Beam and Target Requirements

When selecting the beam-target combination for a Coulomb-excitation study, several points need to be taken into consideration. As discussed in Sect. 2.3, atomic number of the collision partner will strongly influence the cross sections and the excitation pattern. In case of high-precision studies of low-spin excited states, it may be beneficial to limit the influence of higher-lying states on the measured excitation cross section by selecting a light collision partner. Sometimes a series of experiments using a variety of beam-target combinations are necessary to disentangle the role of various excitation paths. Evidently, the energies of γ-ray transitions that are expected to be observed in the target and in the projectile should not overlap, and if it is planned to use target excitation for normalisation, as explained in Sect. 2.5.1.3, the relevant matrix elements, including the spectroscopic quadrupole moment of at least the first excited state, need to be known to high precision.

Moreover, unambiguous identification of collision partners is required, and the mass resolution capabilities of particle detectors used in Coulomb-excitation studies may result in important limitations. As straggling in the target leads to worsening of the energy resolution after Doppler correction, usually targets of 1–2 mg/cm^2 thickness are used, and with the exception of "thick-target" studies introduced in Sect. 2.5.2.1, target thicknesses rarely exceed 4 mg/cm^2. Moreover, integration of the cross sections over a large range of incident energies due to the beam slowing down in a thick target may decrease the sensitivity to higher-order effects, and knowledge of stopping powers is mandatory to properly account for it. In typical experiments using energies measured in silicon detectors to distinguish between scattered beam particles and recoiling target nuclei, the use of targets few mg/cm^2 thick implies a considerable difference between the masses of the collision partners, i.e. almost a factor of 2. This strongly limits the mass of the target nucleus, if heavy projectiles are used. For example, ^{120}Sn was the heaviest target used for recent Coulomb-excitation studies of ^{222,228}Ra at HIE-ISOLDE [82] and $^{182-188}$Hg at ISOLDE [83]. One should note that the CHICO/CHICO2 device provides a much better mass resolution, as discussed in Sect. 2.5.2.

Finally, some limitations are intrinsically linked to the use of radioactive ion beams. In many cases, pure beams are not achievable, and excitation of beam contaminants is also observed. The beam composition needs to be monitored during the experiment and properly accounted for in the data analysis. Moreover, limitations on the half-lives of nuclei that can be delivered as a post-accelerated ISOL beam are due to the trapping and charge-breeding times of typically at least 100 ms that are involved in this technique. One of the most short-lived nuclei studied with low-energy Coulomb excitation was ^{99}Rb with a half-life of 54(4) ms. Even with the shortest trapping and breeding times possible, it was still observed that about 60% of ^{99}Rb ions decayed to ^{99}Sr and ^{99}Y on their way from the primary target to the Coulomb-excitation setup [11]. The same experiment can be used as an example of a successful low-energy Coulomb-excitation study performed at a very low beam intensity of a few thousands of particles per second. One should note, however, that ^{99}Rb is very collective and that some model assumptions were necessary in the analysis, as outlined in Refs. [11, 35]. In general, a rule of thumb is that one needs at least 10^4 pps for integral cross-section measurements and 10^5 pps for differential studies.

2.6   Summary and Outlook

Low-energy Coulomb excitation has been used for over 60 years to determine transition probabilities and spectroscopic quadrupole moments of short-lived excited states in a wide range of nuclei. Over the years, both the theoretical description of the process and the data analysis techniques have been refined, and currently it is possible to perform detailed studies that firmly establish nuclear shapes through the use of quadrupole sum rules. An exciting new era opened recently with the availability of radioactive ion beams, which made it possible to apply this powerful technique to nuclei far from stability. New challenges emerged, related, among others, to a lack of precise spectroscopic information on the nuclei under study that can provide crucial inputs and stringent constraints to analyses of Coulomb-excitation data. For example, information on lifetimes of excited states greatly increases sensitivity of Coulomb-excitation data to spectroscopic quadrupole moments and relative signs of electromagnetic matrix elements. High-precision β-decay measurements are able to bring information on very weak decay branches, which can be important in terms of transition probabilities. In particular for heavy nuclei, significant $E0$ branches for $I \rightarrow I$ transitions can appear that strongly influence the observed decay pattern and should be a subject of complementary conversion-electron studies. As state-of-the-art theoretical calculations are now able to predict properties of nuclei with an unprecedented level of detail, it is crucial to confront them with high-precision experimental data. It can be anticipated that low-energy Coulomb excitation will play a prominent role in future nuclear-structure studies as a part of a combined approach involving a multitude of complementary techniques.

Acknowledgments I would like to thank all those with whom I collaborated on topics related to this chapter, in particular (alphabetically) P.A. Butler, E. Clément, D.T. Doherty,

L.P. Gaffney, P.E. Garrett, A. Görgen, K. Hadyńska-Klęk, M. Komorowska, W. Korten, A. Nannini, P.J. Napiorkowski, M. Rocchini, J. Srebrny, P. Van Duppen, N. Warr and K. Wrzosek-Lipska.

References

1. T. Glasmacher, Lect. Notes Phys. **764**, 27 (2009)
2. D. Cline, Annu. Rev. Nucl. Part. Sci. **36**, 683 (1986)
3. M.P. Fewell, D.C. Kean, R.H. Spear, A.M. Baxter, J. Phys. G: Nucl. Phys. **3**, L27 (1977)
4. M.T. Esat, D.C. Kean, R.H. Spear, M.P. Fewell, A.M. Baxter, Phys. Lett. **72B**, 49 (1977)
5. M.T. Esat, M.P. Fewell, R.H. Spear, T.H. Zabel, A.M. Baxter, S. Hinds, Nucl. Phys. **A362**, 227 (1981)
6. R.H. Spear, T.H. Zabel, M.T. Esat, A.M. Baxter, S. Hinds, Nucl. Phys. **A378**, 559 (1982)
7. W.J. Vermeer, M.T. Esat, R.H. Spear, Nucl. Phys. **A389**, 185 (1982)
8. W.J. Vermeer, M.T. Esat, J.A. Kuehner, R.H. Spear, A.M. Baster, S. Hinds, Phys. Lett. **122B**, 23 (1983)
9. K. Kumar, Phys. Rev. Lett. **28**, 249 (1972)
10. J. Van de Walle, F. Aksouh, F. Ames et al., Phys. Rev. Lett. **99**, 142501 (2007)
11. C. Sotty, M. Zielińska, G. Georgiev et al., Phys. Rev. Lett. **115**, 172501 (2015)
12. E. Clément, M. Zielińska, S. Péru et al., Phys. Rev. C **94**, 054326 (2016)
13. A. Görgen, W. Korten, J. Phys. G: Nucl. Part. Phys. **43**, 024002 (2016)
14. K. Alder, A. Winther, *Electromagnetic Excitation* (North-Holland, Amsterdam, 1975)
15. H. Frauenfelder, R. Steffen, in *Alpha-, Beta and Gamma Spectroscopy*, ed. by K. Siegbahn (North-Holland, Amsterdam, 1965)
16. F. Bosch, H. Spehl, Z. Phys. **A280**, 329 (1977)
17. R. Brenn, H. Spehl, A. Weckherlin, H. Doubt, G. Van Middelkoop, Z. Phys. **A281**, 219 (1977)
18. K. Krane, Nucl. Instr. Meth. **98**, 205 (1972)
19. H. Ower, *The Coulex code CLX/DCY* (unpublished)
20. T. Czosnyka, D. Cline, C.Y. Wu, Bull. Am. Phys. Soc. **28**, 745 (1983). GOSIA User's Manual. http://slcj.uw.edu.pl/en/gosia-code/
21. A. Winther, J. de Boer, A computer program for multiple coulomb excitation, in *Coulomb Excitation* ed. by K. Alder, A. Winther (Academic Press, New York/London, 1966)
22. D.C. Radford, C. Baktash, J.R. Beene et al., Nucl. Phys. **A746**, 83c (2004)
23. J.M. Allmond, D.C. Radford, C. Baktash et al., Phys. Rev. C **84**, 061303(R) (2012)
24. J.M. Allmond, A.E. Stuchbery, A. Galindo-Uribarri et al., Phys. Rev. C **92**, 041303(R) (2015)
25. R. Kumar, M. Saxena, P. Doornenbal et al., Phys. Rev. C **96**, 054318 (2017)
26. R. Kumar, P. Doornenbal, A. Jhingan et al., Phys. Rev. C **81**, 024306 (2010)
27. D. Rosiak, M. Seidlitz, P. Reiter et al., Phys. Rev. Lett. **121**, 252501 (2018)
28. A. Illana, M. Zielińska et al., submitted to Phys. Rev. C
29. K. Hadyńska-Klęk, P.J. Napiorkowski, M. Zielińska et al., Phys. Rev. Lett. **117**, 062501 (2016)
30. K. Hadyńska-Klęk, P.J. Napiorkowski, M. Zielińska et al., Phys. Rev. C **97**, 024326 (2018)
31. S. Akkoyun, A. Algora, B. Alikhani et al., Nucl. Instr. Meth. Phys. Res. **A668**, 26 (2012)
32. M. Zielińska, K. Hadyńska-Klęk, EPJ Web Conf. **178**, 02014 (2018)
33. N. Amzal, P.A. Butler, D. Hawcroft et al., Nucl. Phys. **A734**, 465 (2004)
34. J. de Boer, J. Eichler, *Advances in Nuclear Physics* (Plenum, New York, 1968)
35. M. Zielińska, L. Gaffney, K. Wrzosek-Lipska et al., Eur. Phys. J. **A 52**, 99 (2016)
36. A.M. Hurst, P.A. Butler, D.G. Jenkins et al., Phys. Rev. Lett. **98**, 072501 (2007)
37. J. Heese, K.P. Lieb, L. Lühmann et al., Z. Phys. **A325**, 45 (1986)
38. J. Ljungvall, A. Görgen, M. Girod et al., Phys. Rev. Lett. **100**, 102502 (2008)
39. K. Wrzosek-Lipska, L. Próchniak, M. Zielińska et al., Phys. Rev. C **86**, 064305 (2012)
40. J. Srebrny, D. Cline, Int. J. Mod. Phys. E 20, 422 (2011)
41. W. Andrejtscheff, P. Petkov, Phys. Lett. **B 329**, 1 (1994)
42. M. Rocchini, K. Hadyńska-Klęk, A. Nannini et al., Phys. Rev. C **103**, 014311 (2021)

43. J. Henderson, Phys. Rev. C **102**, 054306 (2020)
44. T. Schmidt, K.L.G. Heyde, A. Blazhev, J. Jolie, Phys. Rev. C **94**, 014302 (2017)
45. A. Poves, F. Nowacki, Y. Alhassid, Phys. Rev. C **101**, 054307 (2020)
46. C.Y. Wu, D. Cline, T. Czosnyka et al., Nucl. Phys. **A 607** 178 (1996)
47. S. Paschalis, I.Y. Lee, A.O. Macchiavelli et al., Nucl. Instrum. Methods Phys. Res. **A 709**, 44 (2013)
48. C.Y. Wu, D. Cline, A. Hayes et al., Nucl. Instrum. Methods Phys. Res. **A 814**, 6 (2016)
49. A.D. Ayangeakaa, R.V.F. Janssens, C.Y. Wu et al., Phys. Lett. **B 754**, 254 (2016)
50. A.D. Ayangeakaa, R.V.F. Janssens, S. Zhu et al., Phys. Rev. Lett. **123**, 102501 (2019)
51. Y. Toh, C.J. Chiara, E.A. McCutchan et al., Phys. Rev. C **87**, 041304(R) (2013)
52. E. Clément, A. Görgen, W. Korten et al., Phys. Rev. C **75**, 054313 (2007)
53. M. Bender, P. Bonche, P.-H. Heenen, Phys. Rev. C **74**, 024312 (2006)
54. M. Girod, J.-P. Delaroche, A. Görgen, A. Obertelli, Phys. Lett. **B 676**, 39 (2009)
55. T.R. Rodriguez, Phys. Rev. C **90**, 034306 (2014)
56. E. Clément, M. Zielińska, W. Korten et al., Phys. Rev. Lett. **116**, 022701 (2016)
57. B. Elbek, C.K. Bockelman, Phys. Rev. **105**, 657 (1957)
58. J.A. Brown, F.D. Becchetti, J.W. Jänecke et al., Phys. Rev. Lett. **66**, 19 (1991)
59. P.E. Garrett, M. Zielińska, A. Bergmaier et al., submitted to Phys. Rev. C.
60. A. Faessler, Nucl. Phys. **A 85**, 653 (1966)
61. A. Arima, T. Otsuka, F. Iachello, I. Talmi, Phys. Lett. **B 66**, 205 (1977)
62. N. Pietralla, P. von Brentano, A.F. Lisetskiy, Prog. Part. Nucl. Phys. **60**, 225 (2008)
63. L. Coquard, N. Pietralla, T. Ahn et al., Phys. Rev. C **80**, 061304(R) (2009)
64. L. Coquard, N. Pietralla, G. Rainovski et al., Phys. Rev. C **82**, 0234317 (2010)
65. A. Goasduff, D. Mengoni, F. Recchi et al., Nucl. Instrum. Methods Phys. Res. **A 1015**, 165753 (2021)
66. M. Rocchini, K. Hadyńska-Klęk, A. Nannini et al., Nucl. Instrum. Methods Phys. Res. **A 971**, 164030 (2020)
67. J. Mierzejewski, J. Srebrny, H. Mierzejewski et al., Nucl. Instrum. Methods Phys. Res. **A 659**, 84 (2011)
68. Y. Toh, M. Oshima, T. Hayakawa et al., Rev. Sci. Instrum. **73**, 47 (2002)
69. M. Zielińska, T. Czosnyka, J. Choiński et al., Nucl. Phys. **A 712**, 3 (2002)
70. A.N. Ostrowski, S. Cherubini, T. Davinson et al., Nucl. Instrum. Methods Phys. Res. **A 480**, 448 (2002)
71. A.M. Hurst, C.Y. Wu, J.A. Becker et al., Phys. Lett. **B 674**, 168 (2009)
72. E. Lunderberg, J. Belarge, P.C. Bender et al., Nucl. Instrum. Methods Phys. Res. **A 885**, 30 (2018)
73. S. Hellgartner, D. Mücher, K. Wimmer et al., submitted to Phys. Lett. B.
74. S. Hellgartner, Ph.D. Thesis, Technische Universität München, 2015
75. V. Bildstein, R. Gernhäuser, T. Kröll et al., Eur. Phys. J. **A 48**, 85 (2012)
76. M.W. Simon, D. Cline, C.Y. Wu et al., Nucl. Instrum. Methods Phys. Res. **A 452**, 205 (2000)
77. R.C. Pardo, G. Savard, R.V.F. Janssens, Nucl. Phys. News **26**, 5 (2016)
78. D.T. Doherty, J.M. Allmond, R.V.F. Janssens et al., Phys. Lett. **B 766**, 334 (2017)
79. J. Pakarinen, J. Ojala, P. Ruotsalainen et al., Eur. Phys. J. **A 56**, 149 (2020)
80. W. Rother, A. Dewald, G. Pascovici et al., Nucl. Instrum. Methods Phys. Res. **A 654**, 196 (2011)
81. T. Konstantinopoulos, M. Axiotis, A. Lagoyannis et al., HNPS Adv. Nucl. Phys. **19**, 33 (2020)
82. P.A. Butler, L.P. Gaffney, P. Spagnoletti et al., Phys. Rev. Lett. **124**, 042503 (2020)
83. N. Bree, K. Wrzosek-Lipska, A. Petts et al., Phys. Rev. Lett. **112**, 162701 (2014)

Ab Initio Approaches to Nuclear Structure

3

Robert Roth

Abstract

Ab initio nuclear structure theory has experienced a phase of ground-breaking developments over the past decade. Compared to the situation in the early 2000s, we now have a rich variety of powerful and complementary tools that connect the underlying theory of the strong interaction to nuclear structure observables. This enables us to describe a much larger domain of nuclei and observables with controlled and quantified theoretical uncertainties—in the ab initio spirit. In this lecture we provide a pedagogical introduction into the ab initio toolbox with a focus on basis expansion approaches, particularly on configuration interaction methods, like the no-core shell model, and decoupling approaches, like the in-medium similarity renormalization group.

3.1   Introduction

The landscape of methods that define the state of the art in ab initio nuclear structure theory has been completely transformed since the early 2000s. New and innovative many-body schemes have been developed that radically expand the boundaries of what is possible computationally. At the same time, the connection to the underlying theory of the strong interaction has been strengthened through the use of effective field theories for the construction of nuclear interactions.

In this lecture we provide an introduction to ab initio nuclear structure theory with a focus on basis expansion methods. This is really meant to be a lecture and not

R. Roth (✉)
Institut für Kernphysik, Technische Universität Darmstadt, Darmstadt, Germany

Helmholtz Forschungsakademie Hessen für FAIR, Darmstadt, Germany
e-mail: robert.roth@physik.tu-darmstadt.de

S. M. Lenzi, D. Cortina-Gil (eds.), *The Euroschool on Exotic Beams, Vol. VI*,
Lecture Notes in Physics 1005, https://doi.org/10.1007/978-3-031-10751-1_3

a review article. We will present the material in a pedagogical manner, focusing on systematics, clarity, and consistency. We will explore the basic theoretical concepts and their interrelations and not so much the multitude of applications and results. We will not attempt to cover the field of ab initio theory completely—given the wealth of recent developments, this would fill a complete volume of this lecture series.

3.2   The Big Picture

We start by formulating the general nuclear structure problem as a quantum many-body problem. Already the first steps in this formulation imply specific assumptions, e.g., on the effective degrees of freedom for the theoretical description. This, in turn, has consequences for the formulation of the relevant interactions and, thus, the effective Hamiltonian governing the structure and dynamics of the system. Finally, the quantum many-body problem has to be solved, which in our case amounts to the solution of a many-particle Schrödinger equation. In this section, we go through these steps and establish the basic language and notation. We will also give an overview of the different classes of approaches to the many-body problem and address the meaning of the term "ab initio."

Constituents We consider nuclei as quantum multi-particle systems, composed of nucleons as effective degrees of freedom. It is understood that nucleons themselves have a complicated substructure, being bound states of quarks governed by the complicated quark-gluon interactions of quantum chromodynamics (QCD). However, we do not wish to resolve this underlying layer of microphysics.

This seemingly obvious choice (from the perspective of low-energy nuclear structure physics) has profound consequences. From the beginning we decide to work with an *effective theory* that has a limited range of validity resulting from the choice of the effective degrees of freedom. By construction the nucleons in our effective theory are point-like particles; they are inert and have no internal structure. This is obviously not the full truth—nucleons are extended objects with a typical root-mean-square radius of their charge distribution (for the proton) of around 0.8 fm, and they have internal excitations, e.g., the Δ resonances at about 300 MeV of excitation energy. For the effective theory of point-nucleons, this implies (i) a limited range of applicability and (ii) the need to account for corrections to the relevant observables.

In a proper effective theory, these points should follow in a systematic and transparent fashion from the formulation of the theory. The range of validity should be clearly defined from the outset, and the theory should provide a consistent way to construct the corrections to observables resulting from the unresolved physics. One of the significant advances in nuclear structure theory over the past decade addresses exactly this point—with the advent of chiral effective field theory, the step from QCD to the world of point-nucleons has become a well-defined procedure. Nuclear structure calculations have matured from a model to an effective theory.

Interactions The simplifications with respect to the degrees of freedom necessarily entail complications with respect to the interactions among them. The effective interaction of point-nucleons also has to encapsulate the complicated quark-gluon dynamics at the level of QCD that is not resolved in our description. The dynamics of quarks and gluons creates the net interaction that two or more nucleons experience, and this net effect has to be mimicked by the effective interaction among point-nucleons in the effective theory.

This situation is not unique to nuclear structure physics. We find a similar scenario in molecular physics when describing a system of atoms. Consider, e.g., a system of two or more neutral ^{4}He atoms. A simple effective theory could assume point-like ^{4}He particles, not resolving the complicated internal structure of the atoms. The effective interaction between the point-atoms has to capture all the underlying dynamics of the atoms; this is the famous van der Waals interaction in atomic and molecular physics. The mechanism behind these forces can be understood intuitively. At large distances the electrically neutral atoms do not affect each other, and there is no interaction. Only at short distances, the mutual polarization of their electron distributions induces a net interaction. A similar mechanism is at work when color-neutral baryons come close enough so that their quark-gluon distributions overlap.

In atomic physics it is possible to compute the residual interaction among the atoms from solving the Schrödinger equation for the multi-electron two-atom problem. In principle this can also be done in QCD, specifically in lattice QCD simulations. Research along these lines is on the way [1–3] and has shown how difficult this problem is. For the time being, we have to resort to effective field theories based on QCD for a description of the nuclear interaction [4]. Chiral effective field theories (EFT) have become a foundation of modern nuclear structure theory.

Many-Body Problem Having defined the constituents and their interactions, we are now in the position to formulate the basic equation that governs the structure and dynamics of the many-body system, the Schrödinger equation. At this point we restrict ourselves to a non-relativistic description of the many-body problem—relativistic effects might enter as specific corrections in the Hamiltonian, but we will not attempt a fully relativistic treatment.

In most cases we are interested in stationary properties of nuclei, and we do not need to address the explicit time evolution of the many-body problem. Thus, the central equation we have to deal with is the stationary Schrödinger equation, in other words, the eigenvalue problem of the Hamiltonian:

$$\hat{H} |\Psi_n\rangle = E_n |\Psi_n\rangle , \tag{3.1}$$

where $\hat{H}$ is the Hamiltonian of the A-nucleon system, E_n are the energy eigenvalues, and $|\Psi_n\rangle$ are the corresponding eigenvectors. We will use the representation-independent Dirac notation throughout this lecture. The discrete index $n = 0, 1, 2, \ldots$ already implies that we are concerned with the discrete part of the spectrum of the Hamiltonian, i.e., the bound states of the nucleus. We will not discuss the continuous part of the spectrum, i.e., the domain of nuclear scattering, reactions, and resonances in this lecture.

For a more detailed look at the eigenvalue problem, we first have to define the Hilbert space in which we are working. We consider a system of A indistinguishable nucleons with spin and isospin degrees of freedom. A typical basis for the description of the single-nucleon degrees of freedom in a finite nucleus consists of a spatial part, which encodes position and momentum information, a spin part, and an isospin part. For the spin part, we simply use the eigenstates of the single-particle spin operators $\hat{s}$ and $\hat{s}_z$ with quantum numbers $s = \frac{1}{2}$ and $m_s = \pm\frac{1}{2}$. Analogously for the isospin, we use eigenstates of $\hat{t}$ and $\hat{t}_3$ with quantum numbers $t = \frac{1}{2}$ and $m_t = \pm\frac{1}{2}$. For this lecture we will limit ourselves to spherical basis sets $|nlm_l\rangle$ in the spatial degrees of freedom, labeled by a generic radial quantum number n and orbital angular momentum quantum numbers l and m_l. We will couple orbital angular momentum and spin to obtain total angular momentum quantum numbers j and m. The coupled single-particle states thus read:

$$|p\rangle = |nljmm_t\rangle = |n(l\tfrac{1}{2})jm\rangle \otimes |\tfrac{1}{2}m_t\rangle . \qquad (3.2)$$

We will use the collective indices $p, q, \ldots$ to label single-particle basis states throughout this lecture.

When proceeding to the many-body system, we have to take the permutation antisymmetry of the states for a system of identical fermions into account. The simplest way to construct a basis of the A-body antisymmetric Hilbert space $\mathcal{H}_A$ uses antisymmetrized product states—so-called Slater determinants. Starting from a complete single-particle basis $\{|p\rangle\}$, we select A different single-particle states $|p_1\rangle, \ldots, |p_A\rangle$ and construct product states for all possible permutations of the single-particle indices. Summing over all these permutations with appropriate signs defines an *antisymmetrized product state* or *Slater determinant*:

$$|p_1 \ldots p_A\rangle = \frac{1}{\sqrt{A!}} \sum_{\pi} \mathrm{sgn}(\pi) \hat{P}_{\pi} |p_1\rangle \otimes \ldots \otimes |p_A\rangle . \qquad (3.3)$$

Here, $\hat{P}_{\pi}$ is the permutation operator, which rearranges the single-particle indices according to the permutation π, and $\mathrm{sgn}(\pi)$ indicates the signum or parity of the permutation π. The prefactor is chosen such that the many-body states are normalized provided that the single-particle states are normalized. Note that our notation does not explicitly indicate the antisymmetric character of the A-body states $|p_1 \ldots p_A\rangle$—antisymmetry is always implied, and we will never go back to simple product states.

The set of all antisymmetrized product states generated from an orthonormal single-particle basis automatically provides an orthonormal basis of the antisymmetric A-nucleon Hilbert space $\mathcal{H}_A$. This is very convenient.

Another very convenient aspect of this basis is the formalism of second quantization. We can define creation operators $\hat{a}_p^\dagger$ and annihilation operators $\hat{a}_p$ that add or remove particles to or from a given Slater determinant $|p_1...p_A\rangle$, automatically yielding a normalized antisymmetrized $(A+1)$ or $(A-1)$-particle state, respectively. We can even construct a complete A-body Slater determinant starting from the zero-body vacuum state $|0\rangle$ through the application of a chain of creation operators:

$$|p_1...p_A\rangle = \hat{a}_{p_1}^\dagger \cdots \hat{a}_{p_A}^\dagger |0\rangle . \tag{3.4}$$

The complications of antisymmetry are now hidden in the anti-commutation relations of fermionic creation and annihilation operators. Finally, creation and annihilation operators can be used to represent any operator, e.g., the components of the Hamiltonian, in an elegant way—we will make heavy use of this later on.

Basis Expansion, Truncation, and Convergence Why did we go through these basic elements from many-body quantum mechanics? Well, they prompt a simple and powerful strategy for the solution of the many-body Schrödinger equation. This strategy can be summarized under the label *basis expansion* and is at the heart of all methods discussed in this lecture.

Assume we have constructed an orthonormal basis $|\Phi_\nu\rangle$ of the A-nucleon Hilbert space $\mathcal{H}_A$, e.g., the antisymmetrized product states $|\Phi_\nu\rangle = |\{p_1...p_A\}_\nu\rangle$ discussed before. We can immediately use this basis to transfer the abstract, representation-independent eigenvalue problem (3.1) into a specific representation in the $|\Phi_\nu\rangle$ basis. We can expand the eigenstates $|\Psi_n\rangle$ in this basis:

$$|\Psi_n\rangle = \sum_\nu C_\nu^{(n)} |\Phi_\nu\rangle \tag{3.5}$$

with expansion coefficients $C_\nu^{(n)}$. Furthermore, we can multiply Eq. (3.1) from the left with all possible basis vectors $\langle\Phi_\nu|$ and insert the above expansion of the eigenstates to obtain a coupled system of algebraic equations:

$$\sum_{\nu'} \langle\Phi_\nu| \hat{H} |\Phi_{\nu'}\rangle \, C_{\nu'}^{(n)} = E_n \, C_\nu^{(n)} \qquad \forall \nu , \tag{3.6}$$

which can be conveniently cast into a matrix equation:

$$\begin{pmatrix} \langle\Phi_1| \hat{H} |\Phi_1\rangle & \langle\Phi_1| \hat{H} |\Phi_2\rangle & \langle\Phi_1| \hat{H} |\Phi_3\rangle & \cdots \\ \langle\Phi_2| \hat{H} |\Phi_1\rangle & \langle\Phi_2| \hat{H} |\Phi_2\rangle & \langle\Phi_2| \hat{H} |\Phi_3\rangle & \cdots \\ \langle\Phi_3| \hat{H} |\Phi_1\rangle & \langle\Phi_3| \hat{H} |\Phi_2\rangle & \langle\Phi_3| \hat{H} |\Phi_3\rangle & \cdots \\ \vdots & \vdots & \vdots & \ddots \end{pmatrix} \cdot \begin{pmatrix} C_1^{(n)} \\ C_2^{(n)} \\ C_3^{(n)} \\ \vdots \end{pmatrix} = E_n \begin{pmatrix} C_1^{(n)} \\ C_2^{(n)} \\ C_3^{(n)} \\ \vdots \end{pmatrix} . \tag{3.7}$$

So the Schrödinger equation or the abstract eigenvalue problem of the Hamiltonian translates naturally into a standard matrix eigenvalue problem. The input for this are the many-body matrix elements of the Hamiltonian $\langle \Phi_\nu | \hat{H} | \Phi_{\nu'} \rangle$ in the basis of choice, and the output are the energy eigenvalues E_n and the eigenvectors $(C_1^{(n)}, C_2^{(n)}, C_3^{(n)}, ...)^{\mathrm{T}}$, which contain the expansion coefficients.

Seems like a pretty straightforward numerical exercise to solve this eigenvalue problem. The only problem is that the matrix is infinite dimensional. Already the single-particle basis $|p\rangle$ and the single-nucleon Hilbert space $\mathcal{H}_1$ are infinite dimensional, and the A-nucleon basis $|p_1...p_A\rangle$ and the A-nucleon Hilbert space are even more so.

The way out of this misery are *truncations*. We discard many-body basis states based on a specific truncation criterion and, in this way, reduce the infinite-dimensional basis to a finite set of states. This finite set spans the so-called *model space* $\mathcal{M}_A$, which is a subspace of $\mathcal{H}_A$. There are many ways to define physics-motivated truncation criteria, and we will dive into this in Sect. 3.5.1. For the time being and as a simple example, assume that the single-particle basis $|p\rangle$ is truncated to a finite set—in case $|p\rangle$ is built from a harmonic oscillator basis, we could simply limit the principal oscillator quantum number $e = 2n + l \le e_{\max}$ with the truncation parameter $e_{\max}$ and discard all single-particle states with $e > e_{\max}$. This renders the A-body basis finite and defines the full configuration interaction scheme, as discussed later.

Imposing a basis truncation implies a departure from the exact solution of (3.1). The solution of the eigenvalue problem in the finite model space is only an approximation to the exact Schrödinger equation. However, it is a very controlled approximation, since the full problem and the exact solution are formally recovered if we include more and more states into the model space and effectively remove the truncation again, i.e., if we consider $e_{\max} \to \infty$ and thus $\mathcal{M}_A \to \mathcal{H}_A$. Obviously, we cannot do this in practical calculations, but we can explore the dependence of the solution and of all observables on the truncation. Ideally we would observe *convergence*, i.e., the observables becoming independent of the truncation for sufficiently large truncation parameters and model spaces. Whether or not we are able to reach convergence before the matrix is getting intractably large will be a central question for later.

Conclusions—Ab Initio So far, we have approached nuclear structure theory from a bird's-eye view. However, we have already encountered the key concepts and difficulties of many nuclear structure methods: basis choice, model space truncation, convergence, and uncertainty quantification. We have seen enough to discuss the notorious qualifier "ab initio" that is used as a quality label for many recent nuclear structure calculations. There is no agreed-upon definition of what qualifies a method as ab initio, and we will not attempt to provide a rigorous definition. However, we will mention some aspects relevant for the use of the term.

First, none of the nuclear structure methods qualifies as "ab initio" or "from first principles" from the perspective of QCD—there are promising attempts to describe light nuclei in lattice QCD [1–3], but more work is needed to provide quantitative results. The baseline for the many-body solutions are Hamiltonians rooted in QCD but using nucleons as effective degrees of freedom. Today, chiral EFT provides the most systematic way to establish the connection to QCD. However, the chiral EFT construction of interactions in itself uses truncations and ad hoc assumptions, which affect the nuclear observables. Furthermore, chiral EFT interactions come with parameters, the low-energy constants, that have to be fitted to experimental data. The selection of data and the experimental uncertainties of the data itself influence the predicted nuclear observables as well.

Second, all methods for the solution of the many-body problem contain some level of approximation. For basis expansion methods, the truncation to a finite model space introduces such an approximation, and there might be additional truncations and approximations involved. These truncations affect the observables and, there-fore, necessitate a systematic convergence analysis. The many-body method has to provide systematic control parameters that govern the transition to a formally exact solution—for basis expansion methods, this is the transition from the model space to the full Hilbert space. We have to use these control parameters to demonstrate that observables are sufficiently converged, i.e., sufficiently independent of the truncation. Doing this is not straightforward, particularly if multiple truncations and approximations are involved—the convergence with respect to all of them has to be addressed; otherwise the calculation is merely an uncontrolled approximation.

Third, the truncations and approximations introduced at the level of the Hamilto-nian and in the many-body solution are the source of systematic theory uncertainties. Even though we strive for convergence with respect to all truncations, we will hardly ever reach complete convergence. Therefore, the remaining systematic uncertainties in the theoretical predictions have to be quantified. This uncertainty quantification should be done within the theoretical framework used for the calculation, e.g., by exploring the convergence behavior with respect to truncations and by translating this into an uncertainty estimate for each and every observable. In addition, there might be statistical uncertainties inherited from experimental data used to calibrate the theory, e.g., for the fit of the low-energy constants in the interaction. Some many-body methods also produce additional statistical uncertainties through some statistical sampling process. For a long time, nuclear theory has been (and often still is) remarkably unconcerned about its uncertainties—a complete discussion of all sources of uncertainties leading to qualified error bars (not just guesswork) should be part of all ab initio calculations. If not, the label "ab initio" is not justified.

3.3   Hamiltonian

The foundation of any ab initio calculation is the Hamiltonian $\hat{H}$. As discussed in the previous section, it is not trivial to write down the Hamiltonian for the nuclear many-body problem. We have to develop an effective theory framework to construct the interactions that enter into the Hamiltonian—this theory framework will be chiral

EFT. Before addressing this, we start with a discussion of the symmetries of the Hamiltonian and the consequences for the many-body eigenstates.

3.3.1 Intrinsic Hamiltonian

Nuclei are self-bound systems. Unlike the electrons in an atom that are trapped in the Coulomb field of the central nucleus, the nucleons are held together solely by their mutual interactions. This seemingly trivial fact already has important consequences; it implies a number of symmetries.

From the fact that all the relevant interactions are intrinsic, i.e., act only among the nucleons of the system, we can conclude that the properties of the nucleus, e.g., the binding or excitation energies, are invariant under basic spatial transformations. These energies must not change if we place the nucleus at a different position in space, rotate the nucleus as a whole, or give the nucleus a non-vanishing total momentum. Therefore, we expect the Hamiltonian to exhibit these symmetries as well: translational invariance, rotational invariance, and invariance under momentum boosts, i.e., the Galilean symmetries.

For the construction of the operators for the two- and multi-nucleon interactions ($\hat{V}_{NN}$, $\hat{V}_{3N}$, etc.), these symmetries are taken into account explicitly. In addition, we have to pay attention to the kinetic energy operator $\hat{T}$. Naively, we might write the kinetic energy in the A-body system as a sum of single-nucleon kinetic energy operators:

$$\hat{T} = \sum_{i=1}^{A} \frac{1}{2m} \hat{\mathbf{p}}_i^2 \, , \tag{3.8}$$

where m is the average nucleon mass—in most ab initio calculations, the mass difference between proton and neutron is not included. This operator is not invariant under momentum boosts; it contains the kinetic energy associated with the center-of-mass motion of the nucleus. We have to subtract the operator for the center-of-mass kinetic energy $\hat{T}_{cm}$ to arrive at the *intrinsic kinetic energy*:

$$
\begin{aligned}
\hat{T}_{int} = \hat{T} - \hat{T}_{cm} &= \sum_{i=1}^{A} \frac{1}{2m} \hat{\mathbf{p}}_i^2 - \frac{1}{2Am} \left(\sum_{i=1}^{A} \hat{\mathbf{p}}_i \right)^2 \\
&= \sum_{i<j}^{A} \frac{1}{2m} (\hat{\mathbf{p}}_i - \hat{\mathbf{p}}_j)^2 = \sum_{i=1}^{A} \frac{1}{2m(A-1)} \hat{\mathbf{p}}_i^2 + \sum_{i<j}^{A} \frac{1}{2m} \hat{\mathbf{p}}_i \cdot \hat{\mathbf{p}}_j \, .
\end{aligned}
\tag{3.9}
$$

The last two expressions show two practical forms of the intrinsic kinetic energy that are being used in nuclear structure calculations—they are equivalent at the operator level. However, within an approximate many-body scheme, the results obtained with the two forms might differ [5].

With this we obtain the *intrinsic Hamiltonian*, which obeys all the Galilean symmetries:

$$\hat{H} = \hat{T}_{\text{int}} + \hat{V}_{\text{NN}} + \hat{V}_{\text{3N}} + \cdots . \tag{3.10}$$

All the many-body approaches discussed in the following will use this type of Hamiltonian. However, this does not guarantee that the solutions and observables in a truncated many-body calculation also exhibit these symmetries. We will come back to this point.

3.3.2 Practitioners' View on Chiral EFT

For constructing the two- and many-nucleon interaction operators, most ab initio approaches resort to chiral EFT. There are many excellent reviews on chiral EFT for nuclear interactions [6–12], and we invite the reader to explore these sources.

We will look at chiral EFT from the perspective of a user, i.e., a many-body practitioner that needs interactions as input for the solution of the many-body problem. Even from this vantage point, there are important aspects in the fabric of chiral EFT we need to be aware of:

- *Chiral order*: Chiral EFT is built on an expansion in a small parameter and the organization of contributions in powers of this small parameter, which is called power counting. The small parameter Q is a ratio of the typical momentum scale in the system P and the breakdown scale of the EFT Λ_χ and is of the order $Q = P/\Lambda_\chi \approx 1/3$. The expansion of the interaction is truncated at some finite power in Q. We will call this the *chiral order* and consider interactions at leading order (LO) corresponding to Q^0, next-to leading order (NLO) with Q^2, next-to-next-to leading order (N^{2}LO) with Q^3, and so on.
- *Many-body forces*: Starting from N^{2}LO, chiral EFT predicts contributions that correspond to irreducible three-nucleon (3N) interactions. Starting from N^{3}LO irreducible four-nucleon interactions emerge. The fact that these many-body forces emerge at higher orders in the expansion indicates that they are expected to be successively weaker. It is a great success of chiral EFT that these terms emerge naturally, in a systematic fashion and in a coherent theoretical framework.
- *Regulator scheme and scale*: Present chiral EFT interactions use a cutoff regularization of divergencies, which is implemented though momentum-dependent cutoff functions. Different types of momenta in the two- and few-nucleon system can be used to formulate the regulator functions (relative momenta vs. momentum transfer). We can even formulate cutoff functions in coordinate space or hybrid schemes that use different regulators for different terms. In addition to the regulator scheme, the cutoff scale Λ can be chosen in a certain range.

- *Fitting strategy*: For given chiral order, regulator scheme, and scale, a number of low-energy constants (LECs) associated with the contact terms have to be determined, typically by a fit to experimental data. There are different strategies to approach this parameter fit. One option is to fix all LECs that appear at the level of two-body interactions in the two-body system, all additional LECs that appear in three-body interactions using three-body observables, and so on. Another option is to fix all LECs simultaneously to the selection of observables in a range of different nuclei. We could also do something in between, fixing the two-body LECs using two-body observables and the others using a selection of many-body observables.
- *Electroweak operators*: Coupling photons and weak interaction bosons to the constituents of a chiral EFT formulation gives access to electromagnetic and weak interaction operators that become relevant for the computation of electromagnetic or weak transitions in nuclei. They can be derived consistently in chiral EFT, which is another great advantage of this approach.

All these aspects are active fields of debate and research today. The different options on chiral order, many-body forces, regulator scheme and scale, and strategy for LEC determination lead to a growing collection of nuclear interactions from chiral EFT that are available for nuclear structure calculations [13–23]. On top of these choices, there are more fundamental questions regarding, e.g., the specific choice of degrees of freedom or the power counting and renormalizability, that require further research on the EFT side [24, 25].

For many-body practitioners and those who compare ab initio results, it is important to understand that there is no "single" or "best" chiral EFT interaction. There will always be a variety of different realizations that are conceptually equally valid but might yield different many-body predictions. Even if all other variables are eliminated, chiral EFT interactions will always come at different chiral orders, and the convergence of this expansion has to be explored.

3.3.3 Uncertainty Quantification

The different choices at the chiral EFT level provide an opportunity to systematically quantify uncertainties in the theoretical description of the Hamiltonian and to propagate these uncertainties to the many-body observables [26].

The paradigm that the theory uncertainties resulting from the modeling of the Hamiltonian should be quantified has entered ab initio nuclear structure theory only recently. The tools and protocols for this uncertainty quantification (UQ) are still in their infancy. Because of the many different design choices and truncations involved in the construction of chiral interactions, there is no complete protocol yet.

The simplest starting point for an UQ protocol is the convergence of an observable obtained with an increasing order of the chiral expansion. For the moment we assume that the many-body calculations are precise, i.e., we do not consider additional uncertainties due to the many-body approach. Assuming that an observable $X^{(n)}$ at chiral order n follows the same power series expansion in the small parameter Q as the interaction, we can try to estimate the remainder of the truncated series based on the behavior of the finite number of terms we have access to. One can use a simple heuristic scheme [27–29] based the differences $\Delta X^{(n)} = X^{(n)} - X^{(n-1)}$ in the observable in subsequent lower orders, scaled with expansion parameter to estimate the remainder, e.g., through $\delta X^{(n)} = \max\{Q \Delta X^{(n)}, Q^2 \Delta X^{(n-1)}, \ldots\}$. There are more elaborate schemes using Bayesian statistics to model the distribution of the remainders [30–32], e.g., the pointwise model presented in [33] that is being used routinely.

Applying these schemes is simple. However, in order to assess the chiral truncation uncertainty, e.g., at order N^3LO, we have to compute the observable for all orders up to N^3LO. Therefore, the computational cost for predicting the observable with uncertainties is four times that of computing just the observable. This is a recurring theme—a systematic uncertainty quantification (in contrast to guesswork) increases the total computational cost significantly.

Another starting point for UQ is the propagation of the statistical uncertainties related to the LEC determination into many-body observables. This requires the construction of sets of interactions (at each chiral order) that sample the LECs as to reproduce the fit observables within their experimental uncertainties. This has been explored for N^2LO interactions, e.g., in [34]. Obviously, propagating statistical uncertainties requires an even larger number of many-body calculations or emulations of such many-body calculations.

3.4 Preconditioning

Now that we have specified the Hamiltonian, we can start to work on the solution of the Schrödinger equation. The general strategy for this was laid out in Sect. 3.2. The critical element in all applications of basis expansion methods is convergence of the observables with model space size. This eventually limits the range of applicability of specific many-body schemes, because the computational cost grows dramatically with model space size and eventually defines the largest feasible basis size. If a specific observable for a given nucleus does not reach a sufficient level of convergence within these model space limits, then an accurate prediction will not be possible. So some obvious questions are as follows: Are there ways to accelerate the convergence of a given basis expansion approach? Is it possible to precondition the eigenvalue problem such that smaller model spaces are sufficient to reach convergence? Can this be done without modifying the physics outputs of such a calculation? Luckily, the answer to all these questions is: yes!

The most direct way to precondition the many-body problem consists in a transformation of the Hamiltonian itself, and we will discuss this option in Sects. 3.4.1 and 3.4.2. Another way to accelerate the convergence in the context of a basis expansion approach is the choice of an optimized single-particle basis, as explored in Sect. 3.4.3. Finally, we can simplify the numerical treatment with an approximate

inclusion of many-body forces through normal-ordering techniques discussed in
Sect. 3.4.4.

3.4.1 Unitary Transformations, Pre-diagonalization, and Correlations

The most efficient way to precondition the many-body problem in a basis expansion
approach is a transformation of the Hamiltonian itself with the aim to accelerate the
convergence. We want to arrive at converged observables in smaller model spaces,
the smaller the better, because this will allow us to tackle heavier nuclei. However,
the transformation should not change the results of the many-body calculation—all
observables should be invariant under the transformation.

A general class of transformations that formally guarantees the invariance of
observables are unitary transformations. Assume a unitary operator $\hat{U}$ with $\hat{U}^\dagger \hat{U} = \hat{1} = \hat{U}\hat{U}^\dagger$. The unitary transformation of the Hamiltonian is then given by:

$$\hat{\tilde{H}} = \hat{U}^\dagger \hat{H} \hat{U} .\tag{3.11}$$

A key property of unitary transformation is that they do not change the spectrum of
the transformed Hamiltonian, i.e., all eigenvalues are invariant. This can be shown
starting from the eigenvalue equation for $\hat{H}$ and by inserting $\hat{U}\hat{U}^\dagger$ and multiplying
with $\hat{U}^\dagger$:

$$\hat{H} |\Psi_n\rangle = E_n |\Psi_n\rangle \;\Leftrightarrow\; \hat{U}^\dagger \hat{H} \hat{U} \hat{U}^\dagger |\Psi_n\rangle = E_n \hat{U}^\dagger |\Psi_n\rangle \;\Leftrightarrow\; \hat{\tilde{H}} |\tilde{\Psi}_n\rangle = E_n |\tilde{\Psi}_n\rangle ,\tag{3.12}$$

where we have introduced the transformed eigenstates $|\tilde{\Psi}_n\rangle = \hat{U}^\dagger |\Psi_n\rangle$. By solving
the eigenvalue problem of $\hat{\tilde{H}}$, we obtain the same energy eigenvalues E_n with the
transformed eigenstates. Considering other observables obtained from the energy
eigenstates, e.g., through expectation values of an operator $\hat{O}$, we can perform a
similar calculation inserting $\hat{U}\hat{U}^\dagger$ twice:

$$O_n = \langle\Psi_n| \hat{O} |\Psi_n\rangle = \langle\Psi_n| \hat{U}\hat{U}^\dagger \hat{O} \hat{U}\hat{U}^\dagger |\Psi_n\rangle = \langle\tilde{\Psi}_n| \hat{\tilde{O}} |\tilde{\Psi}_n\rangle\tag{3.13}$$

with the transformed operator

$$\hat{\tilde{O}} = \hat{U}^\dagger \hat{O} \hat{U} .\tag{3.14}$$

Thus, also expectation values and matrix elements can be computed using the
transformed eigenstates $|\tilde{\Psi}_n\rangle$ together with the consistently transformed operator
$\hat{\tilde{O}}$.

In summary, many-body observables are invariant under unitary transformations. When solving the many-body problem with the transformed Hamiltonian, we only have to make sure that observables are evaluated with consistently transformed operators. Two questions remain: (1) What does the unitary transformation have to do to the Hamiltonian to accelerate model space convergence? (2) How to formulate and implement such a transformation?

Imagine the matrix representation of the Hamiltonian in a large (infinite) many-body basis, where all regions of the matrix are populated with non-zero matrix elements. If we now truncate the basis to a small (finite) model space, and only solve the eigenvalue problem of the small matrix, the eigenvalues and eigenvectors will change as compared to the full matrix. This could be avoided if the full matrix would have a block-diagonal structure, i.e., if the model space would form one block and the rest of the Hilbert space would form the other block and all the matrix elements connecting the two blocks would be zero. Solving the eigenvalue problem in the small model space would reproduce part of the spectrum of the full matrix.

This block diagonalization or block decoupling idea with respect to the actual model space of the many-body calculation, the so-called P-space, and the excluded part of the Hilbert space, the so-called Q-space, is at the heart of the Okubo-Lee-Suzuki (OLS) transformation [35, 36]. In OLS a similarity transformation is constructed explicitly from the formal decoupling condition, i.e., the requirement that the transformed Hamiltonian should not connect P- and Q-space. We will not go into the formalism, but just remark that, by construction, the resulting transformed Hamiltonian will depend on the nucleus, single-particle basis, and model space. This results in a non-trivial convergence behavior as function of model space size, which makes an uncertainty quantification difficult.

It would be advantageous to construct a transformation that performs a pre-diagonalization in a more generic sense, independent of the specific nucleus and model space, with a transformed interaction that is *universal* and can be employed in any basis expansion approach. Such a method will not provide a perfect block decoupling of a model space, but it will nonetheless accelerate the convergence and provide a regular convergence behavior that obeys the variational principle.

A first method that implemented generic unitary transformations in this spirit is the unitary correlation operator method (UCOM) [37–39]. Here we explicitly design the operator for the unitary transformation, guided by the structure of the interaction and the physics of correlations induced in the many-body states. The UCOM concept highlights the intimate connection between decoupling and correlations.

As a reminder, correlations are a property of many-body states that distinguish them from states of a system of non-interacting, independent particles. The eigenstates of a system of non-interacting fermions are Slater determinants, i.e., the basis states we typically use in our basis expansion approaches. A strongly correlated state can only be represented by a superposition of a huge number of Slater determinants. Looking at the structure of the Hamilton matrix, a Hamiltonian with many strong off-diagonal matrix elements will induce more correlations, because these off-diagonal matrix elements are generating the admixture of the corresponding basis states to the eigenstates. A Hamiltonian with fewer off-diagonal matrix elements will generate less admixtures, weaker correlations, and faster convergence. An extensive discussion of UCOM and the concept of correlations can be found in [39].

Over the past decade, a new method has essentially replaced OLS, UCOM, and similar approaches in many applications. The reason for its success is the simplicity and flexibility of the underling formulation. This method is the similarity renormalization group, which will be discussed in detail in the following sections.

3.4.2 Similarity Renormalization Group

3.4.2.1 General Idea

The similarity renormalization group (SRG) transformation is the most elegant and versatile way to implement a unitary transformation to pre-diagonalize the Hamiltonian. It goes back to Wegner [40, 41] as well as Glazek and Wilson [42], and was adopted in nuclear structure physics in 2007 [43] and has thrived since then [39, 44–46].

We start by formulating a continuous unitary transformation of the initial Hamiltonian $\hat{H}$:

$$\hat{H}(\alpha) = \hat{U}^\dagger(\alpha)\, \hat{H}\, \hat{U}(\alpha), \tag{3.15}$$

using a unitary transformation operator $\hat{U}(\alpha)$, which depends on a continuous parameter α, the so-called *flow parameter*. The unitarily transformed Hamiltonian $\hat{H}(\alpha)$ now also depends on the flow parameter. For $\alpha = 0$ we define an initial condition requiring that $\hat{U}(\alpha = 0) = \hat{1}$ so that the evolved Hamiltonian coincides with the initial Hamiltonian $\hat{H}(\alpha = 0) = \hat{H}$.

Formally, we can take the derivative of Eq. (3.15) with respect to the flow parameter α, which leads to:

$$\frac{d}{d\alpha}\hat{H}(\alpha) = \left(\frac{d}{d\alpha}\hat{U}^\dagger(\alpha)\right)\hat{H}\,\hat{U}(\alpha) + \hat{U}^\dagger(\alpha)\,\hat{H}\left(\frac{d}{d\alpha}\hat{U}(\alpha)\right)$$

$$= \left(\frac{d}{d\alpha}\hat{U}^\dagger(\alpha)\right)\hat{U}(\alpha)\,\hat{H}(\alpha) + \hat{H}(\alpha)\,\hat{U}^\dagger(\alpha)\left(\frac{d}{d\alpha}\hat{U}(\alpha)\right), \tag{3.16}$$

where we have inserted the unitarity relation $\hat{U}^\dagger(\alpha)\,\hat{U}(\alpha) = \hat{1}$ to recover the transformed Hamiltonian $\hat{H}(\alpha)$. We now define the so-called *generator* $\hat{\eta}(\alpha)$ through:

$$\hat{\eta}(\alpha) = -\hat{U}^\dagger(\alpha)\left(\frac{d}{d\alpha}\hat{U}(\alpha)\right). \tag{3.17}$$

From the flow parameter derivative of the unitarity relation $\hat{1} = \hat{U}^{\dagger}(\alpha)\,\hat{U}(\alpha)$, we find that the generator is an anti-Hermitian operator, i.e., $\hat{\eta}^{\dagger}(\alpha) = -\hat{\eta}(\alpha)$. Combining all this leads to the final form of the SRG *flow equation*:

$$\frac{d}{d\alpha}\hat{H}(\alpha) = [\hat{\eta}(\alpha),\,\hat{H}(\alpha)]. \tag{3.18}$$

The steps from Eq. (3.15) to this flow equation are very general and do not rely on any specifics of the transformation. The difference between the flow equation (3.18) and the direct transformation (3.15) lies in inputs needed to evaluate the transformed Hamiltonian: For the direct transformation, we have to specify the unitary operator $\hat{U}(\alpha)$; for the flow equation, we only need to define the generator $\hat{\eta}(\alpha)$.

There are many different ways to define the generator $\hat{\eta}(\alpha)$. The simplest and most intuitive choice goes back to Wegner [40, 41] and is based on the commutator of the transformed Hamiltonian $\hat{H}(\alpha)$ with its "diagonal part" $\hat{H}^{d}(\alpha)$:

$$\hat{\eta}_{W}(\alpha) = [\hat{H}^{d}(\alpha),\,\hat{H}(\alpha)] = [\hat{H}^{d}(\alpha),\,\hat{H}^{od}(\alpha)]. \tag{3.19}$$

Identifying the diagonal and off-diagonal parts of the Hamiltonian requires a matrix representation of the Hamiltonian with respect to some specific basis. One can identify the diagonal part with the strict diagonal of the Hamilton matrix, or one can use more general band or block-diagonal structures to identify $\hat{H}^{d}(\alpha)$. The off-diagonal part then follows via $\hat{H}^{od}(\alpha) = \hat{H}(\alpha) - \hat{H}^{d}(\alpha)$.

Irrespective of the specific choice, the flow equation will suppress the off-diagonal part of the Hamiltonian throughout the flow evolution. If the Hamiltonian has reached a perfect diagonal form, then $\hat{H}^{od}(\alpha) = 0$, the generator (3.19) vanishes, and the flow evolution stops—this defines the fix point of the SRG flow evolution.

This is the most important aspect of the SRG, it provides a simple and elegant way to pre-diagonalize the Hamiltonian with respect to a specific basis. The choice of the generator defines which basis this is and how exactly the pattern of diagonal and off-diagonal pieces should look like. This makes the whole approach very flexible and intuitive. It turns out that it is much simpler to construct a generator that drives a specific, physics-motivated pre-diagonalization than to formulate the corresponding unitary operator directly.

As indicated earlier, instead of thinking in terms of diagonalization, one can also think in terms of *decoupling*. Any suppression of off-diagonal matrix elements entails a decoupling of certain parts of the basis from the rest of the basis. For a generator that drives the Hamiltonian toward a block-diagonal structure, the individual blocks would eventually decouple, and it would suffice to solve the eigenvalue problem for an individual block to recover a part of the exact spectrum.

3.4.2.2 Consistent Observables

When using a unitarily transformed Hamiltonian (3.15) in a subsequent many-body calculation, we have to make sure that the operators of all other observables of interest are transformed consistently (cf. Sect. 3.4.1). Thus, for a generic observable $\hat{O}$, we have to evaluate the transformation:

$$\hat{O}(\alpha) = \hat{U}^\dagger(\alpha)\,\hat{O}\,\hat{U}(\alpha)\,. \tag{3.20}$$

We can use the same steps as for the Hamiltonian to convert this explicit unitary transformation into flow equation:

$$\frac{d}{d\alpha}\hat{O}(\alpha) = [\hat{\eta}(\alpha),\,\hat{O}(\alpha)]\,, \tag{3.21}$$

with the initial condition $\hat{O}(\alpha = 0) = \hat{O}$. Note that the anti-Hermitian generator $\hat{\eta}(\alpha)$ has to be the same as in the flow equation of the Hamiltonian. Since the generator necessarily contains the evolved Hamiltonian, we have to solve the flow equation for $\hat{H}(\alpha)$ as well. To handle this numerically, the two flow equations (3.18) and (3.21) are solved simultaneously as a coupled system of differential equations of twice the size. If there is only one other observable of interest, then this is usually no problem; however, it is getting tedious if several operators have to be transformed.

A more elegant way is the construction of the unitary transformation $\hat{U}(\alpha)$ itself. We can recast equation (3.17), by multiplying from the left with the unitary operator $\hat{U}(\alpha)$ into a differential equation for $\hat{U}(\alpha)$:

$$\frac{d}{d\alpha}\hat{U}(\alpha) = -\hat{U}(\alpha)\,\hat{\eta}(\alpha)\,. \tag{3.22}$$

This differential equation again involves the generator which depends on the evolved Hamiltonian and, therefore, has to be solved simultaneously with the flow equation (3.18). However, once this is done, we have a representation of the unitary operator $\hat{U}(\alpha)$ which can be used to explicitly transform any other operator (including the Hamiltonian) using (3.20).

In a numerical setting where (3.22) is integrated stepwise starting from the initial condition $\hat{U}(\alpha = 0) = \hat{1}$, there is also the option to compute the Hamiltonian $\hat{H}(\alpha)$ entering the generator from an explicit transformation (3.15) with the $\hat{U}(\alpha)$ obtained in the previous integration step. In this way we only need to handle one differential equation.

There is a third way to handle the general transformation of operators, the so-called Magnus expansion [47, 48]. It is similar to the solution of the differential equation for $\hat{U}(\alpha)$, but this time we first parametrize the unitary operator $\hat{U}(\alpha)$ in terms of an anti-Hermitian Magnus operator $\hat{\Omega}(\alpha)$ through:

$$\hat{U}(\alpha) = \exp(-\hat{\Omega}(\alpha)) \tag{3.23}$$

with $\hat{\Omega}(\alpha = 0) = 0$. One might be tempted to think that the Magnus operator is the same as the generator $\hat{\eta}(\alpha)$—it is not. Since the generator $\hat{\eta}(\alpha)$ depends on α itself and does not commute with itself for different values of the flow parameter, the formal integration of (3.22) does not simply yield an exponential of the generator. This is exactly what the Magnus expansion takes care of.

The Magnus operator can be obtained from the following differential equation:

$$\frac{d}{d\alpha}\hat{\Omega}(\alpha) = \sum_{k=0}^{\infty} \frac{B_k}{k!}[\hat{\Omega}(\alpha), \hat{\eta}(\alpha)]_k \,, \tag{3.24}$$

where B_k are the Bernoulli numbers and $[\hat{X}, \hat{Y}]_k = [\hat{X}, [\hat{X}, ...[\hat{X}, \hat{Y}]]]$ are the k-fold nested commutators, where k gives the number of $\hat{X}$ factors, i.e., $[\hat{X}, \hat{Y}]_0 = \hat{Y}$, $[\hat{X}, \hat{Y}]_1 = [\hat{X}, \hat{Y}]$, etc. We will not be able to handle the infinite series on the right-hand side exactly, and we have to truncate this series in finite order. The good news is that no matter how bad this truncation for $\hat{\Omega}(\alpha)$ is, it will not destroy the unitarity of the transformation. Another technical benefit is that for typical applications, the differential equation (3.24) is numerically easier to handle (less stiff) and thus more efficient to solve (fewer and larger steps).

Once we have obtained $\hat{\Omega}(\alpha)$, we still have to evaluate the unitary transformation of the observables through (3.20) and (3.23). This can be done via another expansion, the Baker-Campbell-Hausdorff series:

$$\hat{O}(\alpha) = \exp(+\hat{\Omega}(\alpha))\,\hat{O}\,\exp(-\hat{\Omega}(\alpha)) = \sum_{n=0}^{\infty} \frac{1}{n!}[\hat{\Omega}(\alpha), \hat{O}]_n \tag{3.25}$$

using nested commutators. Again, we will have to truncate this infinite series in numerical applications, but typically this series converges rather quickly. Note that this truncation might destroy the unitarity of the transformation.

It might seem that the Magnus approach only causes complications, and, indeed, whenever a direct construction of $\hat{U}(\alpha)$ through (3.22) is possible, this will be the preferred method. However, the Magnus expansion will have important applications in an advanced version of SRG that will be discussed in Sect. 3.6.1.

3.4.2.3 Free-Space Similarity Renormalization Group

So far, we have discussed a generic version of the SRG at the operator level. The Hamiltonian in this discussion is the Hamiltonian for the A-body system, and the basis used to identify diagonal and off-diagonal parts is the full A-body basis, as introduced in Sect. 3.2, or at least an A-body basis in a huge model space. Therefore, we have gained nothing regarding to the computational complexity of the problem— it is more challenging to solve a coupled system of differential equations for the matrix elements in a huge model space than to solve the eigenvalue problem of this matrix. We have to use the SRG evolution in a different setting to really gain something. There are two such beneficial settings that we will discuss in this lecture, the free-space SRG and the in-medium SRG.

The free-space SRG is built on two design choices: (i) the use of a generator that yields a basis-independent transformation and implements a more generic decoupling idea and (ii) the evaluation of the SRG flow equations in few-body spaces, typically $A = 2$ and $A = 3$, and a subsequent embedding of the evolved operators into A-body space via a cluster expansion.

Let us first discuss the generator that is most widely used in the free-space SRG. To construct a universal Hamiltonian for use in a wide array of many-body methods for different nuclei, we adopt a more generic concept of pre-diagonalization or decoupling—and the concept that comes to mind is the decoupling of energy or momentum scales. We could use the diagonal of the Hamiltonian in the eigenbasis of the total momentum or the kinetic energy operator. Or even simpler, we can use the kinetic energy operator directly, instead of the diagonal part of the Hamiltonian. This leads to the SRG generator that was introduced in [43, 49] and is widely used in nuclear physics:

$$\hat{\eta}_T(\alpha) = \frac{m^2}{\hbar^4} \, [\hat{T}_{\text{int}}, \hat{H}(\alpha)], \qquad (3.26)$$

where $\hat{T}_{\text{int}}$ is the intrinsic kinetic energy (3.9) and the prefactor including the nucleon mass m is chosen such that the flow parameter α has units [length^4]. For this specific generator, it is reasonable to associate the flow parameter with a momentum scale λ, using the relation $\lambda = \alpha^{-1/4}$.

Using this generator, the SRG flow equations will drive the Hamiltonian toward a band-diagonal form in momentum representation. In other words, it decouples low-momentum states from high-momentum states, since the matrix elements far off the diagonal in a momentum representation are suppressed. This suppression will also be effective in other basis representations, e.g., the harmonic oscillator basis; we will illustrate this in Sect. 3.4.2.5.

3.4.2.4 Cluster Expansion and Cluster Truncation

Despite the simplicity of the equations, we cannot evaluate the flow equations at the general A-body level. However, we can evaluate them in few-body systems, typically $A = 2$ and $A = 3$, and reconstruct the evolved operator in A-body space from this in an approximate way. The formal background of this procedure is the *cluster expansion* and a *cluster truncation*.

We can decompose the transformed Hamiltonian $\hat{H}(\alpha)$ for the A-body system into irreducible k-body operators $\hat{H}^{[k]}$:

$$\hat{H}(\alpha) = \hat{H}^{[1]}(\alpha) + \hat{H}^{[2]}(\alpha) + \hat{H}^{[3]}(\alpha) + \cdots + \hat{H}^{[A]}(\alpha). \qquad (3.27)$$

Each k-body operator of the cluster expansion can be written in its second-quantized form with k-body matrix elements $H_{q_1...q_k}^{p_1...p_k}(\alpha) = \langle p_1...p_k | \hat{H}^{[k]}(\alpha) | q_1...q_k \rangle$ and a product of k creation and k annihilation operators:

$$\hat{H}^{[k]}(\alpha) = \frac{1}{(k!)^2} \sum_{p_1,...,p_k} \sum_{q_1,...,q_k} H_{q_1...q_k}^{p_1...p_k}(\alpha) \, \hat{a}_{p_1}^\dagger \cdots \hat{a}_{p_k}^\dagger \hat{a}_{q_k} \cdots \hat{a}_{q_1} \qquad (3.28)$$

To work out the individual terms of the cluster expansion, i.e., the matrix elements $H_{q_1 \ldots q_k}^{p_1 \ldots p_k}(\alpha)$ for $k = 1, 2, 3, \ldots$, we simply perform the SRG transformation of the Hamiltonian in Hilbert spaces of increasing particle number A. For $A = 1$ the SRG transformation does not do anything, since the generator does not have a one-body contribution, so the one-body part of the transformed Hamiltonian equals the initial one-body part. For $A = 2$ we get a non-trivial transformation of the Hamiltonian. After subtraction of the previously obtained one-body part embedded into two-body space, this yields the irreducible two-body part. For $A = 3$ we again get a transformed Hamiltonian that, after subtraction of the previously determined one- and two-body contributions, yields the irreducible three-body part [39, 45, 50].

This scheme continues up to the A-body level. Even if the initial Hamiltonian only has up to three-body terms, because we only include up to 3N interactions, the transformed Hamiltonian contains induced terms beyond the three-body level. These induced multi-particle contributions formally have to be included to warrant the unitarity of the transformation and to benefit from the exact unitary equivalence discussed in Sect. 3.4.1. Induced multi-particle interactions are the price to pay for the improved convergence with the transformed Hamiltonian—a clear case of the no-free-lunch theorem.

For the SRG transformation, it is easy to see how induced multi-particle terms emerge. Consider the SRG flow equation (3.18) with the free-space SRG generator (3.26). Now we solve the flow equation as an initial value problem in a simplistic Euler-type approach, i.e., we assume a small step $\Delta\alpha$ in the flow parameter and use a two-point finite difference form to approximate the derivative of the Hamiltonian. After simple rearrangements we get for one Euler step:

$$\hat{H}(\alpha + \Delta\alpha) \approx \hat{H}(\alpha) + \Delta\alpha \frac{m^2}{\hbar^4}[[\hat{T}_{\text{int}}, \hat{H}(\alpha)], \hat{H}(\alpha)] . \qquad (3.29)$$

For simplicity we assume that the initial Hamiltonian $\hat{H}(\alpha = 0)$ only consists of a two-body term, and we recall that the intrinsic kinetic $\hat{T}_{\text{int}}$ can also be written as a two-body operator. We can now use a general property of commutators: The commutator of an n-body and an m-body operator yields contributions up to $(n + m - 1)$-body operators. Thus, the commutator of two two-body operators produces up to three-body operators, and the nested commutator of three two-body operator generates up to four-body operators. As a consequence, a single Euler step from $\alpha = 0$ to $\alpha = \Delta\alpha$ induces up to four-body terms in the Hamiltonian. And the many small Euler steps needed to reach a finite flow parameter will induce multi-particle terms of arbitrary particle rank.

Obviously, we cannot keep all induced multi-particle interactions in practical calculations. First of all, it is computationally not possible to solve the SRG flow equations for larger A—we can routinely handle the evolution in three-body space, but already the evolution in four-body space is not fully tractable at present. Second, even if the all induced multi-particle terms would be available, their inclusion in the final many-body calculation would be prohibitive. Already the step from two-body to three-body interactions significantly increases the computational complexity, and some approaches are still not able to include three-body interactions without additional approximation.

Therefore, we have to truncate the cluster expansion at finite particle numbers and by this introduce an approximation at the level of the Hamiltonian. Remember that already the initial Hamiltonian constructed within chiral EFT is subject to truncations with respect to chiral order and multi-particle interactions, and the truncation of the cluster expansion can be viewed in the same context. The present state of the art is to include the chiral interactions up to the three-body level and to truncate the evolved Hamiltonian also at the three-body level [45, 50, 51].

In this situation it is important to quantify the uncertainties resulting from the cluster truncation. Luckily, the SRG offers a handy tool to assess these truncation uncertainties without explicitly calculating the next order of the expansion. We can use the continuous flow parameter α as a diagnostic. Because the un-truncated SRG evolution is unitary and preserves the spectrum and the observables, any dependence of the energy eigenvalues or other observables on the flow parameter in a converged many-body calculation using truncated operators signals the impact of discarded multi-particle terms in the cluster expansion [45, 50, 51].

3.4.2.5 Example: SRG Evolution in Three-Body Space

As an illustration of the free-space SRG transformation, we consider the three-body system, and corresponding illustrations for the simpler two-body system can be found in the literature [39, 44].

In order to numerically solve the SRG flow equation for the Hamiltonian in three-body space, we need an appropriate basis to convert the operator flow equation (3.18) into a set of coupled differential equations for the matrix elements. We will use a harmonic oscillator basis in the relative coordinates of the three-particle system, the so-called Jacobi coordinates. The antisymmetrized states of the relative harmonic oscillator basis, which will be discussed in more detail in Sect. 3.4.3.1, can be written as $|Ei J^\pi M, T M_T\rangle$, with a three-particle principal quantum number E and a collective quantum number i that encapsulates the orbital angular momentum and spin degrees of freedom [50].

The matrix elements of the initial chiral NN+3N interaction in this basis for the quantum numbers of the triton, i.e., $J^\pi = 1/2^+$, $T = 1/2$, $M_T = -1/2$, are depicted in the top-left panel of Fig. 3.1. The rows and columns are spanned by the quantum numbers (E, i), where the apparent block structure results from sections of the basis with fixed E. The lower-left panel shows the lowest eigenvalues resulting for the numerical solution of the eigenvalue problem of the corresponding Hamilton matrix, truncated to $E \leq N_{max}$—as will be discussed later, this corresponds to a Jacobi no-core shell model calculation for the triton ground state. We observe that the matrix has strong off-diagonal matrix elements and that the ground-state energy needs large model spaces with $N_{max} \gtrsim 16$ to converge. The next two columns show the matrix elements and energies for the SRG-evolved interaction for flow parameter $\alpha = 0.04\,\mathrm{fm}^4$ and $0.16\,\mathrm{fm}^4$, respectively. The matrices clearly show the suppression of off-diagonal matrix elements with increasing flow parameter—note that although

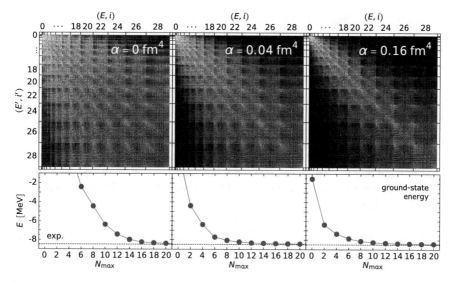

Fig. 3.1 Three-body matrix elements (top panels) and ground-state energy convergence (bottom panels) for the triton with initial and SRG-evolved NN+3N interactions (Modified from [50])

we use the kinetic energy generator (3.26) here, the pre-diagonalization is also evident in a harmonic oscillator basis. The corresponding plots for the ground-state energies show a much faster convergence, and now $N_{max} \approx 8$ is sufficient to reach the same level of convergence as reached at $N_{max} \approx 16$ for the initial interaction. The converged energy is the same in all cases, since the SRG transformation includes all three-body terms.

3.4.3 Single-Particle Basis

The convergence behavior of basis expansion calculations also depends on the choice of the underlying single-particle basis. We can try to optimize this basis with respect to global properties of the nucleus, e.g., its spatial size. Obviously, the convergence will deteriorate if we choose basis sets spanning length scales that are completely different from the intrinsic length scale of the nucleus. We will discuss three types of single-particle bases for the description of finite nuclei: the harmonic oscillator basis, the Hartree-Fock basis, and a specific variant of a natural orbital basis.

3.4.3.1 Harmonic Oscillator Basis

The harmonic oscillator (HO) is the default basis for any type of localized many-body system, simply because the basis functions are analytically known and there are many special relations for the HO basis that are of critical importance for practical calculations.

When we talk about the HO basis, we refer to the eigenstates of a single particle in a spherical harmonic oscillator potential characterized by an oscillator frequency Ω or an oscillator length $a = \sqrt{\hbar/(m\Omega)}$. The single-particle HO Hamiltonian reads:

$$\hat{h}_{\text{HO}} = \frac{1}{2m}\,\hat{\mathbf{p}}^2 + \frac{m\Omega^2}{2}\,\hat{\mathbf{x}}^2 \,. \tag{3.30}$$

The analytic solution of the eigenvalue problem of this Hamiltonian can be found in any textbook on quantum mechanics. We exploit spherical symmetry and introduce orbital angular momentum quantum numbers l and m_l plus an addition radial quantum number n, so that the eigenstates are characterized as $|nlm_l\rangle$. The energy eigenvalues are given by $\epsilon_{nl} = \hbar\Omega(2n+l+3/2) = \hbar\Omega(e+3/2)$ with the principal quantum number $e = 2n+l$. The spectrum is equidistant with a fixed energy spacing $\hbar\Omega$ between adjacent single-particle levels.

There is one feature that makes the HO unique and directly results from the fact that the HO Hamiltonian is a quadratic form in position and momentum. If we consider two classical particles of mass m with positions $\mathbf{x}_1$ and $\mathbf{x}_2$, then we can introduce a relative coordinate $\mathbf{r} = \mathbf{x}_1 - \mathbf{x}_2$ and a center-of-mass coordinate $\mathbf{X} = \frac{1}{2}(\mathbf{x}_1 + \mathbf{x}_2)$. We can extend this to the momenta of the two particles $\mathbf{p}_1$ and $\mathbf{p}_2$ and define a relative momentum $\mathbf{q} = \frac{1}{2}(\mathbf{p}_1 - \mathbf{p}_2)$ and a total or center-of-mass momentum $\mathbf{P} = \mathbf{p}_1 + \mathbf{p}_2$. We can do the same with the position and momentum operators in quantum mechanics, and, because of our choice of prefactors, the canonical commutation relations between position and momentum operators also hold for the relative and center-of-mass operators.

We can transfer this to the (classical or quantum) Hamiltonian of a system of two non-interacting particles in a HO potential and write it either using the single-particle operators or the relative and center-of-mass operators. A simple calculation shows that:

$$\begin{aligned}
\hat{h}_{\text{HO},1} + \hat{h}_{\text{HO},2} &= \frac{1}{2m}\,\hat{\mathbf{p}}_1^2 + \frac{m\Omega^2}{2}\,\hat{\mathbf{x}}_1^2 + \frac{1}{2m}\,\hat{\mathbf{p}}_2^2 + \frac{m\Omega^2}{2}\,\hat{\mathbf{x}}_2^2 \\
&= \frac{1}{2\mu}\,\hat{\mathbf{q}}^2 + \frac{\mu\Omega^2}{2}\,\hat{\mathbf{r}}^2 + \frac{1}{2M}\,\hat{\mathbf{P}}^2 + \frac{M\Omega^2}{2}\,\hat{\mathbf{X}}^2 = \hat{h}_{\text{HO,rel}} + \hat{h}_{\text{HO,cm}}.
\end{aligned} \tag{3.31}$$

with the reduced mass $\mu = m/2$ and the total mass $M = 2m$. In the second line, we have identified a HO Hamiltonian $\hat{h}_{\text{HO,rel}}$ in the relative quantities and a HO Hamiltonian $\hat{h}_{\text{HO,cm}}$ for the center-of-mass quantities. The first line tells us that a tensor product of two single-particle HO states, i.e., $|n_1 l_1 m_{l1}\rangle \otimes |n_2 l_2 m_{l2}\rangle$, will be an eigenstate of this two-body Hamiltonian with an energy eigenvalue $\hbar\Omega(e_1 + e_2 + 3)$. The second line tells us that a tensor product of HO eigenstates for the relative motion $|n_{\text{rel}} l_{\text{rel}} m_{l,\text{rel}}\rangle$ and the center-of-mass motion $|n_{\text{cm}} l_{\text{cm}} m_{l,\text{cm}}\rangle$ is also an eigenstate with eigenvalue $\hbar\Omega(e_{\text{rel}} + e_{\text{cm}} + 3)$. Thus we have two different eigenbasis sets for the same Hamiltonian, spanning the same two-particle Hilbert space with the same degenerate subspaces for the two-body principal quantum number $E_2 = e_1 + e_2 = e_{rel} + e_{cm}$. Therefore, there has to be a unitary transformation connecting the states of the two bases within each of the degenerate subspaces. This basis transformation is the celebrated Talmi or Talmi-Moshinsky-Smirnov transformation [52–54]:

$$|n_{rel}n_{cm}[l_{rel}l_{cm}]LM_L\rangle = \sum_{n_1,n_2,l_1,l_2} \langle\langle n_1n_2, l_1l_2; n_{rel}n_{cm}, l_{rel}l_{cm}; L\rangle\rangle \, |n_1n_2[l_1l_2]LM_L\rangle ,$$

(3.32)

where the sum is restricted to $e_{rel} + e_{cm} = e_1 + e_2$. For convenience we have coupled the two orbital angular momenta in each basis to total orbital angular momentum L and M_L. The transformation coefficients $\langle\langle n_1n_2, l_1l_2; n_{rel}n_{cm}, l_{rel}l_{cm}; L\rangle\rangle$ are the so-called Moshinsky coefficients or harmonic oscillator brackets [55].

Why is this important? Well, we use this transformation all the time in practical calculations; here are a few examples:

- *Computation of NN matrix elements*: For the many-body calculation, we need two-body matrix elements with respect to antisymmetrized product states, as they appear, e.g., in the second-quantized form of a two-body operator. For the computation of the matrix elements of the chiral NN interaction or for the SRG evolution, a relative two-body basis is much more convenient. We can exploit spherical symmetry and the center-of-mass part of the basis separates, which drastically reduces the number of matrix elements. Therefore, we first compute all the relative HO matrix elements, perform the SRG-transformation in the relative HO basis, and in the end use the Talmi transformation to compute the matrix elements in terms of single-particle quantum numbers for use in the subsequent many-body calculation.

- *Computation of 3N matrix elements*: This is essentially the same story as for the NN interaction, only more complicated since we have to work with three particles. We need an extension of the relative and center-of-mass coordinates for the three-body system, which leads to the so-called Jacobi coordinates. This construction of the Jacobi coordinates translates into a corresponding hierarchical nesting of relative HO quantum numbers and the corresponding Talmi transformations. We refer the reader to [50] for a detailed discussion.

- *Center-of-mass separation*: We can formally extend the idea of the Jacobi coordinates and the corresponding relative HO states and associated Talmi transformation to the many-body level. For an A-body model space spanned by all Slater determinants of HO single-particle states up to a maximum total quantum number $\sum_i e_i \leq E_{A,max}$, there is an alternative basis of relative and center-of-mass HO states, connected through an A-body Talmi transformation. The relative HO basis allows for an explicit separation of the center-of-mass state of the A-body system from the intrinsic state. Therefore, also the Slater determinant basis in an $E_{A,max}$-truncated space allows for an exact center-of-mass separation. We will come back to this point in Sect. 3.5.3.

All of this is unique to the HO, and, therefore, we will always use the HO basis at certain stages of the calculation. However, there is also a dark side, related to the asymptotic behavior of the HO wave functions. The potential term in the HO Hamiltonian grows quadratically with $x = |\mathbf{x}|$, and, as a result, the HO coordinate space wave functions fall off with a Gaussian $e^{-x^2/(2a^2)}$ behavior. This is unrealistic for a self-bound system. If a localized many-particle system is bound by finite-range attractive interaction between the particles, then we can invoke a schematic mean-field-type picture. The average interaction of a particle with all the others will resemble a potential well, which goes to zero in the exterior. The bound single-particle wave functions for such a mean-field potential will fall off exponentially and not like a Gaussian. If we use the HO basis for a many-body calculation, then we have to correct for the unrealistic asymptotic behavior by superpositions of many basis states in order to build up the exponential asymptotic. This will slow down the model space convergence, particularly for weakly bound states and halo nuclei, which show a prominent exponential tail in their density distribution.

3.4.3.2 Hartree-Fock Basis

An obvious candidate for a more suitable set of single-particle states is the Hartree-Fock (HF) basis. The HF approximation itself is covered in many textbooks [56,57], so we keep this discussion brief.

The HF method provides a variational approximation for the ground state of the A-body system assuming a trial state that consists of a single Slater determinant. The variational degrees of freedom are the single-particle states that enter into the trial state. The HF equations that determine the single-particle basis simply result as Euler-Lagrange equations from the minimization of the expectation value of the many-body Hamiltonian with the Slater determinant trial state. This simple picture holds for closed-shell nuclei, where all magnetic substates of the highest j-shell are occupied. For the purpose of constructing a basis in open-shell nuclei, we can use a simple constrained HF scheme with an equal-filling approximation for the partially occupied shell.

The HF single-particle basis definitely has advantages. It is constructed from a variational calculation for the nucleus and the Hamiltonian under consideration, using a single Slater determinant that will automatically be a basis state in the subsequent many-body calculation. We can view the subsequent calculation as an expansion around this variational optimum. Global properties of the ground state that are accessible already in the simplified mean-field picture are built into the many-body basis.

However, it also has some problems. Strictly speaking, the variational approach only constrains the energetically lowest single-particle states that are occupied in the HF ground state. Higher-lying single-particle states are only determined through technical constraints on the density matrix and orthogonality. Furthermore, the HF potential really resembles a potential well with a finite number of bound states with negative single-particle energies. In addition there is a continuous spectrum of solutions at positive energies, representing unbound single-particle states. These unoccupied and unbound states depend on the specific way in which we solve the HF equations. Since our Hamiltonian is specified in terms of HO matrix elements, we will represent the HF single-particle states in an HO basis expansion with a truncation with respect to the principal HO quantum number e requiring $e \leq e_{max}$ [58]. The underlying HO basis depends on the oscillator frequency Ω, and this dependence carries over particularly to the unoccupied and unbound single-particle states.

Another limitation of the HF basis optimization is the fact that it does not account for correlations in the many-body state. The HF ground state is a single Slater determinant without any correlations, and it is typically far above the exact ground-state energy. Even for SRG-evolved Hamiltonians, HF typically recovers only half of the binding energy. For a bare chiral Hamiltonian, the ground state might not even be bound at the HF level. This mismatch influences the structure of the single-particle states; they are optimized for a badly approximated ground state.

3.4.3.3 Natural Orbital Basis

There is a way to inform the single-particle basis about the correlated ground state of the system and this way involves so-called *natural orbitals*. In general, natural orbitals are single-particle states that result as eigenstates of the one-body density matrix.

We can start from a highly correlated many-body state $|\Psi\rangle$ and compute the one-body density matrix with respect to a specific single-particle basis, e.g., the HO basis, via:

$$\rho_{pq} = \langle\Psi|\hat{a}_p^\dagger\hat{a}_q|\Psi\rangle , \tag{3.33}$$

where the creation and annihilation operators of the second quantization are defined with respect to the HO single-particle basis (or any other computational reference basis). We can now solve the matrix eigenvalue problem of the one-body density matrix, which yields eigenvectors that define the natural orbital single-particle basis and the eigenvalues the mean occupation numbers of the natural orbital states.

The natural orbitals inherit the angular momentum, spin, and isospin structure of the reference basis and only differ in the radial wave functions, because a scalar one-body density matrix exhibits a corresponding block structure. So just like in the spherical or constrained HF case, the new single-particle basis can be expressed via a simple basis transformation with respect to the radial quantum numbers:

$$|\nu(l\tfrac{1}{2})jm; \tfrac{1}{2}m_t\rangle = \sum_n C_{n\nu}^{(ljm_t)} |n(l\tfrac{1}{2})jm; \tfrac{1}{2}m_t\rangle , \tag{3.34}$$

where ν indicates the natural orbital basis and n the initial HO basis, and the expansion coefficients $C_{n\nu}^{(ljm_t)}$ are given by the eigenvectors of the density matrix.

The interesting question is as follows: how to construct the correlated many-body state $|\Psi\rangle$ that determines the density matrix? Typically, we will consider an approximation for the ground state of the system for the construction of the natural orbitals, although one could also use a density matrix defined for a mixture of states. In principle one can perform a preparatory ab initio calculation, e.g., in a configuration interaction framework, to extract a proxy for the ground state. This has be done [59] but is computationally very expensive.

A much simpler way to construct an approximation for the correlated ground state is many-body perturbation theory [60]. We can start from a HF calculation and add low-order perturbative corrections on top of the unperturbed HF ground state:

$$|\Psi\rangle = |\Psi^{(0)}\rangle + |\Psi^{(1)}\rangle + |\Psi^{(2)}\rangle + \dots \quad \text{with} \quad |\Psi^{(0)}\rangle = |\text{HF}\rangle . \tag{3.35}$$

The perturbative corrections $|\Psi^{(i)}\rangle$ $(i > 0)$ account for the most important beyond-HF correlations. Inserting this into the definition of the density matrix and keeping terms up to the second order in the perturbation yields:

$$\rho \approx \rho^{(00)} + \rho^{(02)} + \rho^{(20)} + \rho^{(11)} , \tag{3.36}$$

where $\rho^{(00)}$ denotes the unperturbed HF density matrix and

$$\rho_{pq}^{(02)} = \rho_{qp}^{(20)} = \langle \Psi^{(0)}| \hat{a}_p^\dagger \hat{a}_q |\Psi^{(2)}\rangle , \quad \rho_{pq}^{(11)} = \langle \Psi^{(1)}| \hat{a}_p^\dagger \hat{a}_q |\Psi^{(1)}\rangle . \tag{3.37}$$

Explicit expressions for these density matrix elements can be found in [60]. With small computational effort, large single-particle spaces can be used to evaluate the perturbative corrections, much larger than the model spaces of the subsequent many-body solution. Thus, the basis can be informed about correlation effects in a very large model space and effectively supply this information for the many-body treatment in a smaller space.

We will use the natural orbital basis in connection with different many-body approaches and compare the performance of the different basis sets in Sect. 3.5.3.

3.4.4 Normal Ordering of Many-Body Interactions

A less obvious preconditioning consists of a rearrangement of the Hamiltonian, the so-called normal ordering. For some of the many-body methods discussed later, normal ordering is a necessary step to formulate the basic working equations of the method in an efficient way. On top of this, normal ordering can be used to define an approximation for the inclusion of many-body interactions, which presents a major simplification for all many-body methods.

Normal ordering is the simple process for rearranging the order of creation and annihilation operators in the second-quantized form of the Hamiltonian (or any other operator). Let us start with a generic Hamiltonian for the many-body system containing up to three-body operators:

$$\hat{H} = \hat{H}^{[1]} + \hat{H}^{[2]} + \hat{H}^{[3]} . \tag{3.38}$$

As in Sect. 3.4.2.4, $\hat{H}^{[k]}$ represents the k-body part of the Hamiltonian. In the second quantization, it can be written as:

$$\hat{H}^{[k]} = \frac{1}{(k!)^2} \sum_{p_1,\dots,p_k} \sum_{q_1,\dots,q_k} H^{p_1\dots p_k}_{q_1\dots q_k} \, \hat{a}^\dagger_{p_1} \cdots \hat{a}^\dagger_{p_k} \, \hat{a}_{q_k} \cdots \hat{a}_{q_1} \tag{3.39}$$

with the shorthand notation $H^{p_1\dots p_k}_{q_1\dots q_k} = \langle p_1\dots p_k| \hat{H}^{[k]} |q_1\dots q_k\rangle$ for the matrix elements of the k-body part. This is the standard form of an operator in the second quantization with the creation operators to the left of the annihilation operators. For later reference, we call this the *vacuum normal-ordered form* of the Hamiltonian. Here, vacuum refers to the vacuum state $|0\rangle$, the only state in the zero-particle Hilbert space, and from the basics of the second quantization, we remember that $\hat{a}_q |0\rangle = 0$ for all q and thus $\langle 0| \hat{a}^\dagger_p \cdots \hat{a}_q |0\rangle = 0$ and also $\langle 0| \hat{H}^{[k]} |0\rangle = 0$.

Now we start to reinterpret and reshuffle things. Assume a Slater determinant $|\Phi\rangle$, which represents a simplistic approximation for the ground state of a closed-shell system, as obtained, e.g., in a HF calculation. We call this specific Slater determinant the *reference state*. Acting with the annihilation operator $\hat{a}_q$ on the reference state $|\Phi\rangle$ can lead to different results, depending on whether the single-particle state $|q\rangle$ is occupied or unoccupied in the reference state. Using an index convention to distinguish states i, j that are occupied in $|\Phi\rangle$, the *hole states*, from states a, b that are unoccupied in $|\Phi\rangle$, the *particle states*, we find:

$$\hat{a}_i |\Phi\rangle \neq 0 \,, \qquad \hat{a}_a |\Phi\rangle = 0 \,, \qquad \hat{a}^\dagger_i |\Phi\rangle = 0 \,, \qquad \hat{a}_a |\Phi\rangle \neq 0 \,. \tag{3.40}$$

Comparing this to the behavior of annihilation and creation operator applied to the vacuum state $|0\rangle$, we observe that $\hat{a}_a$ and $\hat{a}^\dagger_i$ seem to behave like annihilation operators with respect to $|\Phi\rangle$. Turning this around, $|\Phi\rangle$ behaves like a vacuum state with respect to the set $\hat{a}_a$ and $\hat{a}^\dagger_i$ of annihilation operators. So let us take Eq. (3.39) and rearrange the product of creation and annihilation operators such that at these reinterpreted annihilation operators $\hat{a}_a$ and $\hat{a}^\dagger_i$ are to the right of the reinterpreted creation operators $\hat{a}^\dagger_a$ and $\hat{a}_i$. This defines normal ordering with respect to the reference state.

Let us convert this into notation. We start from the Hamiltonian $\hat{H}$ and rewrite it in terms of the normal-ordered Hamiltonian $\hat{H}_N$:

$$\hat{H} = \langle \Phi| \hat{H} |\Phi\rangle + \hat{H}_N \tag{3.41}$$

and the expectation value of $\hat{H}$ with respect to the reference state $|\Phi\rangle$ given by:

$$\langle \Phi| \hat{H} |\Phi\rangle = \sum_i H^i_i + \frac{1}{2} \sum_{ij} H^{ij}_{ij} + \frac{1}{6} \sum_{ijk} H^{ijk}_{ijk} \,. \tag{3.42}$$

The normal-ordered Hamiltonian $\hat{H}_N$ again contains a one-body, two-body, and three-body part:

$$\hat{H}_N = \hat{H}_N^{[1]} + \hat{H}_N^{[2]} + \hat{H}_N^{[3]} , \qquad (3.43)$$

each normal-ordered with respect to the reference state. We denote the normal-ordered k-body part as:

$$\hat{H}_N^{[k]} = \frac{1}{(k!)^2} \sum_{p_1,\dots,p_k} \sum_{q_1,\dots,q_k} \tilde{H}_{q_1\dots q_k}^{p_1\dots p_k} \{ \hat{a}_{p_1}^\dagger \cdots \hat{a}_{p_k}^\dagger \hat{a}_{q_k} \cdots \hat{a}_{q_1} \} , \qquad (3.44)$$

where the curly brackets enclosing the string of creation and annihilation operators indicate that this product is arranged in normal order with respect to the reference state, i.e., that particle annihilation and hole creation operators are to the right of particle creation and hole annihilation operators. Since the sums over the single-particle indices p_i and q_i run over particle and hole states, the curly brackets indicate that the proper normal order is established for each term of the sum. As a result, we always have $\langle \Phi | \{ \hat{a}_p^\dagger \cdots \hat{a}_q \} | \Phi \rangle = 0$ and $\langle \Phi | \hat{H}_N^{[k]} | \Phi \rangle = 0$, in perfect analogy to the relations for the vacuum state $|0\rangle$ in vacuum normal order.

Obviously, we cannot simply change the order of creation and annihilation operators; there are the fermionic anti-commutation relations that we have to obey. Changing the order of creation and annihilation operators produces an extra term, $\hat{a}_p^\dagger \hat{a}_q = \delta_{pq} - \hat{a}_q \hat{a}_p^\dagger$, which has fewer creation and annihilation operators. Starting from a k-body operator (in vacuum normal order), there will be contributions of lower particle ranks in reference normal order, generated by the extra terms. They show up, in the coefficient $\tilde{H}_{q_1\dots q_k}^{p_1\dots p_k}$ in front of the normal-ordered operators:

$$\tilde{H}_q^p = H_q^p + \sum_i H_{qi}^{pi} + \frac{1}{2} \sum_{ij} H_{qij}^{pij} , \quad \tilde{H}_{rs}^{pq} = H_{rs}^{pq} + \frac{1}{4} \sum_i H_{rsi}^{pqi} , \quad \tilde{H}_{stu}^{pqr} = H_{stu}^{pqr} .$$

$$(3.45)$$

Working out the commutator algebra manually is a tedious job and can be circumvented by the use of Wick's theorem. We will not go into details here and rather refer to the literature [61].

As mentioned already, normal ordering is instrumental for the formulation of some many-body methods, and we will come back to this in Sect. 3.6. It can be used to define in approximation for the multi-nucleon terms in Hamiltonian, specifically for the contribution of the 3N interaction. Looking at the matrix elements in (3.45), we observe that the three-body contributions H_{stu}^{pqr} of the initial Hamiltonian (in vacuum normal order) enter also into the two-body and one-body part of the reference normal-ordered Hamiltonian. The three-body matrix elements in these terms are partially summed over one or two occupied levels, so effectively selected three-body terms are demoted to lower particle rank. This covers already an important part of the three-body effects, and we might omit the residual normal-ordered three-body term in good approximation. This is the so-called normal-ordered two-body (NO2B) approximation [51, 62, 63].

The great advantage of the NO2B approximation is that the many-body method only has to deal with up to two-body terms, which is a significant formal and computational simplification. We can convert the NO2B Hamiltonian back to vacuum normal order and obtain:

$$\hat{H}_{\text{NO2B}} = \frac{1}{6} \sum_{ijk} H_{ijk}^{ijk} + \sum_{pq} \left(H_q^p - \frac{1}{2} \sum_{ij} H_{qij}^{pij} \right) \hat{a}_p^\dagger \hat{a}_q$$

$$+ \frac{1}{4} \sum_{pqrs} \left(H_{rs}^{pq} + \sum_i H_{rsi}^{pqi} \right) \hat{a}_p^\dagger \hat{a}_q^\dagger \hat{a}_s \hat{a}_r , \qquad (3.46)$$

which can be readily used in many-body methods that do not employ normal ordering otherwise (e.g., the no-core shell model).

So far, we have considered a reference state given by a single Slater determinant, a so-called *single-reference* scheme. What about a reference state that is a superposition of multiple Slater determinants? The immediate consequence is that the simple distinction between particle and hole states breaks down. There will be single-particle states that are occupied in some of the determinants making up the reference state and unoccupied in others. Despite this complication, there is a generalization of normal ordering and Wick's theorem [64–66] for these multi-determinantal reference states or *multi-reference* states for short. From a practical point of view, the partitioning of the summations into particle and hole states is replaced by unrestricted sums, and the information on the structure of the reference state is introduced via density matrices. A more detailed discussion of multi-reference normal ordering and the resulting NO2B approximation can be found in [67].

3.5 Diagonalization Approaches

After our extensive preparations, it is now almost trivial to define a first class of many-body methods. Taking the basis expansion idea introduced in Sect. 3.2 literally, we are left with a large-scale matrix eigenvalue problem for the Hamiltonian represented in the A-body basis of choice. The Hamiltonian and the many-body basis make use of the preconditioning methods discussed in the previous section, which are hidden in the A-body matrix elements. This general class of methods is typically identified as configuration interaction (CI) or exact diagonalization approaches. Methods of this class still differ in the way the many-body basis is truncated and how convergence is assessed.

3.5.1 Many-Body Truncations

There are different physics-motivated strategies to truncate the many-body basis and to define the many-body model space. Essentially, different truncation schemes define different many-body methods. We discuss the main contenders, provide a physics motivation for the truncation, and comment on uncertainty quantification.

Full Configuration Interaction Starting from a finite set of single-particle states and building Slater determinants from all possible combinations automatically result in a finite set of A-body basis states spanning the model space of a so-called full configuration interaction (FCI) scheme. In the context of the HO basis, we can use the principal quantum number $e = 2n + l$ to define a single-particle truncation $e \leq e_{\max}$ with a control parameter $e_{\max}$. The model space of this FCI scheme can be characterized as:

$$\mathcal{M}_{\mathrm{FCI}} = \{\text{all Slater determinants } |p_1 p_2 ... p_A\rangle \text{ with } e_{p_i} \leq e_{\max}\} . \tag{3.47}$$

Obviously, in the limit $e_{\max} \rightarrow \infty$, the FCI calculation will approach the exact result. Therefore, it is in principle straightforward to assess the convergence behavior via an explicit variation of $e_{\max}$. We can derive uncertainty estimates or construct extrapolation schemes to deal with incomplete convergence—all of this is facilitated by the fact that a single parameter, $e_{\max}$, controls the truncation.

This truncation is motivated by the assumption that energetically high-lying single-particle states will contribute less to the basis expansion of low-lying eigenstates. This resonates with the discussion of decoupling of low- and high-energy states in Sect. 3.4.2. However, the FCI truncation still allows for configurations where all particles occupy the highest available single-particle states—it is very unlikely that such configuration will appear in the expansion of low-lying eigenstates. Not surprisingly, FCI calculations turn out to be rather inefficient in nuclear structure applications.

Particle-Hole-Truncated Configuration Interaction In addition to the single-particle truncation of FCI, we use a truncation on the number of particle-hole pairs that distinguish a basis state from a specific reference state $|\Phi\rangle$. The reference state represents a specific basis determinant, typically the one with the A energetically lowest single-particle states occupied and can be obtained, e.g., from an HF calculation. All basis states can be classified according to the number of single-particle states that differ from the reference state $|\Phi\rangle$, or, equivalently, the number of particle-hole excitations needed to construct a basis determinant from the reference state.

We can use the creation and annihilation operators of the second quantization to define n-particle-n-hole ($npnh$) excitations on top of the reference state $|\Phi\rangle$:

1p1h : $\qquad |\Phi_i^a\rangle = \hat{a}_a^\dagger \hat{a}_i \, |\Phi\rangle$

2p2h : $\qquad |\Phi_{ij}^{ab}\rangle = \hat{a}_a^\dagger \hat{a}_b^\dagger \hat{a}_j \hat{a}_i \, |\Phi\rangle$

3p3h : $\qquad |\Phi_{ijk}^{abc}\rangle = \hat{a}_a^\dagger \hat{a}_b^\dagger \hat{a}_c^\dagger \hat{a}_k \hat{a}_j \hat{a}_i \, |\Phi\rangle \, ,$ $\hfill (3.48)$

where we again use the index convention with i, j, k being hole states that are occupied in the reference state and a, b, c particle states that are unoccupied in $|\Phi\rangle$. This hierarchy can be easily extended to the A-particle-A-hole level.

Including all levels of this particle-hole hierarchy will simply recover the FCI model space, and we have only structured the basis in a physically useful manner. However, we can introduce an additional, physics-motivated truncation by only keeping basis states up to a specific maximum $npnh$ level and, thus, define the $npnh$-truncated model space of CI($npnh$):

$$\mathcal{M}_{\text{CI}(npnh)} = \{\text{all Slater determinants } |\Phi\rangle, \, |\Phi_i^a\rangle, \, |\Phi_{ij}^{ab}\rangle, \, ...$$

$$\text{up to the } npnh \text{ level with } e_a, e_b, \leq e_{\max}\} \, . \hfill (3.49)$$

Note that we still need the $e_{\max}$ truncation, since the particle-hole truncation alone would not render the model space finite. Thus, we have to handle two parameters that control the truncation of the model space, n and $e_{\max}$. This makes the study of the convergence as well as the quantification of uncertainties more difficult, particularly since the inclusion of the next particle-hole level typically increases the model space dimension by orders of magnitude.

The physics motivation behind the particle-hole hierarchy results from a perturbative consideration. We can consider the reference state as a simple first approximation for the ground state of our system—very much in the spirit of HF—and, thus, as the lowest-order approximation in a perturbative expansion. The first-order perturbative corrections to this state will only include up to 2p2h excitations on top of $|\Phi\rangle$ if the Hamiltonian contains up to two-body operators. This is because the amplitudes of the perturbative correction to the states involve matrix elements $\langle \Phi_{ij...}^{ab...} | \hat{H} | \Phi\rangle$, which vanish beyond the 2p2h level. Second-order corrections to the many-body states will include up to 4p4h excitations, etc. Thus, multi-particle-multi-hole states enter in increasing orders of perturbation theory, and we expect their contribution to be increasingly suppressed.

So far, we have considered the simple situation of a closed-shell system, which is characterized by a unique Slater determinant as a reference state. For an open-shell system, the filling of single-particle states in energetic order results in a partially occupied j-orbit, i.e., not all $(2j + 1)$ magnetic substates are filled. As a result, there are multiple possible reference Slater determinants with the same total unperturbed energy—there is a set of degenerate reference states. This trivial effect has huge consequences; it is at the heart of all the differences between single-reference (closed-shell) and much more complicated multi-reference (open-shell) methods that will be discussed later. At the moment, we are only interested in model space truncations in a CI framework, and there is an easy fix. Instead of counting particle-hole excitations with respect to a specific reference state, which will depend on which of the degenerate reference states we have picked, we count the single-particle states that are above the last (partially) occupied orbit of the degenerate reference states. This truncation is sometimes called T_{max} truncation, where T is the number of single-particle states above the last reference orbit. When applied in a closed-shell situation, the T_{max} truncation is equivalent to an $npnh$-truncation with $n = T_{max}$.

No-Core Shell Model The previous CI truncations are used across different fields of quantum physics and quantum chemistry. The truncation we are discussing now, which define the so-called no-core shell model (NCSM), is more specific to nuclear physics [68, 69]. In its pure version, the NCSM uses an HO single-particle basis in combination with a truncation with respect to the total HO energy of the many-body basis states. This total HO energy is parametrized in terms of a parameter N_{max}, which counts the HO excitation quanta above the lowest-energy HO determinant, i.e., the reference state. Formally, the number of excitation quanta N is obtained from:

$$N = \sum_{i=1}^{A} e_i - E_{ref}. \quad \text{with} \quad E_{ref} = \sum_{i=1}^{A} e_i^{ref}, \quad (3.50)$$

where E_{ref} is the total principal quantum number of the lowest-energy reference configuration whose single-particle state has principal quantum numbers e_i^{ref}. Note that degeneracy of reference determinants in the case of open-shell systems does not pose a problem here. For the HO basis, the number of excitation quanta N can be translated into an HO excitation energy by multiplying with $\hbar\Omega$. The NCSM model space is, thus, defined as:

$$\mathcal{M}_{NCSM} = \{\text{all Slater determinants } |p_1 p_2 ... p_A\rangle \text{ with } N \leq N_{max}\}. \quad (3.51)$$

Like the FCI scheme, the NCSM model space is based on a truncation with respect to an unperturbed energy. However, the NCSM uses the total energy of the many-body basis state, not the single-particle energy. Therefore, high-energy configurations with all particles in high-lying single-particle states are excluded.

Again, we can resort to perturbation theory to motivate this energy truncation. Looking at the corrections to the many-body state predicted in perturbation theory, the contributions of individual states are always scaled by energy denominators, which correspond exactly to the HO excitation energies $N \hbar\Omega$. Thus, with increasing N, the contribution of configurations gets more and more suppressed by the energy denominators. The N_{max} truncation, therefore, discards configurations that are expected to have small contributions based on a perturbative estimate. In this sense, the N_{max} truncation is much more physics-driven than the simple e_{max} truncation.

Nowadays, the NCSM truncation is also used with other single-particle basis sets than the HO, for example, the natural orbitals discussed in Sect. 3.4.3.3. In this case, the N_{max} parameter loses its direct connection to the unperturbed energies and is purely defined on the basis of a pseudo-principal quantum number $e = 2n + l$. Nevertheless, it remains a useful and efficient truncation, as the gross picture of energy scales is still valid. Since the NCSM is the main workhorse among the ab initio CI methods in nuclear physics, we will come back to it in Sect. 3.5.3.

Valence-Space Shell Model Also the traditional valence-space shell model (VSSM) of nuclear physics can be viewed as a specific incarnation of the general configuration interaction idea. It is based on a partitioning of the single-particle basis into three subsets: core states, valence states, and excluded sates. The many-body basis is then characterized by two different truncations: (1) a simple single-particle truncation that eliminates all the excluded single-particle states and (2) a selective truncation that retains only many-body configurations that have all core states filled. As a result, the basis configurations of the many-body model space all have A_c nucleons occupying the same A_c core states, and the states only differ in the assignments of the $A_v = A - A_c$ remaining valence nucleons to the valence single-particle states. We can summarize the VSSM model space as follows:

$$\mathcal{M}_{VSSM} = \{\text{all Slater determinants } |p_1 p_2 ... p_A\rangle \text{ with}$$
$$\{p_1, ..., p_{A_c}\} = \text{core and } \{p_{A_c+1}, ..., p_A\} \subset \text{valence space}\}. \tag{3.52}$$

For shell model practitioners, this definition might sound unfamiliar. Practical VSSM calculations do not work with an A-body Slater determinant basis, but with an A_v-body basis of the same dimension. Since the same core states are occupied in all basis states of the model space, their contribution to all relevant many-body matrix elements can be computed beforehand and absorbed into the remaining valence-space part of the matrix elements. This is a purely technical step akin to the normal ordering of operators discussed in Sect. 3.4.4. The reduction to an effective A_v-body problem does not imply additional approximations; it is an equivalent reformulation of the problem.

Although the VSSM can be viewed as a truncation of the A-body Hilbert space, its practical applications follow a different philosophy than the other CI-type approaches. In all ab initio CI approaches, the truncation parameters are varied in order to assess the convergence toward the full Hilbert space and to extract model space uncertainties. This is not done in typical VSSM calculations that work with a fixed valence space that is governed by computational feasibility and not by convergence considerations. In practice, a systematic variation of core and excluded space is often not possible, as model space dimensions become intractable very quickly. Therefore, VSSM calculations are traditionally performed in a phenomenological setting, with valence-space interactions fitted to nuclear observables, e.g., excitation energies, for nuclei in the respective valence shell. For other observables, e.g., electromagnetic transition strengths and moments, phenomenological corrections in the form of effective charges are introduced. In addition to purely phenomenological interactions, effective interactions derived in a perturbative framework or in a decoupling scheme are being used. In this way, the connection to the underlying Hamiltonian is retained; however, a quantification of model space uncertainties generally is still lacking.

Symmetry Reduction For all the truncations discussed so far, we can take into account symmetries to further reduce the basis dimension. The most important symmetries for our purpose are charge conservation and rotational invariance.

As a consequence of charge conservation, the number of protons and neutrons is fixed, and, thus, the total isospin projection $M_T = Z - N$ is a good quantum number of the nuclear eigenstates. In technical terms, there is a simultaneous eigenbasis of the Hamiltonian and the operator $\hat{T}_3 = \sum_i \hat{t}_{3,i}$ for the three-component of the total isospin. The Slater determinant basis states that span the CI model spaces are also eigenstates to $\hat{T}_3$, and the eigenvalue is the sum of all single-particle isospin projections m_t, cf. Eq. (3.2). Therefore, only basis states with appropriate M_T contribute to the expansion of the eigenstates, and we can discard all other basis states from the model space. Obviously, this is a simple and effective way to reduce the model space dimension without changing the results.

A similar argument applies to the projection M of the total angular momentum $\hat{J}_z = \sum_i \hat{j}_{z,i}$. Due to the rotational symmetry of the problem, the Hamiltonian does not connect states with different M quantum numbers, and the energy spectrum exhibits a degeneracy with respect to M. Therefore, we can choose a specific value of M at the beginning of the calculation and only include basis states with this specific M into the model space. Again, the Slater determinant states are also eigenstates to $\hat{J}_z$ with eigenvalues given by the sum of the single-particle projections m.

One could consider going one step further with rotational symmetry. The Hamiltonian also commutes with the square of the total angular momentum operator $\hat{\mathbf{J}}^2$; therefore, J is also a good quantum number for the nuclear eigenstates. We could focus on a specific value of J and limit the model space to basis states with this value of J. The problem here is that Slater determinants are generally not eigenstates of $\hat{\mathbf{J}}^2$; therefore, a simple basis-state selection is not possible. One can, however, use information on a specific target value for J through the choice of M. Usually one uses $M = 0$ (for even A) or $M = \frac{1}{2}$ (for odd A), because this is compatible with all possible J values. For targeting larger values of J, one can use $M = J$, which will reduce the model space dimension but exclude all eigenstates with $J < M$.

In order to exploit the good J quantum number, one has to use a J-coupled basis. The construction of such a basis is much more involved than the simple Slater determinant basis, and the computation of many-body matrix elements is far from trivial. Therefore, J-coupled basis sets are rarely used in the CI context. Exceptions are J-scheme versions of the valence-space shell model and Jacobi coordinate version of the NCSM.

A final remark: One can approach the question of symmetries from a completely different angle. Instead of trying to incorporate the good quantum number into the basis to reduce the model space dimension, one can break symmetries on purpose to enrich the basis and include specific correlations and later restore the broken symmetries explicitly in order to arrive at the nuclear eigenstate. This symmetry-breaking and restoration strategy is particularly helpful to describe collective correlations, also called static correlations, such as the intrinsic deformation of a nucleus.

3.5.2 Importance Truncation

All ab initio CI approaches are eventually limited by the trade-off between convergence and model space dimension. The model space has to be large enough to reach acceptable convergence of the target observable, but the computational cost for the calculation increases rapidly with model space size. With an increasing particle number A, one quickly faces the situation that the calculation cannot be converged with numerically tractable basis dimensions. However, there is one more trick—the selective removal of basis states from the model space using an adaptive, state-specific, and physics-guided truncation criterion [70, 71].

Assume we target a small number of low-lying eigenstates $|\Psi^{(m)}\rangle$ for $m = 1, ..., M$ in a CI calculation for a specific model space. The full calculation would yield eigenvectors representing the amplitudes $C_\nu^{(m)}$ for the expansion of the target eigenstates in terms of the many-body basis states $|\Phi_\nu\rangle$:

$$|\Psi^{(m)}\rangle = \sum_\nu C_\nu^{(m)} |\Phi_\nu\rangle \, . \tag{3.53}$$

Many of the amplitudes will be very small or vanishing, i.e., the corresponding basis states do not contribute significantly to the target states. If these amplitudes were known beforehand, we could reduce the basis dimension significantly by discarding those basis states.

In order to estimate the amplitudes a priori, we use initial approximations of the target states, the reference states $|\Psi_{\text{ref}}^{(m)}\rangle$, that are typically determined from a previous CI calculation in a smaller model space $\mathcal{M}_{\text{ref}}$:

$$|\Psi_{\text{ref}}^{(m)}\rangle = \sum_{\nu \in \mathcal{M}_{\text{ref}}} C_{\text{ref},\nu}^{(m)} |\Phi_\nu\rangle \, . \tag{3.54}$$

These reference states carry information about the physical properties of the target eigenstates. Guided by first-order multi-configurational perturbation theory, we estimate the amplitudes of the individual basis states $|\Phi_\nu\rangle \notin \mathcal{M}_{\text{ref}}$ in the expansion of the target eigenstate [71]. This first-order perturbative correction for the amplitudes defines the so-called importance measure:

$$\kappa_\nu^{(m)} = -\frac{\langle\Phi_\nu|\,\hat{H}\,|\Psi_{\text{ref}}^{(m)}\rangle}{\Delta\epsilon_\nu}, \tag{3.55}$$

where $\hat{H}$ is the full Hamiltonian and $\Delta\epsilon_\nu$ is the unperturbed HO excitation energy of the basis state $|\Phi_\nu\rangle$ [70, 71].

The importance measure combines information about the properties of the target states, carried by the reference states, about the many-body basis, and about the Hamiltonian. It is the foundation for the definition of a state-dependent adaptive truncation of the model space, the so-called importance truncation (IT). We define the importance-truncated model space $\mathcal{M}_{\text{IT}}(\kappa_{\text{min}})$ spanned by all states of the reference space $\mathcal{M}_{\text{ref}}$ plus all basis states $|\Phi_\nu\rangle \notin \mathcal{M}_{\text{ref}}$ with importance measure $|\kappa_\nu^{(m)}| \geq \kappa_{\text{min}}$ for at least one $m \in \{1, ..., M\}$. The importance threshold κ_{min} provides an additional truncation parameter, which will be varied later on to probe the contribution of the discarded basis states. Note that the importance measure (3.55) is based on the first-order perturbative correction to the states, not on the perturbative correction to the energies. It is, therefore, not biased to an optimal description of energies, but aims at an optimal description of the states and, thus, of all observables.

Depending on the specific flavor of CI, we can use different strategies to set up the importance-truncated model space. We can use iterative schemes, where the IT-space is successively refined and expanded by using improved reference states from a previous IT calculation. For the NCSM a particularly efficient scheme uses the eigenstate of the next smaller N_{max} as reference state for the construction of the IT model space.

The importance threshold κ_{min} acts as an additional truncation parameter of the CI model space, and it is guaranteed that in the limit $\kappa_{\text{min}} \to 0$ we recover the original CI model space. In practical applications, we typically perform calculations for a sequence of importance thresholds κ_{min} and use an extrapolation of the observables to vanishing importance threshold to effectively account for discarded basis states. More details on the practical aspects of IT calculations can be found in [71, 72].

3.5.3 No-Core Shell Model

Since the NCSM is the most commonly used ab initio CI method in nuclear structure physics [68, 69], we will discuss the main components of the calculation and illustrate the convergence behavior in a little more detail.

Setup and Numerics As discussed before, the NCSM uses a basis build from HO single-particle states and is truncated with respect to the maximum number of HO excitation quanta N_{max}. This model space has the unique advantage that the center-of-mass motion can be separated or factorized exactly from the intrinsic state of the system. Therefore, we can use simple tricks like adding an extra HO Hamiltonian for the center-of-mass to remove spurious center-of-mass excitations from the low-lying excitation spectrum. This model space also offers two alternative basis sets to work with: the Slater determinant HO basis and the relative coordinate or Jacobi HO basis, mentioned in Sect. 3.4.3.1. The former defines the standard m-scheme NCSM, which is more universal and applicable to heavier nuclei, and the latter the Jacobi NCSM, which is very efficient for light nuclei with $A \lesssim 6$ [73]. The standard NCSM has the advantage of numerical simplicity, e.g., for the computation of matrix elements, but it requires large basis dimensions. The Jacobi NCSM can realize much smaller basis dimensions for the same N_{max}, because all symmetries, including translational and rotational invariance, are exploited; however, the computation of matrix elements is much more complicated and practically feasible only for light nuclei.

Let us expand on the computational aspects of standard NCSM calculations in large model spaces. Today, advanced NCSM implementations handle model spaces with up to 10^{10} basis states [74, 75], i.e., they tackle the eigenvalue problem of a $10^{10} \times 10^{10}$ matrix, which might seem impossible. Storing this full matrix with single-precision floating-point numbers would require on the order of 10^8 TB of storage. Luckily, the Hamilton matrix is very sparse, and we only need to store the non-zero matrix elements. This is because a Hamiltonian with up to two- or three-body interactions connects only those pairs of basis states that differ by up to two or three single-particle states, respectively. As a result, the sparsity decreases with increasing particle rank of the Hamiltonian, which is why the NO2B approximation discussed in Sect. 3.4.4 can be useful in the NCSM as well.

These huge basis sizes require an efficient computation of many-body matrix elements—a significant fraction of the total runtime goes into the computation of the matrix elements of the Hamiltonian. The next step in the calculation is the solution of the matrix eigenvalue problem. Since we are only interested in a tiny fraction of the eigenstates at the lower end of the spectrum, we can use iterative Krylov subspace methods, like the famous Lanczos or Arnoldi algorithms. Each iteration of these algorithms requires a matrix-vector multiplication, which can be implemented efficiently with sparse matrix storage and even with distributed memory schemes. Only about 10 iterations are needed to converge the lowest eigenvalue, i.e., the ground state, and for about 100 iterations, we typically get the interesting part of the low-lying excitation spectrum. If the matrix elements are computed very efficiently, one can even consider computing them on the fly during the evaluation of matrix-vector products without ever storing them—this shifts characteristics of the calculation from being very storage intensive to being very compute intensive.

As for all CI methods, the solution of the eigenvalue problem yields the energies of the low-lying states and the eigenvectors that contain the coefficients for the basis expansion of the eigenstates in Eq. (3.5). They can be used in a post-processing step to evaluate all other observables that are defined in terms of expectation values

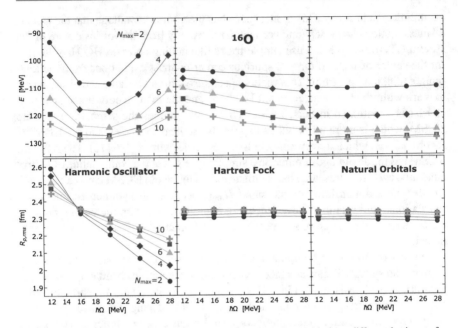

Fig. 3.2 Comparison of the convergence of NCSM calculations with three different basis sets for the ground-state energy and point-proton rms-radius for ^{16}O using an SRG-evolved chiral NN+3N interaction with $\alpha = 0.08\,\mathrm{fm}^4$ (Modified from [60])

or matrix elements with these eigenstates. Note that ground and excited states are always obtained on the same footing.

Convergence and Frequency Dependence The model space size is controlled solely by the truncation parameter N_{max}, and we recover the full Hilbert space in the limit $N_{\mathrm{max}} \to \infty$. The HO basis has an additional parameter, the oscillator frequency $\hbar\Omega$, and all observables have to become independent of $\hbar\Omega$ in the limit of large N_{max}. However, the choice of $\hbar\Omega$ will affect the convergence behavior of the calculation. An example for NCSM calculations with the HO basis is shown in the left-hand panels of Fig. 3.2 for the ground-state energy and the point-proton rms-radius of ^{16}O. The observables are plotted as function of $\hbar\Omega$ with different curves representing different N_{max}. For the ground-state energy, we observe the characteristic monotonic convergence from above—the curves shift downward and flatten out with increasing N_{max}. For the rms-radii, the convergence pattern is completely different; there is no variational principle that warrants monotonic convergence, and we observe that, depending on $\hbar\Omega$, the radius converges from above or below. For both observables there are sweet spots in $\hbar\Omega$ which yield the most rapid convergence. For the energy this is in the range of $\hbar\Omega = 16$–20 MeV and for the rms-radius at 16 MeV, where the radius is practically independent of N_{max}. We can use these optimal values of $\hbar\Omega$ to extract a most converged result and possibly even improve on it via an extrapolation $N_{\mathrm{max}} \to \infty$ with quantified uncertainties. This shows, however, that we have to perform NCSM calculations for multiple values of $\hbar\Omega$ to explore the frequency dependence of each observable.

Basis Optimization What does the convergence look like with the other single-particle basis sets? Of course, with the HF and the natural orbital states discussed in Sect. 3.4.3, we lose the formal properties of the HO, e.g., the exact separation of intrinsic and center-of-mass degrees of freedom. We also do not have the equidistant energy spectrum and the interpretation of $N_{max}\hbar\Omega$ as unperturbed excitation energy anymore, but we can still set up an N_{max}-truncated model space based on the quantum numbers of the single-particle states, using the pseudo-principal quantum number $e = 2n + l$ to evaluate N_{max}. As discussed in Sect. 3.4.3, we still use the HO to represent the single-particle states of these optimized basis sets; therefore, there is still an $\hbar\Omega$ parameter involved.

The NCSM results for ^{16}O with the HF and the natural orbital basis are depicted in the center and right-hand columns of Fig. 3.2. The convergence of the ground-state energy with the HF basis shows an anomalous pattern: the frequency dependence is reduced, but the energies drop almost linearly with N_{max}, and the $N_{max} = 10$ results are well above the corresponding HO energies. This is related to the pathologies of the HF basis discussed in Sect. 3.4.3.2, which is not suitable for NCSM calculations.

The situation is different with the natural orbital basis. As evident from the right-hand panels of Fig. 3.2, energies and radii are, for all practical purposes, independent of $\hbar\Omega$. The N_{max}-convergence of both observables is comparable to the convergence with the HO basis for the optimal $\hbar\Omega$. This is a great result! We do not need to optimize $\hbar\Omega$ anymore; this is done implicitly by the natural orbitals already, because it contains global information on the nucleus, such as its size.

Oxygen Isotopic Chain Using the natural orbital basis, we can attempt large-scale NCSM calculations that reach the limit of particle numbers the NCSM can handle. We consider the oxygen isotopic chain from the neutron-deficient ^{14}O to ^{26}O beyond the neutron dripline, and the results are summarized in Fig. 3.3. Because of the preconditioning, using the natural orbital basis with an SRG-evolved Hamiltonian, we are close to convergence for $N_{max} = 10$, and simple exponential extrapolations can be used to obtain a final energy with an uncertainty estimate. Reaching these large N_{max} parameters for $A \approx 20$ systems requires the IT scheme discussed in Sect. 3.5.2. The right-hand panel of Fig. 3.3 shows the extrapolated ground-state energies for calculations with explicit 3N terms in the Hamiltonian in comparison to the NO2B approximation. We observe a small overestimation of the ground-state energy by 1–2 MeV when using the NO2B approximation, which is acceptable given that the NCSM calculation speeds up by a factor of 10 due to the increased sparsity of the Hamilton matrix.

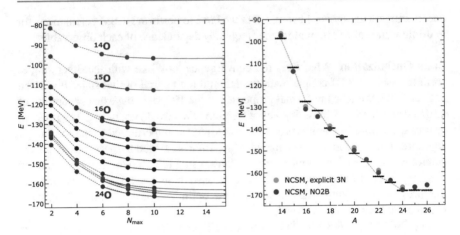

Fig. 3.3 NCSM calculations for the ground-state energies of oxygen isotopes. Left: Convergence of the energy as function of N_{max} using the natural orbital basis together with an exponential extrapolation. Right: Extrapolated ground-state energies obtained with the full 3N interaction and with the NO2B approximation in comparison to experiment

For going much beyond the oxygen isotopes, we have to resort to a different many-body strategy—we will move from diagonalization to decoupling.

3.6   Decoupling Approaches

We have already established the conceptual relation between *diagonalization* and *decoupling* during the discussion of unitary transformation and the SRG in Sect. 3.4.2. Now we transfer this directly to the methods for the solution of the many-body problem. The CI approaches discussed in the previous section use the concept of diagonalization—we construct a matrix representation of the Hamiltonian and solve the matrix eigenvalue problem, which is equivalent to a diagonalization of the matrix. Typically we will only extract a few low-lying eigenvalues and eigenvectors, which can be viewed as a selective diagonalization of a few rows and columns of the matrix. This structure of the matrix could also be viewed as a specific decoupling, i.e., a suppression of the off-diagonal matrix elements that connect the low-lying states with the rest of the model space.

3.6.1   In-Medium Similarity Renormalization Group

The most obvious implementation of the decoupling strategy is the SRG framework discussed in Sect. 3.4.2. We can use the SRG flow equation to drive a continuous decoupling of a selected state or subspace from the rest of the model space. In contrast to the free-space SRG, we now aim at a pre-diagonalization of the Hamiltonian in A-body space. As mentioned earlier, the direct solution of the flow equations for the Hamiltonian in an A-body CI-type matrix representation is neither advantageous nor feasible. Therefore, we combine the SRG flow equation with the normal ordering of the Hamiltonian and a truncation analogous to the NO2B approximation discussed in Sect. 3.4.4. This results in the so-called in-medium similarity renormalization group (IM-SRG) [76–80].

Let us start with the single-reference formulation of normal ordering suitable for closed-shell nuclei. We define a reference state $|\Phi\rangle$ as a single Slater determinant constructed, e.g., in a previous HF calculation or with a natural orbital basis. We convert the relevant operators for the formulation of the SRG flow equation into normal-ordered form with respect to this reference state and truncate after the normal-ordered two-body terms, i.e., the Hamiltonian:

$$\hat{H}(s) = E(s) + \sum_{pq} H_q^p(s)\,\{\hat{a}_p^\dagger \hat{a}_q\} + \frac{1}{4}\sum_{pqrs} H_{rs}^{pq}(s)\,\{\hat{a}_p^\dagger \hat{a}_q^\dagger \hat{a}_s \hat{a}_r\}\ . \tag{3.56}$$

and the anti-Hermitian generator

$$\hat{\eta}(s) = \sum_{pq} \eta_q^p(s)\,\{\hat{a}_p^\dagger \hat{a}_q\} + \frac{1}{4}\sum_{pqrs} \eta_{rs}^{pq}(s)\,\{\hat{a}_p^\dagger \hat{a}_q^\dagger \hat{a}_s \hat{a}_r\}\ . \tag{3.57}$$

The Hamiltonian, the generator, and all their matrix elements are functions of the IM-SRG flow parameter s (we reserve α for the free-space SRG flow parameter). The zero-body part of the Hamiltonian, i.e., the expectation value of the Hamiltonian in the reference state, is denoted by $E(s)$. These normal-ordered operators enter in the IM-SRG flow equation, which looks just like the general SRG flow equation (3.18):

$$\frac{d}{ds}\hat{H}(s) = \left[\hat{\eta}(s), \hat{H}(s)\right]\ . \tag{3.58}$$

After working out the commutator with normal-ordered operators and truncating again after the two-body level, we obtain a system of coupled first-order differential equations for the normal-ordered zero-body, one-body, and two-body matrix elements of the Hamiltonian:

$$
\frac{d}{ds} E = \sum_{pq} \eta_q^p H_p^q (n_p - n_q) + \frac{1}{2} \sum_{pqrs} \eta_{rs}^{pq} H_{pq}^{rs} n_p n_q \bar{n}_r \bar{n}_s
$$

$$
\frac{d}{ds} H_2^1 = \sum_p \left[\eta_p^1 H_2^p + (1 \leftrightarrow 2) \right] + \sum_{pq} (n_p - n_q) \left(\eta_q^p H_{p2}^{q1} - H_q^p \eta_{p2}^{q1} \right)
$$

$$
+ \frac{1}{2} \sum_{pqr} \left[\eta_{pq}^{r1} H_{r2}^{pq} (n_p n_q \bar{n}_r + \bar{n}_p \bar{n}_q n_r) + (1 \leftrightarrow 2) \right]
$$

$$
\frac{d}{ds} H_{34}^{12} = \sum_p \left[\eta_p^1 H_{34}^{p2} - H_p^1 \eta_{34}^{p2} - (1 \leftrightarrow 2) \right] - \sum_p \left[(\eta_3^p H_{p4}^{12} - H_3^p \eta_{p4}^{12}) - (3 \leftrightarrow 4) \right]
$$

$$
+ \frac{1}{2} \sum_{pq} \left[\eta_{pq}^{12} H_{34}^{pq} (1 - n_p - n_q) + (1, 2 \leftrightarrow 3, 4) \right]
$$

$$
- \sum_{pq} (n_p - n_q) \left[(\eta_{p4}^{q2} - \eta_{q3}^{p1} H_{p4}^{q2}) - (1 \leftrightarrow 2) \right] , \tag{3.59}
$$

where we have omitted the flow parameter arguments for brevity. We have introduced single-particle occupation numbers $n_p \in \{0, 1\}$ and $\bar{n}_p = 1 - n_p$ of the reference state to effectively distinguish hole and particle states. Note that the external single-particle indices that enter on the left-hand side are simply denoted by numbers to easily distinguish them from the additional summation indices.

We have not specified the generator $\hat{\eta}(s)$ yet, but we can draw on our discussion in Sect. 3.4.2. The simplest choice for $\hat{\eta}(s)$ is again the Wegner ansatz (3.19):

$$
\hat{\eta}_W(s) = [\hat{H}^d(s), \hat{H}(s)] = [\hat{H}^d(s), \hat{H}^{od}(s)] \tag{3.60}
$$

with a commutator of the diagonal parts $\hat{H}^d(s)$ and off-diagonal parts $\hat{H}^{od}(s)$ of the Hamiltonian. The Wegner generator is not very efficient, and in practical applications, other choices, mainly the imaginary-time and White generators, are being used. We refer to the literature for more details [78–80]. For all generators we still have to decide what we consider as diagonal and off-diagonal; this defines the decoupling pattern.

The single-reference IM-SRG for closed-shell nuclei uses a simple and extreme decoupling pattern. The off-diagonal part of the Hamiltonian is everything that connects the reference state $|\Phi\rangle$ to any other basis state. In the language of the matrix elements $H_p^q(s)$ and $H_{pq}^{rs}(s)$, all matrix elements that connect particle and hole states are considered off-diagonal. Consequently the IM-SRG aims to decouple the reference state from all particle-hole excitations. This is illustrated with an actual IM-SRG evolution for ^{16}O in Fig. 3.4. The top row depicts matrix representations of the Hamiltonian in a particle-hole CI basis with increasing flow parameters marked by triangles. The flow evolution will selectively diagonalize the matrix with respect

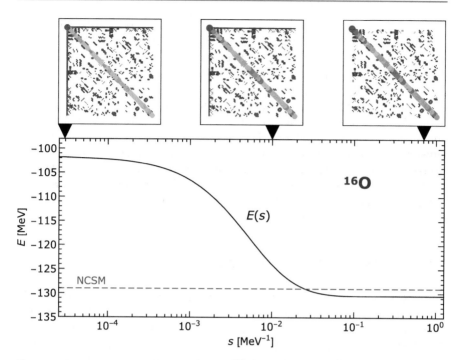

Fig. 3.4 Illustration of an IM-SRG evolution for ^{16}O. The bottom panel shows the flow parameter dependence of the zero-body part $E(s)$ of the Hamiltonian in comparison to the NCSM result for the ground-state energy. The upper panels depict a part of the evolved Hamiltonian in a particle-hole basis representation for different values of s. Each circle indicates a non-vanishing matrix element, and its area encodes the absolute value of the matrix element

to the first row and column with the matrix element $\langle\Phi|\,\hat{H}(s)\,|\Phi\rangle$ on the diagonal. The lower panel shows the evolution of this matrix element, which constitutes zero-body part of the flowing Hamiltonian $E(s) = \langle\Phi|\,\hat{H}(s)\,|\Phi\rangle$. With increasing flow parameter, $E(s)$ first decreases and then stabilizes once the decoupling is achieved. In this decoupled regime, $E(s)$ corresponds to an eigenvalue and directly represents the ground-state energy of the system.

Addressing other ground-state observables, e.g., the rms-radius, requires more work. As discussed in Sect. 3.4.2.2, we have to transform the matrix elements of the observable consistently using the same IM-SRG flow evolution. At this point the Magnus formulation of the flow equations comes in handy (cf. Sect. 3.4.2.2) and is used in many state-of-the-art IM-SRG implementations.

The numerical character of IM-SRG calculations is very different from the CI or NCSM approaches discussed before. We are dealing with an initial value problem for a system of coupled first-order differential equations (3.59) for the matrix elements of the normal-ordered Hamiltonian. Similar to FCI, the single-particle basis has to be truncated to arrive at a finite set of equations. Beyond the m-scheme formulation discussed here, we can use angular momentum-coupled

matrix elements and exploit their symmetries in order to reduce the number of
equations drastically. In this way, large single-particle basis sets become tractable
with moderate computational effort. Note that the particle number A does not
directly affect the dimension of the system of differential equations (3.59), so heavy
nuclei are accessible in principle.

A final comment regarding uncertainties: Multiple truncations are being used
in the IM-SRG framework—the truncation of the single-particle basis, the NO2B
approximation of the initial Hamiltonian, the NO2B truncation of commutator
terms leading to the flow equations, and further truncations at the level of the 3N
matrix elements entering the calculation. All these truncation potentially affect the
results, and it is difficult to quantify them explicitly within the IM-SRG method.
Therefore, comparisons with other many-body approaches, e.g., the NCSM as
shown in Fig. 3.4, are important to gauge the accuracy.

3.6.2 In-Medium No-Core Shell Model

The single-reference IM-SRG formulation is rather limited, and we would certainly
like to address excited states and open-shell nuclei as well in an IM-SRG framework.
One option is to use the single-reference IM-SRG formulation in conjunction with
the VSSM [81, 82]. Here the IM-SRG is used to decouple the valence space from
the closed-shell core and from the excluded space. This approach has been used
successfully in a range of different applications, but it inherits the limitations of the
VSSM discussed in Sect. 3.5.1.

A more powerful option is the use of a multi-reference formulation of the
IM-SRG. It results from the combination of the SRG flow equations with multi-
reference normal ordering, mentioned in Sect. 3.4.4. Instead of being limited
to single-determinant reference states, we can now use much more elaborate
reference states, tailored for open-shell situations. The first formulation of a
multi-reference IM-SRG approach employed reference states from particle number-
projected Hartree-Fock-Bogoliubov calculations, which give access to semi-magic
isotopic chains [83,84]. However, we can go further than this and combine the multi-
reference IM-SRG with reference states from the NCSM. This will lead to a new
hybrid ab initio method, the in-medium no-core shell model (IM-NCSM) [85].

We will not go into the equations for the general multi-reference IM-SRG but
refer to the literature [80, 85]. Compared to the single-reference equations (3.59),
the information on the reference states is encoded in density matrices in addition
to the occupation numbers, which directly result from the multi-reference normal
ordering. For the IM-NCSM, we use an NCSM eigenstate for a small model space,
typically $N_{max} = 0$ or 2 as reference state. The reference state already contains a
significant amount of correlations, and we control the complexity of the reference
state through the size of the model or reference space the state is obtained from.
Since we will be using an angular momentum-coupled formulation of the multi-
reference flow equations, we will limit ourselves to reference states with $J = 0$
leading to scalar density matrices.

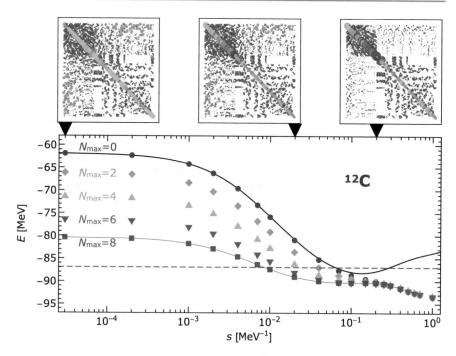

Fig. 3.5 Illustration of an IM-NCSM evolution for ^{12}C. The bottom panel shows the flow parameter dependence of the zero-body part $E(s)$ as well as NCSM results for the ground-state energy obtained with the evolved Hamiltonian in different model spaces. The upper panels depict a part of the evolved Hamiltonian in an NCSM basis representation for different values of s. Each circle indicates a non-vanishing matrix element, and its area encodes the absolute value of the matrix element

The multi-reference IM-SRG flow evolution will decouple the reference space from the rest of the model space. This is illustrated in Fig. 3.5 showing the flow evolution for ^{12}C with an $N_{\max}^{\text{ref}} = 0$ reference space. The upper panels depict a part of Hamiltonian in the NCSM many-body basis at different flow parameters. The upper-left corner of the matrix shows the $N = 0$ sector of the matrix, the rest belongs to the $N = 2$ subspace. We observe that the $N = N_{\max}^{\text{ref}} = 0$ block is decoupled from the rest of the model space throughout the flow evolution, i.e., the off-diagonal blocks are suppressed. As a result of the IM-SRG evolution, we obtain an approximately block-diagonal Hamiltonian, which serves as import for a second NCSM calculation to extract observables. The $N_{\max}$-convergence of the ground-state energies obtained in these NCSM calculations at different points in the flow evolution is depicted by the colored symbols in Fig. 3.5. At small flow parameters, we observe the usual slow convergence of a standard NCSM calculation, with a HF basis in this case. However, with increasing flow parameter, the convergence accelerates up to a point, where all calculations starting from $N_{\max} = 0$ provide the same result. This is the regime where the decoupling of the $N_{\max} = 0$ block is complete, and, therefore, already the diagonalization within this block provides the converged ground-state energy. The converged ground-state energy is stable over a range of flow parameters but starts to change at very large s signaling the impact of induced normal-ordered multi-particle contributions that have been truncated in the flow equations.

The combination of a NCSM calculation for the construction of the reference state, the multi-reference IM-SRG evolution for the coupling of the residual model space, and another NCSM calculation for the extraction of converged observables is very powerful. The final NCSM calculation gives access to exited states and to all relevant observables. Again, the operators for other observables, e.g., radii or electromagnetic properties, have to be transformed consistently in the IM-SRG evolution and can use a Magnus formulation of the multi-reference IM-SRG for this.

The quantification of many-body uncertainties is more difficult in the IM-NCSM than in the NCSM, since multiple truncations are involved. Nevertheless, we can construct uncertainty quantification protocols that use the dependence of observables on the main control parameters of the calculation, i.e., N_{max}^{ref}, N_{max}, and s. To validate these protocols, we can use comparisons to other ab initio methods with known uncertainties. This is illustrated in Fig. 3.6 for the ground-state energies

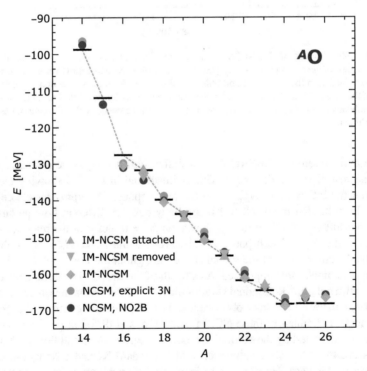

Fig. 3.6 IM-NCSM calculations for the ground-state energies of oxygen isotopes in comparison to the corresponding NCSM results (cf. Fig. 3.3). For odd isotopes the IM-NCSM calculation uses a particle-attached or particle-removed scheme starting from even $J = 0$ reference states. All calculations use an SRG-evolved chiral NN+3N interaction with $\alpha = 0.08$ fm^4

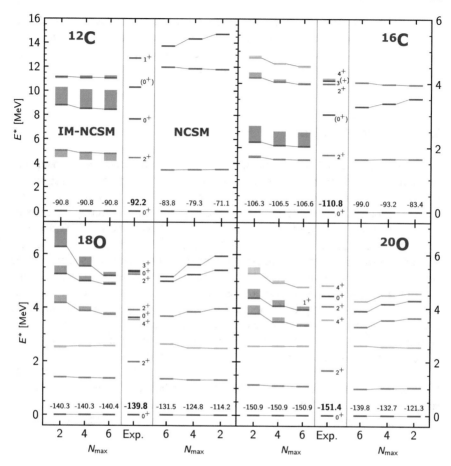

Fig. 3.7 IM-NCSM and NCSM results for the excitation spectra of selected carbon and oxygen isotopes as function of N_{max}. All calculations use an SRG-evolved chiral NN+3N interaction with $\alpha = 0.08\,\text{fm}^4$. The bands shown for the IM-NCSM results provide an indication of the IM-SRG flow parameter dependence (Modified from [85])

of oxygen isotopes in comparison to the NCSM results discussed before. We observe a very good agreement between the two calculations. The small deviations correspond to the expected effect of the NO2B truncation on the order of 2 MeV. A similar comparison for excitation energies is presented in Fig. 3.7, where panel shows the IM-NCSM convergence on the left-hand side and the NCSM convergence of the right-hand side. For those states that are well converged, we again observe good agreement; however, some states converge slowly or are sensitive to the flow parameter of the IM-SRG evolution as indicated by the shaded bands. A prominent example is the 0^+ state in ^{12}C, which has large many-body uncertainties in both NCSM and IM-NCSM.

The IM-NCSM provides access to the same range of observables as the NCSM, for ground excited states and for open and closed-shell systems all on the same footing. Since $N_{\max}$-convergence can be reached in very small model spaces, e.g., $N_{\max} = 0$ or 2, much heavier nuclei up into the calcium region can be described with moderate computational effort.

3.7 Things Left Out

Unfortunately, we could not cover all of the recent developments in ab initio nuclear structure theory, not even in the domain of basis expansion methods. Therefore, we would like to provide a few references to recent review articles for filling these gaps.

In the group of decoupling methods, another prominent and important member is coupled-cluster theory, which shares some aspects with the IM-SRG and is a standard method in many fields of quantum many-body physics and chemistry. We refer to [86] for an overview. Another medium-mass method used in nuclear structure theory is based on propagator theory and known as self-consistent Green's function method. A recent pedagogical review can be found in [87]. A big group apart from the diagonalization and the decoupling approaches are methods built on many-body perturbation theory. There are many different incarnations of perturbation theory, also in hybrid schemes combined with other ab initio methods like the NCSM. A comprehensive overview can be found in [88]. Beyond the basis expansion methods, there is an exciting progress on quantum Monte Carlo methods for finite nuclei, presented, e.g., in [89]. Moreover, lattice EFT methods, merging chiral EFT directly with lattice simulation techniques, have provided exciting results [90]. A nice overall summary of the current state of ab initio methods is provided in [91].

3.8 The Future of Ab Initio Nuclear Structure

Instead of a summary, we provide a brief and somewhat biased outlook into the future of ab initio nuclear structure theory. We are at a point where the focus of the research work in this field is shifting. Over the past decade, the focus was on the development of ab initio frameworks that *make calculations possible*, i.e., that extend the reach of ab initio methods to heavier nuclei, open-shell systems, excited states, electromagnetic observables, etc. These developments will continue, but the focus is shifting to methodological advances that *make calculations precise and accurate*.

Discussing the *precision* of ab initio calculations requires the quantification of all the theory uncertainties that accumulate on the way from the chiral EFT formulation of the interactions to the many-body observables. Throughout the lecture we have discussed individual sources of uncertainties and ways to quantify them within the respective framework. All of this has to be propagated through the whole ab initio toolchain to the eventual observable. Illustrations of how uncertainty quantified ab

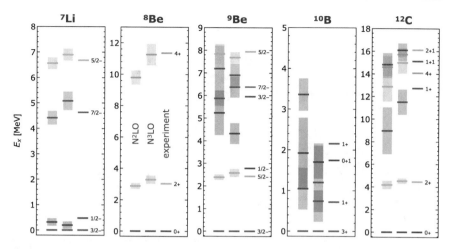

Fig. 3.8 NCSM calculations for the excitation spectra of p-shell nuclei using a family of nonlocal chiral NN+3N interactions up to N^3LO for cutoff $\Lambda = 500$ MeV. The excitation energies at N^2LO and N^3LO are shown with uncertainty bands extracted from the order-by-order behavior of the chiral expansion (Modified from [14])

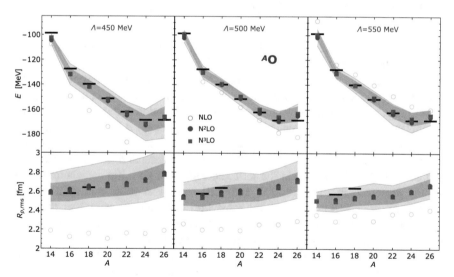

Fig. 3.9 IM-NCSM calculations for the ground-state energies and point-proton rms-radii of even oxygen isotopes using a family of nonlocal chiral NN+3N interactions up to N^3LO for three different cutoffs. Shown are the results at NLO, N^2LO, and N^3LO. The results for the higher order are shown with uncertainty bands that include the chiral truncation uncertainty and the many-body uncertainties (Modified from [14])

initio results will look like are shown in Figs. 3.8, 3.9, and 3.10. We apply the full range of ab initio methods discussed in this lecture, starting from the NCSM for the description of light nuclei and their excitation spectra in Fig. 3.8, via the description

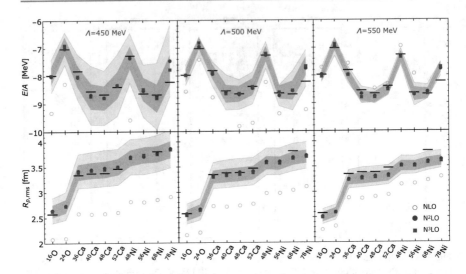

Fig. 3.10 IM-SRG calculations for the ground-state energies and point-proton rms-radii of closed-shell isotopes up to ^{78}Ni using a family of nonlocal chiral NN+3N interactions up to N^{3}LO for three different cutoffs. See Fig. 3.9 for details (Modified from [14])

of beyond p-shell nuclei in the IM-NCSM in Fig. 3.9, to the study of medium-mass closed-shell nuclei in the IM-SRG in Fig. 3.10

All calculations use a family of chiral NN+3N interactions, presented in [14], that allow for a systematic variation of the chiral order and the cutoff and, thus, enable a quantification of uncertainties due to the truncation of the chiral expansion. To this end, the many-body calculations have to be repeated for each chiral order and each cutoff, i.e., the computational cost multiplies. This is the price to pay for assessing the uncertainties related to the input interaction, an aspect that was not addressed quantitatively in the past generations of ab initio calculations. The results presented in Figs. 3.8, 3.9, and 3.10 show that these uncertainties can be sizable; in most cases the chiral truncation uncertainties are larger than the many-body truncation uncertainties. Therefore, increasing the precision of the calculations primarily requires a reduction of the uncertainties associated with the input interactions. Work along these lines is under way with chiral EFT interactions that include conceptual improvements and higher orders [92, 93].

Assessing the *accuracy* of ab initio calculations, i.e., the agreement with experiment, once the precision is acceptable, i.e., once theory uncertainties are small enough to enable meaningful comparisons, is a next step. With the precision of the present calculations, all observables are in agreement with experiment within the estimates uncertainties. With an improved precision, this can change, and we might find discrepancies between theory and experiment, hinting at weak links in the chain of ab initio tools that connect nuclear structure observables to the underlying theory of the strong interaction.

Acknowledgments Supported by the DFG through the Sonderforschungsbereich (SFB) 1245 (Project ID 279384907) and the BMBF through Verbundprojekt 05P2021 (ErUM-FSP T07, Contract No. 05P21RDFNB). Calculations were performed using the Lichtenberg II high-performance computer at the Technische Universität Darmstadt.

References

1. S.R. Beane, W. Detmold, K. Orginos, M.J. Savage, Prog. Part. Nucl. Phys. **66**, 1 (2011)
2. M.J. Savage, Prog. Part. Nucl. Phys. **67**, 140 (2012)
3. T. Inoue, Few Body Syst. **62**, 106 (2021)
4. S. Weinberg, Phys. Lett. B **251**, 288 (1990)
5. H. Hergert, R. Roth, Phys. Lett. B **682**, 27 (2009)
6. U. van Kolck, Prog. Part. Nucl. Phys. **43**, 337 (1999)
7. P.F. Bedaque, U. van Kolck, Ann. Rev. Nucl. Part. Sci. **52**, 339 (2002)
8. E. Epelbaum, Prog. Part. Nucl. Phys. **57**, 654 (2006)
9. E. Epelbaum, H.W. Hammer, U.G. Meissner, Rev. Mod. Phys. **81**, 1773 (2009)
10. R. Machleidt, D.R. Entem, Phys. Rept. **503**, 1 (2011)
11. E. Epelbaum, U.G. Meissner, Ann. Rev. Nucl. Part. Sci. **62**, 159 (2012)
12. H.W. Hammer, S. König, U. van Kolck, Rev. Mod. Phys. **92**, 025004 (2020)
13. Y. Nosyk, D.R. Entem, R. Machleidt, Phys. Rev. C **104**, 054001 (2021)
14. T. Hüther, K. Vobig, K. Hebeler, R. Machleidt, R. Roth, Phys. Lett. B **808**, 135651 (2020)
15. W.G. Jiang, A. Ekström, C. Forssén, G. Hagen, G.R. Jansen, T. Papenbrock, Phys. Rev. C **102**, 054301 (2020)
16. P. Reinert, H. Krebs, E. Epelbaum, Eur. Phys. J. A **54**, 86 (2018)
17. D.R. Entem, R. Machleidt, Y. Nosyk, Phys. Rev. C **96**, 024004 (2017)
18. M. Piarulli, L. Girlanda, R. Schiavilla, R. Navarro Pérez, J.E. Amaro, E. Ruiz Arriola, Phys. Rev. C **91**, 024003 (2015)
19. A. Ekström, G.R. Jansen, K.A. Wendt, G. Hagen, T. Papenbrock, B.D. Carlsson, C. Forssén, M. Hjorth-Jensen, W. Navrátil, W. Nazarewicz, Phys. Rev. C **91**, 051301 (2015)
20. E. Epelbaum, H. Krebs, U.G. Meißner, Phys. Rev. Lett. **115**, 122301 (2015)
21. A. Gezerlis, I. Tews, E. Epelbaum, M. Freunek, S. Gandolfi, K. Hebeler, A. Nogga, A. Schwenk, Phys. Rev. C **90**, 054323 (2014)
22. A. Ekström, et al., Phys. Rev. Lett. **110**, 192502 (2013)
23. D.R. Entem, R. Machleidt, Phys. Rev. C **68**, 041001 (2003)
24. U. van Kolck, Front. in Phys. **8**, 79 (2020)
25. E. Epelbaum, H. Krebs, P. Reinert, Front. in Phys. **8**, 98 (2020)
26. B.D. Carlsson, A. Ekström, C. Forssén, D.F. Strömberg, G.R. Jansen, O. Lilja, M. Lindby, B.A. Mattsson, K.A. Wendt, Phys. Rev. X **6**, 011019 (2016)
27. E. Epelbaum, H. Krebs, U.G. Meißner, Eur. Phys. J. A **51**, 53 (2015)
28. S. Binder, et al., Phys. Rev. C **93**, 044002 (2016)
29. S. Binder, et al., Phys. Rev. C **98**, 014002 (2018)
30. R.J. Furnstahl, N. Klco, D.R. Phillips, S. Wesolowski, Phys. Rev. C **92**, 024005 (2015)
31. J.A. Melendez, S. Wesolowski, R.J. Furnstahl, Phys. Rev. C **96**, 024003 (2017)
32. S. Wesolowski, R.J. Furnstahl, J.A. Melendez, D.R. Phillips, J. Phys. G **46**, 045102 (2019)
33. J.A. Melendez, R.J. Furnstahl, D.R. Phillips, M.T. Pratola, S. Wesolowski, Phys. Rev. C **100**, 044001 (2019)
34. A. Ekström, G. Hagen, Phys. Rev. Lett. **123**, 252501 (2019)
35. S. Okubo, Prog. Theor. Phys. **12**, 603 (1954)
36. K. Suzuki, S.Y. Lee, Prog. Theor. Phys. **64**, 2091 (1980)
37. H. Feldmeier, T. Neff, R. Roth, J. Schnack, Nucl. Phys. A **632**, 61 (1998)
38. H. Hergert, R. Roth, Phys. Rev. C **75**, 051001 (2007)
39. R. Roth, T. Neff, H. Feldmeier, Prog. Part. Nucl. Phys. **65**, 50 (2010)

40. F. Wegner, Ann. Phys. (Leipzig) **3**, 77 (1994)
41. F.J. Wegner, Phys. Rep. **348**, 77 (2001)
42. S.D. Glazek, K.G. Wilson, Phys. Rev. D **48**, 5863 (1993)
43. S.K. Bogner, R.J. Furnstahl, R.J. Perry, Phys. Rev. C **75**, 061001 (2007)
44. S.K. Bogner, R.J. Furnstahl, A. Schwenk, Prog. Part. Nucl. Phys. **65**, 94 (2010)
45. R. Roth, J. Langhammer, A. Calci, S. Binder, P. Navratil, Phys. Rev. Lett. **107**, 072501 (2011)
46. R. Wirth, R. Roth, Phys. Rev. C **100**, 044313 (2019)
47. W. Magnus, Commun. Pure Appl. Math. **7**, 649 (1954)
48. T.D. Morris, N. Parzuchowski, S.K. Bogner, Phys. Rev. C **92**, 034331 (2015)
49. S. Szpigel, R.J. Perry, *Quantum Field Theory—A 20th Century Profile* (Hindustan Book Agency, 2000), chap. The Similarity renormalization group, pp. 59–81
50. R. Roth, A. Calci, J. Langhammer, S. Binder, Phys. Rev. C **90**, 024325 (2014)
51. R. Roth, S. Binder, K. Vobig, A. Calci, J. Langhammer, P. Navratil, Phys. Rev. Lett. **109**, 052501 (2012)
52. I. Talmi, Helv. Phys. Acta **25**, 185 (1952)
53. M. Moshinsky, Nucl. Phys. **13**, 104 (1959)
54. M. Moshinsky, Y.F. Smirnov, *The Harmonic Oscillator in Modern Physics* (Harwood Academic Publishers, Amsterdam, 1996)
55. G.P. Kamuntavicius, R.K. Kalinauskas, B.R. Barrett, S. Mickevicius, D. Germanas, Nucl. Phys. A **695**, 191 (2001)
56. P. Ring, P. Schuck, *The Nuclear Many-Body Problem* (Springer-Verlag, New York, 1980)
57. J. Suhonen, *From Nucleons to Nucleus: Concepts of Microscopic Nuclear Theory*. Theoretical and Mathematical Physics (Springer, Berlin, Germany, 2007)
58. R. Roth, P. Papakonstantinou, N. Paar, H. Hergert, T. Neff, H. Feldmeier, Phys. Rev. C **73**, 044312 (2006)
59. C. Constantinou, M.A. Caprio, J.P. Vary, P. Maris, Nucl. Sci. Tech. **28**, 179 (2017)
60. A. Tichai, J. Müller, K. Vobig, R. Roth, Phys. Rev. C **99**, 034321 (2019)
61. I. Shavitt, R.J. Bartlett, *Many-Body Methods in Chemistry and Physics: MBPT and Coupled-Cluster Theory*. Cambridge Molecular Science (Cambridge University Press, Cambridge, 2009)
62. S. Binder, J. Langhammer, A. Calci, P. Navratil, R. Roth, Phys. Rev. C **87**, 021303 (2013)
63. S. Binder, J. Langhammer, A. Calci, R. Roth, Phys. Lett. B **736**, 119 (2014)
64. W. Kutzelnigg, D. Mukherjee, J. Chem. Phys. **107**, 432 (1997)
65. D. Mukherjee, Chem. Phys. Lett. **274**, 561 (1997)
66. L. Kong, M. Nooijen, D. Mukherjee, J. Chem. Phys. **132**, 234107 (2010)
67. E. Gebrerufael, A. Calci, R. Roth, Phys. Rev. C **93**, 031301 (2016)
68. B.R. Barrett, P. Navratil, J.P. Vary, Prog. Part. Nucl. Phys. **69**, 131 (2013)
69. P. Navratil, S. Quaglioni, I. Stetcu, B.R. Barrett, J. Phys. G **36**, 083101 (2009)
70. R. Roth, P. Navratil, Phys. Rev. Lett. **99**, 092501 (2007)
71. R. Roth, Phys. Rev. C **79**, 064324 (2009)
72. C. Stumpf, J. Braun, R. Roth, Phys. Rev. C **93**, 021301 (2016)
73. P. Navratil, G.P. Kamuntavicius, B.R. Barrett, Phys. Rev. C **61**, 044001 (2000)
74. C. Forssén, B.D. Carlsson, H.T. Johansson, D. Sääf, A. Bansal, G. Hagen, T. Papenbrock, Phys. Rev. C **97**, 034328 (2018)
75. M. Shao, H.M. Aktulga, C. Yang, E.G. Ng, P. Maris, J.P. Vary, Comput. Phys. Commun. **222**, 1 (2018)
76. K. Tsukiyama, S.K. Bogner, A. Schwenk, Phys. Rev. Lett. **106**, 222502 (2011)
77. H. Hergert, S.K. Bogner, S. Binder, A. Calci, J. Langhammer, R. Roth, A. Schwenk, Phys. Rev. C **87**, 034307 (2013)
78. H. Hergert, S.K. Bogner, T.D. Morris, A. Schwenk, K. Tsukiyama, Phys. Rept. **621**, 165 (2016)
79. H. Hergert, S.K. Bogner, J.G. Lietz, T.D. Morris, S. Novario, N.M. Parzuchowski, F. Yuan, Lect. Notes Phys. **936**, 477 (2017)
80. H. Hergert, Phys. Scripta **92**, 023002 (2017)
81. K. Tsukiyama, S.K. Bogner, A. Schwenk, Phys. Rev. C **85**, 061304 (2012)

82. S.R. Stroberg, A. Calci, H. Hergert, J.D. Holt, S.K. Bogner, R. Roth, A. Schwenk, Phys. Rev. Lett. **118**, 032502 (2017)
83. H. Hergert, S. Binder, A. Calci, J. Langhammer, R. Roth, Phys. Rev. Lett. **110**, 242501 (2013)
84. H. Hergert, S.K. Bogner, T.D. Morris, S. Binder, A. Calci, J. Langhammer, R. Roth, Phys. Rev. C **90**, 041302 (2014)
85. E. Gebrerufael, K. Vobig, H. Hergert, R. Roth, Phys. Rev. Lett. **118**, 152503 (2017)
86. G. Hagen, T. Papenbrock, M. Hjorth-Jensen, D.J. Dean, Rept. Prog. Phys. **77**, 096302 (2014)
87. V. Somà, Front. in Phys. **8**, 340 (2020)
88. A. Tichai, R. Roth, T. Duguet, Front. in Phys. **8**, 164 (2020)
89. S. Gandolfi, D. Lonardoni, A. Lovato, M. Piarulli, Front. Phys. **8**, 117 (2020)
90. T.A. Lähde, U.G. Meißner, *Nuclear Lattice Effective Field Theory: An Introduction*, vol. 957 (Springer, Berlin, 2019)
91. H. Hergert, Front. Phys. **8**, 379 (2020)
92. E. Epelbaum, et al., Phys. Rev. C **99**, 024313 (2019)
93. P. Maris, et al., Phys. Rev. C **103**, 054001 (2021)

Nuclear Data and Experiments for Astrophysics 4

Anu Kankainen and Stephane Goriely

Abstract

Nuclear astrophysics aims to understand the origin of elements and the role of nuclear processes in astrophysical events. Nuclear reactions and reaction rates depend strongly on nuclear properties and on the astrophysical environment. Nuclear inputs for stellar reaction rates involve a variety of nuclear properties, theoretical models, and experimental data. Experiments providing data for nuclear astrophysics range from stable ion beam direct measurements to radioactive beam experiments employing inverse kinematics or indirect methods. Many properties relevant for astrophysical calculations, such as nuclear masses and β-decays, have also been intensively studied. This contribution shortly introduces selected astrophysical processes, discusses the related nuclear data needs, and gives examples of recent experimental and theoretical efforts in the field.

4.1 Origin of Elements and Nucleosynthesis Processes

4.1.1 The Composition of the Universe

Our knowledge of the composition of the Universe in general, and of our Solar System in particular, results almost entirely from the analysis of electromagnetic spectra originating from the various observable sources in the Universe, i.e., the

A. Kankainen (✉)
Department of Physics, University of Jyväskylä, Jyväskylä, Finland
e-mail: anu.kankainen@jyu.fi

S. Goriely
Institut d'Astronomie et d'Astrophysique, Université Libre de Bruxelles, Brussels, Belgium
e-mail: sgoriely@astro.ulb.ac.be

galaxies, the interstellar medium, and the stars of all types (including first of all, our sun) but also from the minute portion of matter which is accessible to the human kind, i.e., meteorites, planets (the Earth and the Moon in particular), energetic solar particles, and the galactic as well as extra galactic cosmic rays. Information provided by such sources on the present composition of the Universe can be found in Refs. [1–4].

One of the fundamental developments resulting from the various observations performed for the last decades is found in the determination of the composition of our own Solar System at the time of its formation some 4.6 billion years ago [5–8]. It is principally based on a special class of meteorites called carbonaceous chondrites of type CI1, considered as the most unaffected sample of matter accessible to man and representative of the primitive solar material. The analysis of the solar spectrum is in good agreement with the meteorite analysis and helps in addition to determine the abundances of some volatile elements, such as H, He, C, N, O, and Ne, which cannot be measured in meteorites reliably. In some cases (Ar, Kr, Xe, Hg), it remains difficult to extract accurate abundances from observational data, and some theoretical consideration is then required. From the primitive solar composition, it is possible to understand the differences observed today in the various constituents of the Solar System calling for the numerous physico-chemical and geological processes having taken place for the last 4.6 billion years. If the elementary composition appears relatively diversified among the various constituents of the Solar System, a very high homogeneity of the isotopic composition is found. For this reason, the isotopic composition of the terrestrial matter is generally used to determine the abundance of the nuclides in the Solar System. The resulting abundance distribution is shown in Fig. 4.1.

Figure 4.1 presents some interesting features. In particular, H and He are the most abundant species in the Solar System. In contrast, Li, Be, and B are extremely underabundant in comparison with the neighboring elements. For nuclei heavier than C, the abundances decrease with increasing atomic numbers A. On top of this general decreasing trend, there are superimposed abundance peaks, with the most prominent peak found for Fe. For $A \leq 56$, secondary peaks every multiple of 4 can be seen, while above Fe a large peak is observed around $80 \leq A \leq 90$ and two double peaks at $A = 130$–138 and $A = 195$–208. In the $A \gtrsim 50$ region, the abundances are also characterized by a saw feature. Such features, as well as the other remarkable features seen in Fig. 4.1, have been recognized since the early analysis of these curves as bearing the signature of specific nuclear properties.

For practical reasons, but also to highlight the link between observations and nucleosynthesis models, it is of particular relevance to divide the abundance curve of the elements heavier than iron into three distributions associated with the stable nuclides located at the bottom of the valley of β-stability, on the neutron-rich side of the valley, and on the neutron-deficient side. For even values of A, many isobars can exist; in this case, the stable most neutron-rich isobar is called r-nucleus and the most proton-rich p-nucleus. The s-nuclei are located between these two isobars, (i.e., at the bottom of the valley). When only one isobar exists, it is usually classified as an sr mix nucleus. The actinides are considered as being of r type. This denomination

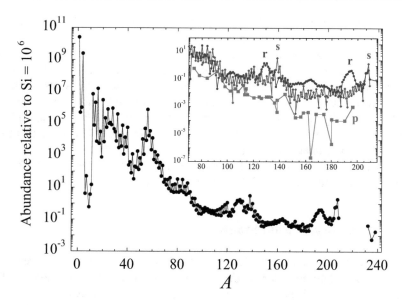

Fig. 4.1 Distribution of isotopic abundances characteristic of our Solar System at the time of formation [7]. The insert shows the decomposition of the Solar System distribution into the s-, r-, and p-abundances for elements heavier than iron [9–11]

is strongly related to the identification of the different mechanisms responsible for the production of the s-, r-, and p-nuclei, i.e., the so-called s-process (for slow), r-process (for rapid), and p-process (for proton). After performing such a nuclear decomposition (see, e.g., [9]), it is found that the double peak structure observed in Fig. 4.1 is now divided into two components, the "heavy" peaks at $A = 138$ and $A = 208$ attributed to the s-process and the "light" r-process peaks at $A = 130$ and $A = 195$ (see Fig. 4.1). The p-nuclei are in contrast about 100–1000 times less abundant than their s and r isobaric counterparts.

Let us finally mention that if the bulk of the Solar System material is found to be of a very high isotopic homogeneity, a small portion of this material ($\lesssim 10^{-4} M_\odot$, where $M_\odot$ is the mass of the Sun) is characterized by a variety of more or less different isotopic compositions. These so-called isotopic anomalies are observed either in meteoritic material which condensed in the solar nebula or in grains probably of circumstellar origins. These grains were formed around stars of various types and survived the protosolar nebula and their inclusion within meteorites. While the Solar System composition illustrated in Fig. 4.1 is considered as resulting from a perfect mix of the ashes produced by a large number of nucleosynthetic events that took place in the Galaxy during the $\sim 10^{10}$ years preceding the Solar System formation, the isotopic anomalies are believed to be caused by a relative small number of events. The analysis of some anomalies due to the in situ radioactive decay of short-lived nuclides (with half-lives of $10^5 \lesssim t_{1/2} \lesssim 10^8$ yr) can even provide severe constraints on the time elapsed between their production and their injection in the Solar System in formation. More information on the isotopic anomalies can be found in the review papers [12–14].

The Solar System is the object of the Universe that provides us with the most complete set of observational data concerning the elements and isotopes abundances. A myriad of information exists nevertheless on the composition of other objects which emphasizes features similar to our solar abundances, as well as a large diversity. Diversity is found not only among objects belonging to different classes but also among objects of the same type. In particular, the abundances observed at the stellar surface can vary with the age of the star, its location in the Galaxy, or its spectral type. Two major effects are found responsible for this abundance diversity: stellar evolution and the chemical evolution of the Galaxy.

4.1.2 Nucleosynthesis Models

One of the most fundamental questions astrophysics tries to answer concerns the present and past composition of the Universe and of its many constituents. The theory of nucleosynthesis aims at identifying the various processes that can be invoked to explain the origin of the nuclides observed in nature, as well as the astrophysical sites capable of providing the conditions required for these processes to take place. The works of [15, 16] represent milestone in this field.

Nuclear reactions represent the fundamental ingredients of all nucleosynthesis models. Two major classes of nuclear reactions are invoked: the thermonuclear reactions and the non-thermal transformations also known as spallation reactions. Thermonuclear reactions took place at the level of the primordial or cosmological (Big Bang) level as well as inside the stars all along the galactic evolution up to date. On the other hand, spallation reactions are important in diluted and cold medium, as the interstellar medium, through the interaction with galactic cosmic rays (GCRs), and at the surface of stars or in their surroundings through interaction with energetic stellar particles [17].

The primordial Big Bang nucleosynthesis (BBN) is responsible for the bulk He content of the Universe as well as for the synthesis of some other nuclei, like D, ^{3}He, and ^{7}Li. All the other nuclides, as well as a fraction of the galactic ^{7}Li, and maybe ^{3}He, result from thermonuclear reactions taking place inside the stars. The only exceptions concern the ^{6}Li, Be, and B nuclei for which spallation reactions from the nuclear interaction of GCRs (accelerated CNO nuclei) with the interstellar medium (mainly protons and α-particles) are invoked [17].

In stars, the thermonuclear reactions can be induced by charged particles (proton or α-particles) or neutrons. In the former case, the reactions mainly take place on light or medium heavy nuclei $A \lesssim 60$–70, since the reactions involving heavier species are not probable enough (because of the too high Coulomb barrier) to play a significant role in realistic stellar environments (cf. Sect. 4.2). The importance of the charged particle-induced reactions is twofold. First, they are fundamental for the energy production enabling the star to counterbalance its energy loss (energetic equilibrium), and second, they locally modify the stellar content where they take place. The neutron-induced reactions are obviously not restricted to species lighter than Fe, since no Coulomb barrier exists in this case. However, these reactions never contribute to the nuclear energy production.

The origin of most of the elements lighter than those of the Fe group has been explained, mainly thanks to the direct link between their nucleosynthesis and the energetic evolution of stars [18–20]. However, the synthesis of nuclei heavier than Fe is far from being well understood at the present time. The major mechanisms called for to explain the production of the heavy nuclei are the slow neutron-capture process (or s-process), occurring during hydrostatic stellar burning phases, the rapid neutron-capture process (or r-process) believed to develop during the explosion of a star as a supernova or during the coalescence of two-neutron stars (NSs) or a NS and a black hole (BH) in a binary system, and the p-process occurring in core-collapse supernova (CCSN) or Type Ia supernovae (SNIa). Recently, an intermediate neutron-capture process (or i-process) has also been proposed to explain observed abundances in low-metallicity stars. More information on these four nucleosynthesis processes is given in Fig. 4.2 and below.

4.1.2.1 The s-Process
For the last decades, an extremely intense amount of work has been devoted to the s-process of nucleosynthesis called to explain the origin of the stable nuclides heavier than iron located at the bottom of the valley of nuclear stability [22–25].

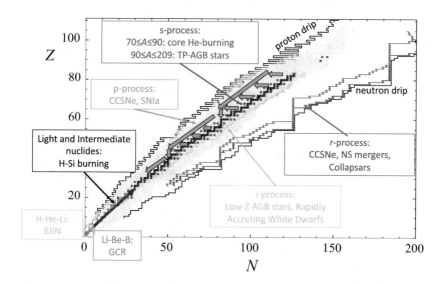

Fig. 4.2 Schematic representation in the (N, Z) plane of the different astrophysical sites responsible for the synthesis of the stable nuclides. The nucleosynthetic contributions by BBN and by GCR are also displayed. The open black squares correspond to stable or long-lived nuclei, and the open yellow squares to the nuclei with experimentally known masses [21]. Nuclei with neutron or proton separation energies tending to zero define the neutron or proton "drip lines" (solid black lines), as predicted from a mass model. More details can be found in [4] (Modified from Ref. [4])

Even though the observation of the radioactive Tc in stellar envelopes clearly proves that the s-process takes place during hydrostatic burning phases of a star, it remains difficult to explain the origin of the large neutron concentrations required to produce s-elements. Two nuclear reactions are suggested as possible neutron sources, i.e., $^{13}C(\alpha,n)^{16}O$ and $^{22}Ne(\alpha,n)^{25}Mg$. These reactions could be responsible for a large production of neutrons during given burning phases, namely, the core He-burning of massive stars (heavier than 10 $M_\odot$) and the shell He-burning during the thermal asymptotic giant branch (AGB) instabilities well-known as thermal pulses (TP) of low- and intermediate-mass stars (lower than typically 10 $M_\odot$).

As reviewed in great detail by Karakas and Lattanzio [24], the s-process in AGB stars is thought to occur in their He-burning shell surrounding an inert C-O core, either during recurrent and short convective TP episodes or in between these pulses. A rather large diversity of s-nuclide abundance distributions are predicted to be produced. A fraction of the synthesized s-nuclides (along with other He-burning products) could then be dredged up to the surface shortly after each pulse. In low-mass AGB stars (less than 3 $M_\odot$), it is generally considered that the necessary neutrons for the development of the s-process are mainly provided by $^{13}C(\alpha,n)^{16}O$, which can operate at temperatures around $(1\sim1.5) \times 10^8$ K. The efficiency of this mechanism is predicted to be the highest in stars with metallicities [Fe/H] lower than solar ([Fe/H] $\ll$ 0). The astrophysical models underlying the thermal pulse scenario are still quite uncertain, in particular in the description of the mechanisms that could be at the origin of the neutron production. The neutron production in these locations depends sensitively on the mechanism of proton ingestion into underlying He-rich layers in amounts and at temperatures that allow the operation of the $^{12}C(p,\gamma)^{13}N(\beta^+)^{13}C(\alpha,n)^{16}O$, while the production of ^{14}N by $^{13}C(p,\gamma)^{14}N$ is inefficient enough to avoid the hold-up of neutrons by the ^{14}N neutron poison. TP-AGB models including empirical diffusive overshoot have been relatively successful to explain such a partial mixing of protons from the H-rich envelope into the C-rich layers during the third dredge-up [24, 25], but it remains difficult to model such mixing mechanisms in common one-dimensional models.

Massive stars, and more specifically their He-burning cores and, to some extent, their C-burning shells, are also predicted to be s-nuclide producers through the operation of the $^{22}Ne(\alpha,n)^{25}Mg$ reaction. This neutron source can indeed be active in these locations that are hotter than the He shell of AGB stars. In addition, ^{22}Ne burning can also be activated in the C-burning shell of massive stars. Many calculations performed in the framework of realistic stellar models come to the classical conclusion that this site is responsible for a substantial production of the $70 \lesssim A \lesssim 90$ s-nuclides and can in particular account for the Solar System abundances of these species. It has also been shown that rotation can significantly increase the efficiency of the s-process, especially at low metallicity [26–28]. Because of the rotational mixing operating between the H-shell and He-core during the core helium burning phase, the abundant ^{12}C and ^{16}O isotopes in the convective He-burning core are mixed within the H-shell, boosting the CNO cycle and forming primary ^{14}N that finally leads to the synthesis of extra ^{22}Ne, hence an increased neutron production.

4.1.2.2 The r-Process

The r-process of stellar nucleosynthesis is called for to explain the production of the stable (and some long-lived radioactive) neutron-rich nuclides heavier than iron that are observed in stars of various metallicities, as well as in the Solar System. Reviews can be found in Refs. [4, 11, 29].

Nuclear physics-based and astrophysics-free r-process models of different levels of sophistication have been constructed over the years. They all have their merits and their shortcomings. The ultimate goal was to identify realistic sites for the development of the r-process. For long, the core-collapse supernova of massive stars has been envisioned as the privileged r-process location. One- or multi-dimensional spherical or aspherical explosion simulations in connection with the r-process nucleosynthesis are reviewed in Refs. [4, 11, 29]. Progress in the modeling of type II supernovae and γ-ray bursts has raised a lot of excitement about the so-called neutrino-driven wind environment. However, until now a successful r-process cannot be obtained ab initio without tuning the relevant parameters (neutron excess, entropy, expansion timescale) in a way that is not supported by the most sophisticated existing models [30, 31]. Although these scenarios remain promising, especially in view of their potential to contribute to the galactic enrichment significantly, they remain affected by large uncertainties associated mainly with the still incompletely understood mechanism responsible for the supernova explosion and the persistent difficulties to obtain suitable r-process conditions in self-consistent dynamical explosion and NS cooling models [30, 32, 33]. In particular, a subclass of CCSNe, the so-called collapsars, corresponding to the fate of rapidly rotating and highly magnetized massive stars and generally considered to be at the origin of observed long γ-ray bursts, could be a promising r-process site [34–36]. The production of r-nuclides in these events may be associated with jets predicted to accompany the explosion or with the accretion disk formed around a newly born central BH [37].

Since early 2000s, special attention has been paid to NS mergers as r-process sites following the confirmation by hydrodynamic simulations that a non-negligible amount of matter could be ejected from the system. Newtonian [38], conformally flat general relativistic [39, 40], as well as fully relativistic [41, 42] hydrodynamic simulations of NS-NS and NS-BH mergers with microphysical equations of state have demonstrated that typically some $10^{-3} M_\odot$ up to more than $0.1 M_\odot$ can become gravitationally unbound on roughly dynamical timescales due to shock acceleration and tidal stripping. Also the relic object (a hot, transiently stable hypermassive NS followed by a stable supermassive NS, or a BH-torus system), can lose mass through outflows driven by a variety of mechanisms [40].

Simulations of growing sophistication have confirmed that the ejecta from NS mergers are viable strong r-process sites up to the third abundance peak and the actinides. The r-nuclide enrichment is predicted to originate both from the dynamical (prompt) material expelled during the NS-NS or NS-BH merger phase and from the outflows generated during the post-merger remnant evolution of the relic BH-torus system. The resulting abundance distributions are found to reproduce very well the Solar System distribution, as well as various elemental distributions observed in low-metallicity stars [29]. In addition, the ejected mass of r-process material, combined with the predicted astrophysical event rate (around $10\,\mathrm{My}^{-1}$ in the Milky Way) can account for the majority of r-material in our Galaxy. A further piece of evidence that NS mergers are r-nuclide producers indeed comes from the very important 2017 gravitational wave and electromagnetic observation of the kilonova GW170817 [43, 44].

Despite the recent success of nucleosynthesis studies for NS mergers, the details of r-processing in these events are still affected by a variety of uncertainties, both from the nuclear physics and astrophysics point of view. In particular, it has been shown that weak interactions may strongly affect the composition of the dynamical ejecta and thus the efficiency of the r-process [42, 45–47].

The r-process nucleosynthesis is also important for understanding the origin of the radionuclides that could be used to estimate an approximate age of the Galaxy, the so-called radio-cosmochronometers. The stellar production of heavy elements requires a detailed knowledge not only of the astrophysical sites and physical conditions in which the processes take place but also of the chemical evolution of the Galaxy.

4.1.2.3 The i-Process

The s- and r-processes introduced very early in the development of the theory of nucleosynthesis have to be considered as the end members of a whole class of neutron-capture mechanisms. Supported by some observations that were difficult to reconcile solely with a combination of the s- and r-processes, a process referred to nowadays as an intermediate or i-process has been put forth, with neutron concentrations in the approximate 10^{12} to 10^{16} neutrons/cm^3 range. The mechanism envisaged to be responsible for this production is the ingestion of protons in He- and C-rich layers, leading to the production of ^{13}C through ^{12}C(p,γ)^{13}N(β^+)^{13}C followed by a substantial production of neutrons through ^{13}C(α,n)^{16}O. This is analogous to the mechanism already considered to be active in TP-AGB stars (Sect. 4.1.2.1), but the higher neutron concentrations are expected to result from the very low metallicity of the considered stars and the activation of ^{13}C(α,n)^{16}O in convective regions at higher temperatures (typically $\sim 2.5 \times 10^8$ K).

Various numerical simulations have been proposed to host i-process conditions. These include the proton ingestion during core He flash in very low-metallicity low-mass stars, during the thermal pulse phase of massive AGB (super-AGB) stars of very low metallicity, during the post-AGB phase ("final thermal pulse"), during rapid accretion of H-rich material on white dwarfs, or during shell He-burning in massive very low-metallicity population II or III stars. While the contribution of the i-process to the global galactic enrichment and more particularly to our Solar System remains unclear, it is needed to explain the heavy element patterns observed in peculiar stars, several carbon-enhanced metal-poor (CEMP) stars with simultaneous presence of s-elements and Eu (so-called CEMP-r/s) stars, as well as Sakurai's object V4334 Sgr. More information can be found in Refs. [48–53].

4.1.2.4 The p-Process

The p-process of stellar nucleosynthesis is aimed at explaining the production of the stable neutron-deficient nuclides heavier than iron that are observed in the Solar System and up to now in no other galactic location (for a review see [10]). Various scenarios have been proposed to account for the bulk p-nuclide content of the Solar System, as well as for deviations ("anomalies") with respect to the bulk p-isotope composition of some elements discovered in primitive meteorites. In contrast to the s- and r-processes calling for neutron captures to explain the production of heavy elements, the p-isotopes are produced by photodisintegration reactions on already synthesized s- and r-nuclei. These photoreactions involve (γ,n), (γ,p), and (γ,α) reactions at stellar temperatures of the order of 2–3×10^9 K.

The p-nuclides are mostly produced in the final explosion of a massive star ($M \gtrsim 10\ M_\odot$) as a CCSN or in pre-explosive oxygen-burning episodes [10]. The p-process can develop in the O-Ne layers of the massive stars explosively heated to peak temperatures ranging between 1.7 and 3.3×10^9 K [54, 55]. The seeds for the p-process are provided by the s-process that develops before the explosion in these stellar mass zones. In this way, as explained above, the O-Ne layers that experience the p-process are initially enriched in $70 \lesssim A \lesssim 90$ s-nuclides.

SNIa have also been suggested as a potential site for the p-process. The p-process nucleosynthesis possibly accompanying the deflagration or delayed detonation regimes has been mainly studied in 1D simulations [56, 57] and shown to give rather similar overabundances as CCSN models [10, 58]. However, the predicted SNIa p-nuclide yields suffer from large uncertainties affecting the adopted explosion models as well as the s-seed distributions, detailed information on the composition of the material that is pre-explosively transferred to the white dwarf being missing.

Despite the fact that p-nuclei can be produced consistently with solar ratios over a wide range of nuclei in such scenarios, there remain deficiencies in a few regions, most particularly in the Mo-Ru region where the p-isotopes are strongly underproduced. This fact motivates the search for alternative or additional ways to produce these nuclides. In particular, proton capture and photodisintegration processes in helium star cataclysmics have been suggested as a promising nucleo-synthesis source [59]. Such an object is made of a carbon-oxygen white dwarf with sub-Chandrasekhar mass ($M < 1.4 M_\odot$) accumulating a He-rich layer at its surface. An alternative site proposed to explain the origin of the Mo and Ru p-nuclei is the p-rich neutrino-driven wind in CCSNe where antineutrino absorptions in the proton-rich environment produce neutrons that are immediately captured by neutron-deficient nuclei [60].

4.2 Nuclear Physics Aspects of Nucleosynthesis

4.2.1 Nuclear Reactions of Astrophysical Interest

In a given astrophysical location, two factors dictate the variety of nuclear reactions that can act as energy producers and/or as nucleosynthetic agents. The abundances of the reactants have obviously to be high enough, and the lifetimes of the reactants

against a given nuclear transmutation have to be short enough for this reaction to
have time to operate during the evolutionary timescale of the astrophysical site
under consideration. The probability of a thermonuclear reaction in an astrophysical
plasma is strongly dependent on some specific properties of this plasma. In this
respect, two key guiding features are the distribution of the energies of the reacting
partners and the reaction cross section at a given energy. First, the reacting nuclei
are, locally at least, in a state of thermodynamic equilibrium. In such conditions, all
nuclear species obey a Maxwell-Boltzmann distribution of energies, from which it
is easily inferred that the relative energies $E_{cm} = \frac{1}{2}\mu v^2$ of the reaction partners also
obey such a distribution (where v is the relative velocity between the interacting
nuclei 1 and 2 and $\mu = m_1 m_2/(m_1 + m_2)$ their reduced mass). While in laboratory
experiments, the target nuclei (T) are typically at rest and the projectiles (P) impinge
into the target nuclei at a certain laboratory energy E_{lab}, in stellar environments
the relative energy is more relevant. Therefore, laboratory experiments should be
expressed as a function of the center-of-mass energy $E_{cm} = [M_T/(M_P + M_T)] \times E_{lab}$, where M_P and M_T refer to the atomic masses of the projectile and the target
(at rest), respectively.

Second, the reaction cross section between charged nuclei is dominated by the
probability of penetration of the Coulomb barrier of the interacting nuclei. As a
result, the effective reaction rate is obtained by integrating the strongly energy-
dependent reaction cross sections over the whole Maxwell-Boltzmann energy range.
The resulting integrant exhibits a strong maximum, generally referred to as the
Gamow peak. It is centered on the "most effective energy" given by

$$E_0 = 0.1220(Z_1^2 Z_2^2 \mu)^{1/3} T_9^{2/3} \text{ (MeV)} \tag{4.1}$$

where Z_1 and Z_2 are the proton numbers, μ the reduced mass (in units of u), and T_9
the temperature T expressed in GK (10^9 K).

The Gamow peak is characterized by a width approximated by

$$\Delta = 4(E_0 k_B T/3)^{1/2} = 0.2368(Z_1^2 Z_2^2 \mu)^{1/6} T_9^{5/6} \text{ (MeV)} \tag{4.2}$$

where k_B is the Boltzmann constant [19].

The reactions thus mostly occur in the approximate window from $E_0 - n\Delta$ to
$E_0 + n\Delta$ ($n = 2$–3), assuming the possible role of resonances is small. For this
reason, the energy range of astrophysical relevance for reactions between charged
particles is largely above the thermal energy $k_B T$ and much lower than the Coulomb
barrier. For these reasons, the sequence of hydrostatic burning episodes is characterized by a limited number of reactions between nuclei with increasing charges, from
H-burning to Si-burning, and the charged-particle-induced thermonuclear reactions
of relevance concern mainly the capture of protons or α-particles which offer the
lowest Coulomb barriers. A limited number of fusion reactions involving heavy
ions (^{12}C, ^{16}O) are also of great importance.

The considerations above lead to the most effective energy E_0 in the case of
reactions between charged particles but do not apply to neutron captures in view of

the absence of Coulomb barriers. In this case it can be shown that the most effective energy is of the order of $k_B T$. It has also to be noted that, in contrast to reactions involving charged reactants, the captures of neutrons do not contribute to the energy budget of a star, but are essential players in the synthesis of nuclides heavier than iron through the s-, i-, and r-processes (see Sects. 4.1.2.1–4.1.2.2).

In non-explosive conditions, like in the quiescent phases of stellar evolution which take place at relatively low temperatures, most of the reactions of interest concern stable nuclides. Even so, the experimental determination of their charged-particle-induced cross sections faces enormous problems and represents a real challenge [19]. This relates directly to the smallness of the cross sections due to the fact that E_0 lies well below the Coulomb barrier. As a consequence, the cross sections can dive into the nanobarn to picobarn range.

In explosive situations, the temperatures are typically higher than in the non-explosive cases. The corresponding increase of the effective energies E_0 gives rise to a higher probability of penetration of the Coulomb barriers and consequently larger cross sections. The price to pay to reach this higher energy domain is huge, however. The nuclear flows indeed depart from the bottom of the valley of nuclear stability and involve more or less unstable nuclei, sometimes all the way very close to the nucleon drip lines (see Fig. 4.3).

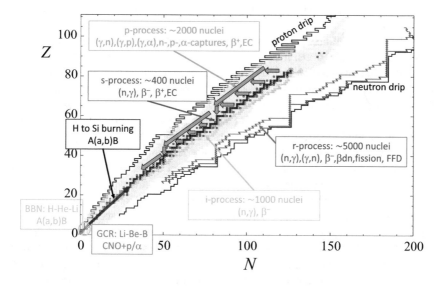

Fig. 4.3 Schematic representation in the (N, Z) plane of the different nuclear data needs for nucleosynthesis applications. The open black squares correspond to stable or long-lived nuclei, and the open yellow squares to the nuclei with experimentally known masses [21]. Nuclei with a neutron or proton separation energies tending to zero define the neutron or proton "drip lines" (solid black lines), as predicted from a mass model. See text for more details and Ref. [4] (Modified from Ref. [4])

For β-decays as well as reaction rates, thermally populated nuclear excited states can contribute to the effective stellar rates. The population of the ith excited state with an excitation energy E_i at temperature T can be derived as

$$P_i = \frac{g_i \exp(-E_i/k_B T)}{\sum_i g_i \exp(-E_i/k_B T)}, \tag{4.3}$$

where $g_i = 2J_i + 1$ is the statistical weight and J_i the spin of the state i. The denominator is called the partition function G. Often a normalized partition function $G^{\text{norm}} = G/g_0 = 1/P_0$ is used to describe the thermal excitations. If $G^{\text{norm}} = 1$, only the ground state is populated ($P_0 = 1$). The thermalization effect is especially noticeable in the case of endothermic reactions on targets with low-lying excited states from which the exit particle channels are greatly favored with respect to the ground state due to restrictions imposed by spin conservation selection rules. The 0^+ isomeric state at 228 keV in ^{26}Al is a good example in this respect. It is much shorter-living, with $t_{1/2} = 6.3460(8)$ s, than the 5^+ ground state with $t_{1/2} = 7.17 \times 10^5$ y. The effective lifetime of ^{26}Al decreases by many orders of magnitude when moving from 0.2 GK to 1.0 GK due to the thermal excitations populating the isomer [61–64]. In many astrophysical conditions, some isomers may not be thermally populated and act as a separate species with respect to the ground state. The role of astrophysically important isomers has been recently discussed, e.g., in Ref. [65].

In stellar environments, target nuclei at high temperatures have typically no or only a few bound electrons. Instead, they are surrounded by a sea of free electrons. This ionization gives rise to various effects. It has first the obvious effect of reducing the probability of capture of bound electrons but opens the possibility to capture free electrons from the surrounding continuum. A less trivial consequence of ionization relates to the possible development of the process of "bound-state β-decay," for which the emitted electron is captured in an atomic orbit previously vacated (in part or in total) by ionization. In addition, the reaction rates for charged particle reactions are different from the rates of bare nuclei due to the electron screening. The screening is also present in the laboratory experiments where target nuclei are surrounded by bound atomic electrons. Hence, the measured reaction rates have to be corrected for the electron screening effect to obtain reaction rates between bare nuclei. Finally, in stellar plasmas, a specific electron screening correction has to be applied and can drastically affect the cross sections for bare nuclei [66, 67]. This correction arises because of the ability of a nucleus to polarize its stellar surroundings. As a result, the Coulomb barrier seen by the reacting nuclei is modified in such a way that the tunneling probability, and consequently the reaction rate, increases over its value in vacuum conditions. Different formalisms have been developed depending on the ratio of the Coulomb energy of reacting nuclei to the thermal energy. Weak screening applies if this ratio is well below unity, while a strong screening is obtained when this ratio is well in excess of unity. In this case, a very large increase of the reaction rates is predicted. The limiting situation of strong screening is reached when solidification of the stellar plasma leads to the special pycnonuclear regime [66, 67]. In this case, the reactions are not governed by temperature like in the thermonuclear regime, but instead by lattice vibrations in dense Coulomb solids. This limiting regime can be approached e.g. at the high densities and low temperatures prevailing in white dwarfs.

4.2.2 Data Needed for the Various Nucleosynthesis Processes

Strong, weak, and electromagnetic interaction processes play an essential role in nuclear astrophysics. As shown in Fig. 4.3, a very large amount of nuclear information is necessary in order to model the various nucleosynthesis processes. These concern the decay properties of a large variety of light to heavy nuclei between the proton and neutron drip lines, including the β-decay or electron capture rates as well as α-decay or spontaneous fission probabilities for the heavy species. For the nuclei lighter than iron, most of the reactions involved during the BBN or the H- to Si-burning stages concern the capture of protons and α-particles at relatively low energies (far below 1 MeV for neutrons and the Coulomb barrier for charged particles). A limited number of fusion reactions involving heavy ions (^{12}C, ^{16}O) are also of direct impact during C and O-burning phases. The nuclear data needed to explain the Li-Be-B nucleosynthesis is quite different since it mainly involves spallation reactions between CNO nuclei accelerated at high energies interacting with the interstellar H and He. A review of the relevant reactions and the precision at which they are needed can be found in Ref. [68].

In addition to reaction rates, some nuclear structure properties, in particular the nuclear mass, may play a key role in nucleosynthesis applications. More specifically, if the r-process nucleosynthesis takes place at sufficiently high temperatures T and high neutron densities N_n, the neutron captures and their inverse photodisintegrations become much faster than β^- decays [69]. In this case, a $(n, \gamma) \rightleftarrows (\gamma, n)$ equilibrium may be established, and the abundances within each isotopic chain determined by the Saha equation (see, e.g., Ref. [10]):

$$\frac{N(A+1, Z)}{N(A, Z)} = N_n \left(\frac{h^2}{2\pi \mu k_B T} \right)^{3/2} \frac{2J(A+1, Z) + 1}{(2J(A, Z) + 1)(2J_n + 1)}$$
$$\times \frac{G^{\mathrm{norm}}(A+1, Z)}{G^{\mathrm{norm}}(A, Z)} e^{Q_{n,\gamma}/(k_B T)}, \tag{4.4}$$

where $Q_{n,\gamma} = [m(A, Z) + m_n - m(A+1, Z)]c^2$ is the Q-value for a neutron capture on nucleus (A, Z) or, in other words, the neutron separation energy S_n of the nucleus $(A+1, Z)$, $Q_{n,\gamma}(A, Z) = S_n(A+1, Z)$. Equation 4.4 highlights the importance of nuclear masses in defining the r-process path at a given time. In NS merger models, the r-process may take place at relatively low temperatures [39], and, at some point, the neutron captures will freeze out, so that the $(n, \gamma) \rightleftarrows (\gamma, n)$ equilibrium is not expected to be established all along the irradiation time. In this case, the abundances cannot be determined simply using Eq. 4.4 but become sensitive to the neutron capture and photoneutron reaction rates. Nuclear masses remain, however, key in estimating the competition between neutron captures, photoneutron emissions, and β-decays.

Fission may also play an important role during the r-process nucleosynthesis though the exact role played by fission on r-abundance distribution strongly depends on the hydrodynamic modeling of the initial neutron richness found in the astrophysical plasma. In astrophysical sites characterized by a large initial neutron richness, e.g., in NS-BH mergers, fission may play a fundamental role, more particularly by (*i*) recycling the matter during the neutron irradiation (or if not, by allowing the possible production of superheavy long-lived nuclei, if any); (*ii*) shaping the r-abundance distribution in the $110 \leq A \leq 170$ mass region at the end of the neutron irradiation; (*iii*) defining the residual production of some specific heavy stable nuclei, more specifically Pb and Bi, but also the long-lived cosmochronometers Th and U; and (*iv*) heating the environment through the energy released [40, 70–72]. In addition to spontaneous fission, neutron-induced and β-delayed fission processes are important for the r-process. In the neutron-induced fission, the additional energy required to overcome the fission barrier is provided by neutrons. In the β-delayed fission mode, the β-decay may lead to an excited state with an excitation energy E_x close to the fission barrier height B_f in the daughter nucleus.

Although important effort has been devoted in the last decades to measure reaction cross sections or nuclear structure properties of astrophysical interest (see Sect. 4.3), experimental data only covers a minute fraction of the whole set of data required for nucleosynthesis applications. Reactions of interest often concern unstable or even exotic (neutron-rich, neutron-deficient, superheavy) species for which no experimental data exist. Given applications (in particular, the nucleosynthesis of elements heavier than iron) involve a large number (thousands) of unstable nuclei for which many different properties have to be determined. Finally, the energy range for which experimental data is available is restricted to the small range reachable by present experimental setups. To fill the gaps, only theoretical predictions can be used, as discussed in Sect. 4.4.

4.3   Nuclear Astrophysics with Radioactive Beams

In order to model various nucleosynthesis processes (see Sect. 4.1.2), different types of nuclear data are needed as discussed in Sect. 4.2.2. For lighter nuclei, the key reactions concern proton and alpha captures. For quiescent hydrogen and helium burning, the relevant temperatures are on the order of 10–100 MK, corresponding to center-of-mass energies less than (or around) 100 keV, far below the Coulomb barriers. As a result, the reaction cross sections for the relevant proton and alpha captures are very low. This poses several challenges. Typically the experiments have not yet reached the relevant energy region but provide cross sections at higher energies, requiring extrapolations down to the relevant energies. Natural background is a major limiting factor for the experiments. Therefore, many direct measurements for stellar burning are nowadays carried out in underground laboratories or other low-background locations. A recent review [73] summarizes the status of these experiments. Here we focus on experiments employing radioactive beams for nuclear astrophysics. Free neutrons are radioactive with a half-life of around 10 mins, but we will not discuss experiments involving neutron beams, which are very important, for example, for the s-process. The status of experiments utilizing neutron beams for astrophysics has been reviewed, for example, in Refs. [73, 74].

4.3.1 Nuclear Reactions in Inverse Kinematics with Radioactive Beams

Many reactions on radioactive nuclei are usually easier to study in inverse kinematics with a radioactive beam on a stable target. As an example, proton-capture reactions can be studied with a radioactive beam on a hydrogen target instead of using normal kinematics, i.e., a proton beam on a radioactive target. For shorter-lived nuclei, inverse kinematics is the only option available. The same applies to other reactions involving radioactive nuclei.

Let us consider the reaction $^{26}\text{Al}(p, \gamma)^{27}\text{Si}$ as an example. This reaction is relevant for the abundance of the cosmic γ-ray emitter ^{26}Al and the observation of its 1809-keV γ-rays with space-based telescopes, such as INTEGRAL [75]. Due to the relatively long half-life of ^{26}Al, $t_{1/2} = 7.17 \times 10^5$ y, a study in normal kinematics is also feasible. The reaction was investigated using proton beams with laboratory energies from 170 keV to 1.5 MeV in the 1980s [76]. Later, it was revisited using the DRAGON recoil separator at TRIUMF and employing a radioactive ^{26}Al beam with laboratory energies of 5.226 MeV and 5.122 MeV [77]. There, ^{26}Al was produced with a 70-μA proton beam on a SiC target.

Radioactive beams for inverse kinematics studies can be produced via nuclear reactions, such as fusion evaporation, fragmentation, or fission, but specific beams can be created using long-lived isotopes extracted from radioactive waste [78]. For example, ^{44}Ti ($T_{1/2} = 85$ y) was extracted from the copper beam dump used for the 590-MeV protons at the Paul Scherrer Institute and later utilized in an experiment at ISOLDE (CERN) [79]. Beam intensities up to around 2×10^6 particles per second were delivered and accelerated to 2.1 MeV/u at REX-ISOLDE before impinging into a helium target [79]. The experiment provided an upper limit estimate for the $^{44}\text{Ti}(\alpha, p)^{47}\text{V}$ reaction cross section within the Gamow window. The limit is at least a factor of 2.2(13) lower than given by the Hauser-Feshbach calculation with the NON-SMOKER reaction code. This brings the calculated ^{44}Ti abundances closer to the observations of the ^{44}Ti yields in Cas A [80, 81] and SN1987A [82] supernova explosions.

Studies of proton captures on light- or intermediate-mass nuclei usually focus on the determination of resonance strengths $\omega\gamma$ because the total reaction rate is typically dominated by a few resonances. The resonance strength for a proton-capture reaction on a target nucleus with spin J_T, leading to a resonant state with spin J_{res}, is determined as

$$\omega\gamma = \frac{(2J_{\text{res}} + 1)}{(2J_p + 1)(2J_T + 1)} \frac{\Gamma_p \Gamma_\gamma}{\Gamma_p + \Gamma_\gamma}, \tag{4.5}$$

where J_p is the proton spin (1/2) and the Γ_p and Γ_γ are the proton and gamma partial widths for the resonance, respectively. From Eq. 4.5, it is clear that at low resonance energies, where the probability for the proton emission is still low ($\Gamma_p \ll \Gamma_\gamma$), the resonance strength is almost entirely determined by the proton width Γ_p. It can be written as $\Gamma_p = C^2 S \, \Gamma_{p,sp}$, where $C^2 S$ is the spectroscopic factor of the state and $\Gamma_{p,sp}$ the single-particle proton width obtained, e.g., via shell-model calculations.

Estimates on relevant spectroscopic factors can be obtained using surrogate methods. Instead of proton captures, the relevant states can be explored via (d, n) proton-transfer reactions. Recently, many studies on this topic have been carried out at the National Superconducting Cyclotron Laboratory. For example, the ^{26}Al$(p, \gamma)^{27}$Si reaction was studied using a 30 MeV/u ^{26}Al^{13+} beam on a deuterated polyethylene, $(CD_2)_n$ target. Spectroscopic factors for states close to the proton threshold in ^{27}Si were obtained by comparing the experimentally determined cross sections to the theoretical predictions for the reaction ^{26}Al$(d, n)^{27}$Si [83]. The results were in agreement with a previous $(^3\text{He}, d)$ study [84], supporting the feasibility of the method. The surrogate technique using (d, n) proton-transfer reactions has been applied to the bottleneck reaction in the nova nucleosynthesis, ^{30}P(p, γ) [85], and for the key reaction to bypass the waiting-point nucleus ^{56}Ni in type I X-ray bursts, ^{56}Ni$(p, \gamma)^{57}$Cu [86].

In addition to (d, n) reactions, relevant information on the resonance states in explosive hydrogen burning scenarios, such as novae and type I X-ray bursts, is obtained via many other methods. β-delayed proton and gamma emissions provide data on the gamma and proton widths of the resonance states. However, β-decay selection rules limit the resonant states that are populated. For example, β-decay of ^{31}Cl populates excited states in ^{31}S that further de-excite via gamma and proton emissions. Thus, the resonant states in the reaction ^{30}P$(p, \gamma)^{31}$S can be studied inversely via β-decay. The β-decay studies have, e.g., indicated a strong $3/2^+$ resonance at 6390 keV in ^{31}S [87]. Information on the excitation energies, spins, and parities of the resonance states is also obtained via high precision gamma spectroscopy and transfer-reaction studies, e.g., employing $(^3\text{He}, t)$ reactions. These studies are not limited by such selection rules like β-decay experiments and therefore cover a larger set of states.

4.3.2 Properties of Exotic Nuclei with Radioactive Beams

Many astrophysical processes proceed via exotic radioactive nuclei as discussed in Sect. 4.1.2. Progress in radioactive beam facilities and measurement techniques has opened new possibilities to study, e.g., nuclei relevant for the r- and i-processes traversing through neutron-rich nuclei. For the r-process, many nuclei will remain experimentally inaccessible and require solid nuclear models (see Sect. 4.4). Experimental data are, however, essential for testing the existing nuclear models and their applicability in different regions of the nuclear chart. The following subsections give a brief overview on experimental techniques and recent experimental results on nuclear properties relevant for nuclear astrophysics, in particular for the r-process.

4.3.2.1 Masses of Exotic Nuclides and Related Techniques

Nuclear masses play a key role in the modeling of astrophysical processes. The reaction Q value, i.e., the energy required for, or released in a reaction, is determined by nuclear masses, $Q = \left(\sum_i m_i - \sum_f m_f \right) c^2$, where m_i and m_f are the mass values for the initial and final states of a reaction. The Q values have a strong effect on reaction and decay rates and therefore have to be known rather precisely for accurate nuclear reaction network calculations.

In practice, experiments determine atomic masses $M(A, Z) = m(A, Z) + Z m_e - B_e(A, Z)/c^2$, where $B_e(A, Z)$ is the total electron binding energy for an atom with mass number A, proton number Z, and nuclear mass $m(A, Z)$. The effect of electron binding energy is usually small though it may contribute, e.g., for low-energy resonant captures involving fully stripped atoms in stellar plasma [88]. As the proton number conserves in nuclear reactions, the electron masses cancel out in the estimate of the Q value. Only for β^+ decays, the electron masses need to be taken into account in the Q-value calculation.

4.3.2.2 Penning-Trap Mass Spectrometry

Several mass measurement methods are used to determine masses of radioactive nuclei. Penning-trap mass spectrometry is the most precise technique at the moment. In a Penning trap, ions are confined radially by a strong, homogeneous magnetic field B and axially by a quadrupolar electric field. The ions have three eigenmotions, axial motion with a frequency v_z and two radial motions with the reduced cyclotron v_+ and magnetron v_- frequencies. For an ion with charge q and mass m in an ideal Penning trap, the two radial motions sum up to the cyclotron frequency v_c:

$$v_c = v_+ + v_- = \frac{1}{2\pi} \frac{q}{m} B \qquad (4.6)$$

In reality, there are misalignments, e.g., in the magnetic field axis or imperfections in the quadrupolar electric field. The invariance theorem [89,90] holds even for these more realistic conditions, coupling the three eigenmotions together:

$$v_c^2 = v_+^2 + v_-^2 + v_z^2. \qquad (4.7)$$

Traditionally, Penning traps have utilized the time-of-flight ion cyclotron resonance (ToF-ICR) method [91,92] to determine ion's cyclotron resonance frequency. In this method, the ions are excited using a quadrupolar radiofrequency pulse with a frequency ν_{RF} and a specific amplitude and duration. The ν_{RF} is scanned around the expected ν_c. When $\nu_{RF} = \nu_c$, the ions are in resonance and gain most energy. This results in the shortest time of flight when the ions are extracted from the trap through a strong magnetic field gradient to an ion detector, typically a microchannel-plate (MCP) detector. The magnetic field strength B is determined by performing a similar measurement with a reference ion which has a well-known mass in literature. The ToF-ICR method takes a rather long time as several frequency points have to be measured around the cyclotron resonance frequency in order to fit the resonance curve to the data. An example of a TOF-ICR spectrum is given in the bottom panel of Fig. 4.4. The quadrupolar excitation times range from around 50 ms up to around 1600 ms, and the total measurement cycle typically takes $\approx$400–1200 ms. Therefore, the Penning-trap mass spectrometry is often limited to nuclei with half-lives longer than $\approx$100 ms. However, in specific cases, where the production rates are high enough, measurements of shorter-living nuclides can also be done with the ToF-ICR method.

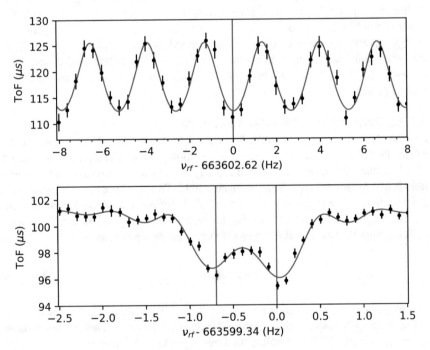

Fig. 4.4 Examples of ToF spectra measured for ^{162}Eu$^+$ ions using a 25-350-25ms (on-off-on) excitation pattern (top) and 1600-ms continuous quadrupolar excitation. The shorter excitation time (top panel) was not sufficient to resolve the low-lying isomeric state from the ground state. In the bottom panel, the cyclotron resonance frequency for the ground state is located at the minimum time of flight indicated with a vertical line at zero. The isomeric state is located at a lower frequency, indicated by the other vertical line. The red curve is a fit to the theoretical lineshape. The fit requires several measured data points (shown in black) around the cyclotron frequency (Reprinted from Ref. [95] with permissions from the American Physical Society)

A slightly higher precision is achieved with the so-called Ramsey method [93, 94], where instead of a continuous quadrupole excitation, time-separated oscillatory fields are applied. In other words, two excitation pulses, each with a rectangular envelope, are applied with a certain time in between when the excitation is off. An example of a Ramsey type of a resonance is given in the upper panel of Fig. 4.4. There, $^{162}Eu^+$ ions have been studied using a 25-350-25ms (on-off-on) excitation pattern (see the top panel of Fig. 4.4). It yields a better precision compared to a 400-ms continuous quadrupolar excitation; however, the resolving power is still not sufficient to resolve the low-lying isomeric state from the ground state. This is achieved with a 1600-ms quadrupolar excitation shown in the bottom panel of Fig. 4.4. It also illustrates how the resolving power of a Penning trap is proportional to the excitation time. The longer the excitation time, the better the resolving power.

The phase-imaging ion cyclotron resonance (PI-ICR) method [96, 97] provides around 40 times better resolving power than the ToF-ICR method. The method is superior in resolving low-lying isomeric states from the ground states, often useful for accurate mass measurements. The frequencies $\nu_\pm$ for the radial ion motions are obtained from the phase $\phi_\pm$ the ion accumulates after time t:$\nu_\pm = \frac{\phi_\pm + 2\pi n}{2\pi t}$, where n is the number of full revolutions ion does during the time t. The phase is determined using a position-sensitive MCP detector. Finally, the cyclotron frequency is computed as a sum of the two radial frequencies (see Eq. 4.6), and the mass is derived from the frequency ratio similar to the ToF-ICR method. The benefit of the PI-ICR method is that every ion counts, i.e., instead of scanning a broad range of frequencies around the cyclotron frequency, every measured ion adds to the phase spot in the 2D image of the ion motion. Figure 4.5 shows an example of a PI-ICR measurement. The PI-ICR method is also applicable to cases with low yields, such as superheavy nuclides [98]. In addition to the ToF-ICR and PI-ICR, Fourier-transform ion cyclotron resonance (FT-ICR) method [99] can be applied in Penning traps; however, it has not yet been widely used for studies of radioactive nuclides due to its complexity. For a recent review on Penning-trap measurements, see, e.g., Refs. [100, 101].

4.3.2.3 Multi-Reflection Time-of-Flight Mass Spectrometers

Multi-reflection time-of-flight (MR-ToF) [102] mass spectrometers offer a faster way to determine masses of exotic nuclides than Penning traps. The method also saves measurement time as several nuclides can be measured at once. The ions injected into a MR-ToF are typically prepared in a radiofrequency quadrupole cooler and buncher at a potential V. They gain a kinetic energy $E_{kin.} = qV = mv^2/2$, where v is the velocity of the ions. As a result, for the same flight path, the flight time t is inversely proportional to the ion's velocity, $t \propto 1/v \propto \sqrt{(m/q)}$, and can be determined as

$$t = a\sqrt{\frac{m}{q}} + b,$$

(4.8)

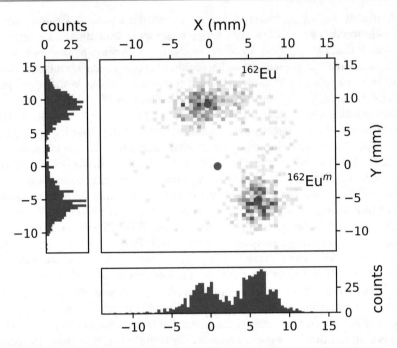

Fig. 4.5 An example of a PI-ICR measurement of ^{162}Eu$^+$ ions. The image of the cyclotron motion of ^{162}Eu$^+$ is magnified and projected onto a position-sensitive detector located outside the Penning trap. The two detected ion spots correspond to the ground and isomeric state of ^{162}Eu. The blue squares show the total number of ions in each bin, darker shading indicating more ions. The red dots show the centers of the cyclotron motion images of the ground and isomeric states and the center of the precision trap. The number of ions projected on the x and y axes is also shown. Positions can be fitted even with a moderate statistics because every ion contributes to the determinations of the spot positions from which the phases, and eventually the cyclotron frequency ratios are determined (Reprinted from Ref. [95] with permissions from the American Physical Society)

where a and b are device-specific parameters. The achieved precision is typically somewhat lower than in Penning-trap mass spectrometry, and the resolving power is not sufficient to resolve low-lying isomeric states (E $\lesssim$ 100 keV). Due to the simple and cost-effective solution for fast and precise mass measurements, MR-ToF mass spectrometers are nowadays widely used in accelerator laboratories around the world. MR-ToF mass spectrometers, e.g., at ISOLDE/CERN [103], at the FRS Ion Catcher in GSI/FAIR [104, 105], and at the TITAN facility in TRIUMF [106], have been utilized for nuclear astrophysics studies. An example of a MR-ToF measurement is shown in Fig. 4.6.

4.3.2.4 Storage Rings

Storage rings have been used for mass measurement of exotic ions for three decades [108]. There are two techniques utilized for mass measurements in storage rings,

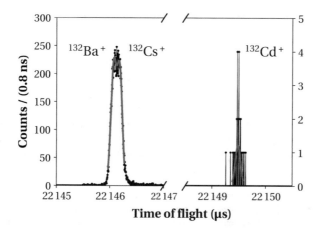

Fig. 4.6 An example of a MR-ToF measurement of ^{132}Cd at ISOLTRAP. Time-of-flight spectrum after 800 revolutions shows the ^{132}Cd$^+$ peak along with isobaric ions (^{132}Ba$^+$ and ^{132}Cs$^+$), used for the calibration together with ^{133}Cs$^+$. Gaussian fits (in red) are also shown. MR-ToF method is suitable for measurements with low statistics (Reprinted from Ref. [107])

but both methods determine the ion's revolution frequency f in the ring:

$$\frac{\delta f}{f} = -\frac{1}{\gamma_t^2}\frac{\delta(m/q)}{(m/q)} + \left(1 - \frac{\gamma^2}{\gamma_t^2}\right)\frac{\delta v}{v}, \qquad (4.9)$$

where $\gamma = 1/\sqrt{1 - (v/c)^2}$ is the Lorenz factor and γ_t is an ion-optical parameter of the storage ring known as the transition energy. In practice, usually revolution times are measured and plotted instead of the revolution frequency.

In the Schottky method, the ions are cooled with electrons to minimize the velocity spread δv. This takes several seconds and limits the use of the Schottky method for shorter-lived nuclei. In the isochronous mass spectrometry (IMS) method, the ions of interest are injected with energies corresponding to $\gamma = \gamma_t$, and no additional cooling is required. The benefit in the IMS method is that a broad variety of ions can be simultaneously measured, and the method is much faster than, e.g., Penning-trap mass spectrometry. An example of an isochronous mass measurement is shown in Fig. 4.7. There are three main storage ring facilities for mass measurements at the moment: Experimental Storage Ring (ESR) [109] at GSI/FAIR, CSRe [110] in Lanzhou and R3 [111] at BigRIPS in RIKEN. Storage rings can also be utilized for reaction cross-sectional measurements for nuclear astrophysics, as exemplified for the ^{96}Ru(p, γ) [112] and ^{124}Xe(p, γ) [113] reactions at the ESR ring.

4.3.2.5 Time of Flight and Magnetic Rigidity

For the most exotic and shortest-lived nuclei, masses can be determined at fragment separator facilities utilizing the relationship between the time of flight and magnetic

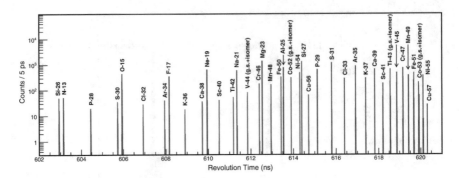

Fig. 4.7 An example of a revolution time spectrum from a storage ring measurement on proton-rich nuclei using isochronous mode at CSRe. The red and blue peaks represent the $T_Z = (N - Z)/2 = -1$ and $T_Z = -1/2$ nuclei, respectively (Reprinted from Ref. [114] with permissions from the American Physical Society)

rigidity $B\rho$:

$$(m/q) = \frac{t}{L} \frac{B\rho}{\gamma}, \tag{4.10}$$

where t is the time of flight and L is the length of the flight path. The dependence of m/q on the time of flight can be calibrated using a set of ions with a well-known mass. The ToF-$B\rho$ technique can only provide a modest precision of several hundreds of keV, but it can be applied also to very short-lived nuclei ($t_{1/2} < \mu s$). Although the lack of precision hinders detailed studies of nuclear structure, general trends and large changes on the mass surface can be detected with the ToF-$B\rho$ method as demonstrated, e.g., in Refs. [115–118].

4.3.2.6 Current Status and Recent Mass Measurements for Nuclear Astrophysics

Table 4.1 summarizes typical precisions achieved for mass-excess values $\Delta = (M(A, Z) - A \cdot u)c^2$ (where u is the atomic mass unit) using different mass measurement techniques and rough half-life limitations or ranges for the experiments. Experimental atomic mass values are evaluated regularly in the so-called atomic mass evaluations (AME). The evaluation takes into account available experimental mass data from various experiments. Experiments provide mass values with respect to other nuclides, e.g., Penning-trap measurements are done with respect to a reference nuclide, and reaction Q-values connect the reactants and products. The AME takes into account all the connections between nuclei and does a least squares optimization of the data, weighted by the experimental uncertainties $\sigma_{exp,i}$ as $w_i = 1/\sigma_{exp,i}^2$ [119]. The optimization yields adjusted mass values tabulated in the AME mass tables. The AME also reveals irregular or anomalous experimental data points deviating from the otherwise smooth mass surface. The most recent

Table 4.1 Different mass measurement techniques, typical precisions achieved for the mass-excess values, and half-life limitations. The precision and half-life limits depend on several factors, such as the production rates and measurement times (statistics). The values are mainly to give an idea of the strengths and weaknesses of the methods

Method	Precisions	Half-lives
ToF-ICR	~0.5–50 keV	$\gtrsim$100 ms
PI-ICR	~0.5–20 keV	$\gtrsim$50 ms
MR-ToF	~20–150 keV	$\gtrsim$10 ms
Schottky MS	~1–50 keV	$\gtrsim$1 s (cooling)
Isochronous MS	~10–200 keV	$\gtrsim$10 μs
ToF-Bρ	~300–500 keV	$\gtrsim$ below 1 μs

AME is AME2020 [21]. The NUBASE evaluations on the ground and isomeric state properties are published together with the AME, with the most recent being NUBASE2020 [120].

Many mass measurements for nuclear astrophysics have been performed recently. For example, masses of 22 neutron-rich rare-earth nuclei have been studied with the JYFLTRAP Penning trap [121], 14 for the first time [95, 122]. The measurements indicated less odd-even staggering in the neutron separation energies than predicted by the commonly used mass models for the r-process calculations (see Sect. 4.4.1). Including the new mass values in the r-process calculations resulted in a smoother trend in the calculated r-process abundances.

The recent precision mass measurements of neutron-rich $^{126-132}$Cd isotopes [107, 123] using the ISOLTRAP Penning trap [107] and its MR-ToF mass spectrometer [103] have reduced the nuclear uncertainties around the second r-process abundance peak. Mass measurements with the MR-ToF mass spectrometer at TITAN [106, 124] and the JYFLTRAP Penning trap [125] have provided new mass data for the first r-process peak region.

4.3.2.7 β-Decay Experiments for Nuclear Astrophysics

β-Decay plays an essential role in neutron-capture processes. The conversion to heavier elements is almost solely done via β^- decays, which compete with neutron captures (and photodisintegrations for high-temperatures environment). As a result, the β-decay half-lives serve as an important input in the nucleosynthesis calculations.

For the r-process, dozens of β-decay half-lives have been recently determined employing fragmentation or in-flight fission of ^{238}U at GSI and at Radioactive Isotope Beam Factory (RIBF) at RIKEN [126–130]. For given magnetic rigidity $B\rho = mv/q$, the fragments are identified based on (i) their energy loss ΔE and (ii) time of flight through the fragment separator. The energy loss is proportional to the proton number Z^2, i.e., heavier elements leave more energy. The energy loss is typically determined using an ionization chamber or stacked silicon detector. The time of flight is usually determined between two scintillator detectors and

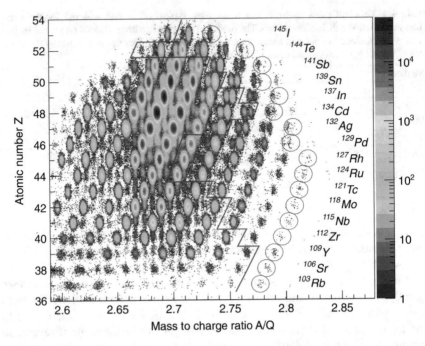

Fig. 4.8 An example of a particle identification (PID) plot. Ions are identified based on their proton number Z and the mass-to-charge ratio. The heaviest studied isotopes are labeled and highlighted by a red circle (Reprinted from Ref. [127] with permissions from the American Physical Society)

is proportional to m/q. A particle identification (PID) plot (see Fig. 4.8 for an example) typically shows the energy loss versus the time of flight but calibrated to show the proton number Z versus A/q.

β-decay half-lives at fragment separator facilities are usually determined by implanting the beam into a stack of silicon detectors and measuring the time difference between the implantation and β^- particles (electrons). During the last decade, the knowledge of the half-lives of neutron-rich nuclei has increased substantially. The measurements at GSI and RIBF have provided around 240 half-life values: around 20 half-lives close to ^{78}Ni [126], 94 in the rare-earth region [130], 110 in the $N = 82$ region [127], and 20 new half-lives in the $N = 126$ region [128, 129]. An example of a β-decay half-life measurement is shown in Fig. 4.9.

β-decays can be studied also at isotope separator on-line (ISOL) facilities, but there the selection of the isotope happens already before the beam arrives at the detector setup. As a result, the experiments focus on one or a few isotopes during a beamtime. On the other hand, β-decay studies at ISOL facilities can be done with very pure beams (see Fig. 4.10). Even isomerically pure beams can be prepared, e.g., by using a Penning trap or selective laser ionization.

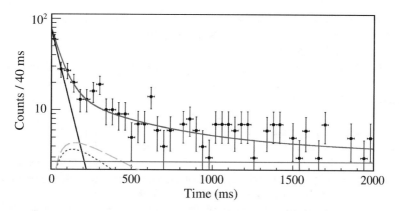

Fig. 4.9 Half-life measurement of ^{79}Ni at RIBF. Time distribution of the β-decay events correlated with the implanted ^{79}Ni ions has been plotted. The fitting function (solid red line) considers the activities of parent nuclei (dashed-dotted black line), β-decay daughter nuclei (fine-dashed blue line), βn-decay daughter nuclei (dashed green line), and a constant background (solid pink line). A half-life of $43.0^{+8.6}_{-7.5}$ ms was determined for ^{79}Ni (Reprinted from Ref. [126] with permissions from the American Physical Society)

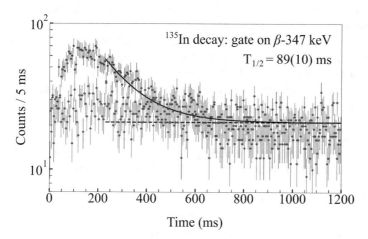

Fig. 4.10 Half-life measurement of ^{135}In at the ISOLDE Decay Station, where the laser-ionized ^{135}In$^+$ beam was accelerated to 40 keV and implanted into an aluminized Mylar tape at the center of the detection setup. The time distribution relative to the proton pulse from the CERN Proton Synchrotron Booster is shown as blue data points for the β-gated 347-keV γ-ray transitions, which belong to the β-delayed neutron daughter ^{134}Sn. The radioactive beam was extracted for period 5–230 ms, followed by the decay. The red data points represent the background (Reprinted from Ref. [131])

β-decays are also essential during the freeze-out phase of the r-process when matter is decaying back to stability. Prior to the freeze-out, the abundance pattern has an odd-even effect due to the odd-even staggering of neutron separation energies. Even-N nuclei are more abundant than their neighboring odd-N nuclei in each isotopic chain. This can also be seen in Eq. 4.4, where $Q_{n,\gamma}$ is higher for a nucleus (A, Z) with an odd neutron number N. During the freeze-out, β-delayed neutron emissions smoothen the abundance pattern.

In β-delayed neutron emission, β-decay of a nucleus (A, Z) will lead to an excited state above the neutron separation energy $S_n(A, Z + 1)$ in the daughter nucleus $(A, Z + 1)$. Since the state is neutron-unbound, it will emit a neutron and lead to a nucleus $(A - 1, Z + 1)$. β-delayed neutron emission (βn) was discovered already in 1939 [132]. Later, also β-delayed two-neutron ($\beta 2n$) [133], three-neutron ($\beta 3n$) [134], and four neutron ($\beta 4n$) [135] decays have been discovered, leading to nuclei $(A - 2, Z + 1)$, $(A - 3, Z + 1)$, and $(A - 4, Z + 1)$, respectively. For the r-process calculations, the β-delayed neutron emission branching ratios are relevant to determine the flow from one mass number to another.

β-delayed neutron branching ratio measurements are nowadays based on ^{3}He counters located in a neutron energy moderator medium, such as polyethylene. The detection of neutrons is based on the reaction ^{3}He$(n, p)^3$H, which releases 764 keV of energy. This is easily detectable and clearly above the noise level. Neutrons are moderated because the cross section for the used detection reaction increases with decreasing neutron energy. The BEta-deLayEd Neutron (BELEN) counter [136] has been designed for the FAIR DESPEC experiment. It has already been utilized in experiments at the ion guide isotope separator on-line (IGISOL) facility [137], where the JYFLTRAP Penning trap was used to select the ions of interest for the β-decay studies. For example, β-delayed two-neutron emission from ^{136}Sb has been studied at IGISOL [138]. More recently, a massive campaign of β-delayed neutron emission measurements has been performed with the BRIKEN (Beta-delayed Neutron Measurements at RIKEN) [139] setup at RIBF. The BRIKEN collaboration has already measured neutron emission probabilities for more than 180 nuclei. In addition to β-delayed neutron emission probabilities, β-delayed neutrons provide a way to determine β-decay half-lives. A recent compilation on β-delayed neutron emission summarizes the current status [140].

4.3.2.8 Neutron-Capture Rates

Neutron-capture rates on radioactive short-lived nuclei are challenging to study. However, many factors affecting the neutron-capture rate calculations can be investigated at radioactive beam facilities. Mass measurements provide data on neutron-capture Q values. The β-Oslo method [141, 142] yields information on level densities and γ-ray strength functions for moderately neutron-rich nuclei. The technique utilizes segmented total absorption γ-ray spectrometers with which both the individual γ-rays and the γ-ray cascade, i.e., the excitation energy, can be determined. In order to efficiently use this method, the β-decay Q value has to be high enough but the neutron-separation energies not too low. This maximizes the range of states that can be detected via β-delayed gamma cascades.

For specific cases, neutron-transfer (d, p) reactions provide information on the key resonance states and spectroscopic factors. For example, single-particle states in ^{133}Sn isotopes have been studied using the ^{132}Sn$(d, p)^{133}$Sn reaction in inverse kinematics [143]. The method has similarities with the (d, n) reactions in inverse kinematics used as a surrogate for proton captures. With more intensive radioactive beams, more possibilities will arrive to study neutron captures; however, the single-particle structure is most pronounced closed to doubly magic nuclei such as ^{132}Sn. Therefore, the method is not as useful for regions far from stability where collectivity is more pronounced.

4.3.2.9 Experiments on Fission

Although fissioning nuclei of r-process interest have not yet been studied experimentally, many experiments provide essential data to test current fission models. The current status of fission and fission experiments has been reviewed in Ref. [144] and fission barriers of superheavy elements in Ref. [145]. In addition to these, there have been many measurements on fission yields for various fissioning systems using a Penning trap (see Sect. 4.3.2.1) as an ion counter (see, e.g., Refs. [146, 147]). The fission yield measurements are useful for testing the predictions from different fission models. They also provide information on isomeric yield ratios in fission.

4.4   Theory for Nuclear Astrophysics

4.4.1   Nuclear Masses

Among the ground-state properties, the atomic mass is obviously the most fundamental quantity. The calculation of the reaction cross section also requires the knowledge of other ground-state properties, such as the deformation, density distribution, or the single-particle-level scheme. When not available experimentally, these quantities need to be extracted from a mass model which aims at reproducing measured masses as accurately as possible, i.e., typically with a root-mean-square (rms) deviation of less than about 0.8 MeV. The importance of estimating all ground-state properties reliably should not be underestimated. For example, the nuclear level densities of a deformed nucleus at low energies (typically at the neutron separation energy) are predicted to be significantly (about 30–50 times) larger than those of a spherical one due principally to the rotational enhancement. An erroneous determination of the deformation can therefore lead to large errors in the estimate of radiative capture cross sections. For this reason, modern mass models not only try to reproduce at the best experimental masses and mass differences but also charge radii, quadrupole moments, giant resonances, fission barriers, shape isomers, infinite nuclear matter properties, . . . [148, 149].

With a view to their astrophysical application in neutron-rich environments, a series of nuclear-mass models have been developed based on the Hartree-Fock-Bogoliubov (HFB) method with Skyrme and contact-pairing forces, together with phenomenological Wigner terms and correction terms for the spurious collective energy within the cranking approximation (see Ref. [150] and references therein); all the model parameters have been fitted to essentially all the experimental mass data. While the first HFB-1 mass model aimed at proving that it was possible to reach a low rms deviation with respect to all experimental masses available at that time, most of the subsequent models were developed to further explore the parameter space widely or to take into account additional constraints. These include in particular a sensitivity study of the mass model accuracy and extrapolation to major changes in the description of the pairing interaction, the spin-orbit coupling, or the nuclear matter properties, such as the effective mass, the symmetry energy, and the stability of the equation of state.

With respect to the 2457 measured masses for $Z, N \geq 8$ nuclei [21], the 32 HFB mass models give an rms deviation ranging between 0.52 MeV for HFB-27 and 0.82 MeV for HFB-1. These rms deviations can be compared to those obtained with other global mass models, such as the Gogny-HFB mass model with the D1M interaction [151] characterized by an rms of 0.81 MeV or the 2012 version of the finite-range droplet model [152] with 0.61 MeV. However, when dealing with exotic nuclei far away from stability, deviations between the HFB mass predictions can become significant, not only in the rigidity of the mass parabola but also in the description of the shell gaps or pairing correlations [153]. The 1σ variance between the 32 HFB mass predictions (with respect to the HFB-24 mass model) is illustrated in Fig. 4.11 where deviations up to about 3 MeV can be found at the neutron drip line for the heaviest species. Such uncertainties can be interpreted as the model uncertainties (due to model defects) inherent to the given HFB model [154]. These model uncertainties have been shown to be significantly larger than the uncertainties associated with local variations of the model parameters in the vicinity of an HFB minimum [153], as estimated using a variant of the Backward-Forward Monte Carlo method [155] to propagate the uncertainties on the masses of exotic nuclei far away

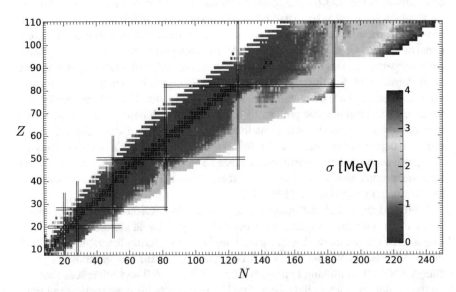

Fig. 4.11 Representation in the (N, Z) plane of the 1σ uncertainty corresponding to the 32 Skyrme-HFB mass models (with respect to HFB-24) for all the 8500 nuclei included in the mass tables from $Z = 8$ up to $Z = 110$

from the experimentally known regions (note that in this method only parameter sets giving rise to masses in reasonable agreement with experiments for all known nuclei are considered).

Many effective interactions have been proposed to estimate nuclear structure properties within the relativistic or non-relativistic mean-field approaches [156]. Except the BSk forces at the origin of the above-mentioned HFB mass models and the D1M interaction at the origin of the Gogny-HFB mass model [151], none of the others have been fitted to a large set of experimental masses. Consequently, their predictions lead to rms deviations typically larger than 2–3 MeV with respect to the bulk of known masses (e.g. masses obtained with the SLy4 force give an rms deviation of the order of 5 MeV [157]). With such a low accuracy, these masses should not be used for applications, such as the r-process nucleosynthesis. Additionally, other global mass models have been developed, essentially within the macroscopic-microscopic approach [152, 158], but this approach remains unstable with respect to parameter variations, as shown in the framework of the droplet model in Ref. [69]. In addition, this approach suffers from major shortcomings, such as the incoherent link between the macroscopic part and the microscopic correction or the instability of the shell correction [148, 149]. For this reason, more fundamental approaches, such as the mean field, are needed for astrophysical applications.

When considering mass models obtained in relatively different frameworks, e.g., the Skyrme-HFB or Gogny-HFB mass models, still large deviations are found in the mass predictions away from the experimentally known region. For example, as shown in Fig. 4.12, deviations up to typically ±5 MeV can be observed for exotic nuclei between HFB-31 [150] and D1M [151] mass predictions, especially around

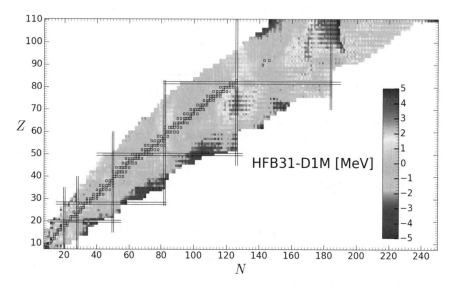

Fig. 4.12 Representation in the (N, Z) plane of the mass differences between HFB-31 [150] and D1M [151] models for all the 8500 nuclei from $Z = 8$ up to $Z = 110$

the $N = 126$ and 184 shell closures. Neutron-capture rates can consequently deviate by three to five orders of magnitude with such mass differences, essentially due to different local variations in the pairing and shell description (see Sect. 4.4.3). Such deviations by far exceed what is acceptable for nucleosynthesis applications. For this reason, further improvements of the mass model are required. These include development of relativistic as well as non-relativistic mean-field models but also the inclusion within such approaches of the state-of-the-art beyond-mean-field corrections, like the quadrupole or octupole correlations by the generator coordinate method [159, 160] and a proper treatment of odd-A and odd-odd nuclei with time-reversal symmetry breaking. Such models should reproduce not only nuclear masses at best but also as many experimental observables as possible. These include charge radii and neutron skin thicknesses, fission barriers and shape isomers, spectroscopic data such as the 2^+ energies, and moments of inertia but also infinite (neutron and symmetric) nuclear matter properties obtained from realistic calculations as well as specific observed or empirical properties of neutron stars, like their maximum mass or mass-radius relations [161, 162].

4.4.2 β-Decay Rates

β-decay rates play a fundamental role in nucleosynthesis in general [4] and more particularly for the r-process nucleosynthesis since they set the timescale of the nuclear flow and consequently of the production of the heavy elements. β^--decay rates have been experimentally determined for 1213 nuclei [120] (see Sect. 4.3.2.7). For the few thousands nuclei missing in r-process nucleosynthesis simulations, only a restricted number of global models is available. These concern the macroscopic gross theory (GT2) [163], the FRDM+RPA [164], the Tamm-Dancoff approximation (TDA) [165], and the relativistic mean field plus QRPA (RMF+RRPA) [166]. Deviations between the predictions of some of these models are illustrated in Fig. 4.13 where ratios larger than a factor of 10 are found in many neutron-rich regions of the (N, Z) plane, especially for heavy or superheavy nuclei.

Here also, more effort needs to be devoted to improve the prediction of β-decay rates, to include consistently not only the contribution of the forbidden transitions [166, 167] but also the deformation effects, the majority of nuclei being deformed [168, 169]. In particular, the first-forbidden transitions have been studied with the finite Fermi system theory [167] and the relativistic QRPA approach [166] but both only for spherical nuclei. Recent studies within the fully self-consistent proton-neutron QRPA model using the finite-range Gogny interaction have now also taken axially symmetric deformations into account [169], but forbidden transition remains to be included and the theory to be applied to systems with an odd number of nucleons. The inclusion of the phonon-phonon coupling has also been shown to give rise to a redistribution of the Gamow-Teller strength and impact the β-decay half-lives of neutron-rich nuclei significantly [170]. Further progress along all these lines will hopefully help improve the predictions. Finally, note that on the basis

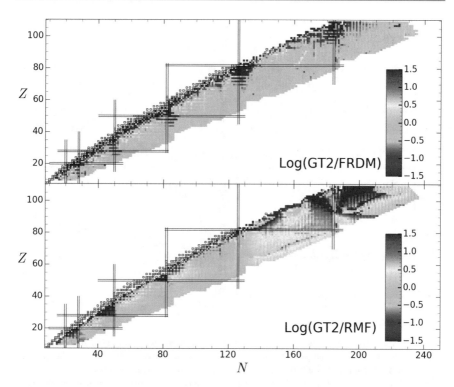

Fig. 4.13 Representation in the (N, Z) plane of the β^--decay rate ratios (in log scale) obtained by three global models. *Upper panel*: Ratio between the HFB-21 + GT2 [163] and the FRDM+RPA rates [164]. *Lower panel*: Ratio between the HFB-21 + GT2 [163] and the RMF+RRPA rates [166]. The open squares correspond to the valley of β-stability. The double solid lines depict the neutron and proton magic numbers

of the β-decay strength, the β-delayed processes, including neutron emission and fission for the heaviest species, also need to be derived [70, 140].

4.4.3 Nuclear Reactions

Most of the low-energy cross-section calculations for nucleosynthesis applications are based on the statistical model of Hauser-Feshbach. Such a model makes the fundamental assumption that the capture process takes place with the intermediary formation of a compound nucleus in thermodynamic equilibrium. The energy of the incident particle is then shared more or less uniformly by all the nucleons before releasing the energy by particle emission or γ-de-excitation. The formation of a compound nucleus is usually justified by assuming that the level density in the compound nucleus at the projectile incident energy is large enough to ensure an average statistical continuum superposition of available resonances. The statistical

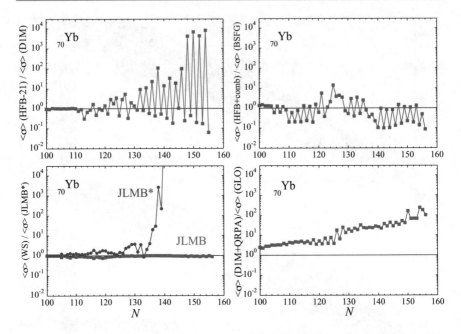

Fig. 4.14 Illustration of some uncertainties affecting the prediction of the radiative neutron-capture rates for the Yb isotopes ($Z = 70$), between the valley of β-stability and the HFB-21 neutron drip line; these include the sensitivity to (*i*) the mass model when using the HFB-21 [183] or D1M [151] models (upper left), (*ii*) the nuclear level densities when using the HFB plus combinatorial [176] or the back-shifted Fermi gas (BSFG) [177] models (upper right), (*iii*) the optical potential adopting a Wood-Saxon (WS) potential [173] or two variants of the microscopic JLMB potentials [174, 175] (lower left), and (*iv*) the γ-ray strength function derived from either the D1M+QRPA [171] or the generalized Lorentzian (GLO) [184] models (lower right). The Maxwellian-averaged rates are estimated within the Hauser-Feshbach statistical model at $T = 10^9$ K

model has proven its ability to predict cross sections accurately for medium- and heavy-mass nuclei. However, this model suffers from uncertainties stemming essentially from the predicted nuclear ingredients describing the nuclear structure properties of the ground and excited states and the strong and electromagnetic interaction properties.

The impact of different input models adopted in the calculation of the reaction rates of astrophysical interest is illustrated in Fig. 4.14. Clear mass models have the strongest impact with deviation reaching four orders of magnitude for the most exotic neutron-rich nuclei. Nuclear level densities are seen to affect rates within typically a factor of 10 with a strong odd-even effect according to the way pairing interaction is treated. The γ-ray strength function may impact the prediction of the rate up to a factor of 100, in particular depending the way the low-energy tail of the giant $E1$ resonance is described, but also the low-energy $M1$ component is included, both for the scissors mode and for the so-called upbend [171, 172]. Finally, the optical potential is known to have a negligible impact in the standard case (e.g. comparing the Woods-Saxon [173] and the microscopic so-called JLMB potential [174] in Fig. 4.14), although a reduction of the imaginary potential may have a drastic impact in reducing the absorption of neutrons by neutron-rich nuclei, as shown when considering the JLMB* potential [175]. More details on our capacity to predict reliably all these ingredients can be found in Refs. [4, 171, 175–178]. A review on the nuclear ingredients of relevance for the description of fission for nucleosynthesis applications and its fundamental role in r-process calculations can be found in Ref. [70].

When the number of available states in the compound nucleus is relatively small, the capture reaction is known to be possibly dominated by direct electromagnetic transitions to a bound final state rather than through a compound nucleus intermediary. It is now well accepted that this direct capture contribution is important and often dominant at the very low energies of astrophysical interest for light or exotic nuclei systems for which few or even no resonant states are available. The direct contribution to the neutron-capture rate can be two to three orders of magnitude larger than the one obtained within the Hauser-Feshbach approach traditionally used in nucleosynthesis applications [178–181]. Significant uncertainties still affect the direct capture predictions. These are related to the determination of the nuclear structure ingredients of relevance, i.e., the nuclear mass, spectroscopic factor, neutron-nucleus interaction potential, and excited level scheme. An important effort will have to be devoted to further improve the prediction of such nuclear inputs within reliable microscopic models, with a special emphasis on the determination of the low-energy excitation spectrum, in particular the spin and parity assignments. The transition from the compound nucleus to the direct capture mechanism, when only a few resonant states are available, also needs to be tackled in a more detailed way, for example, within the Breit-Wigner approach or the so-called high-fidelity-resonance technique [182].

4.5 Summary and Outlook

One of the major issues in modern astrophysics concerns the analysis of the present composition of the Universe and its various constituting objects. Nucleosynthesis models aim to explain the origin of the different nuclei observed in nature by identifying the possible processes able to synthesize them. Though the origin of most of the nuclides lighter than iron is now quite well understood, the synthesis of the heavy elements (i.e., heavier than iron) remains obscure in many respects, from the astrophysics as well as nuclear physics point of views. As far as nuclear physics is concerned, strong, weak, and electromagnetic interaction processes play an essential role in nucleosynthesis processes.

Radioactive beam facilities have provided new ways to explore key reactions for the nucleosynthesis of lighter elements. Many reactions have become available for studies in inverse kinematics using radioactive beams at astrophysically relevant energies. Surrogate reactions or β-delayed gamma and particle spectroscopy have provided information on the properties of key resonance states. On the other hand, low-background facilities located underground have opened new possibilities to study reactions relevant for quiescent hydrogen and helium burning at or near astrophysically relevant energies.

For the synthesis of heavier elements, experiments at radioactive beam facilities have extended our knowledge of exotic nuclei and their properties, which serve as relevant inputs, for example, for the r-process calculations. Mass measurement techniques, such as Penning traps, MR-ToF mass spectrometers, and storage rings, have recently provided mass values for dozens of new neutron-rich nuclei. β-decay studies have yielded information on β-decay half-lives for several dozens of neutron-rich nuclei. Results on β-delayed neutron branchings obtained with the BRIKEN detector setup are coming and provide a major step forward in the knowledge of neutron-rich nuclei. Neutron-capture rates have been probed for specific nuclei using the β-Oslo method to determine the level densities and γ-ray strength functions. Several experiments have studied nuclear fission and fission yields, providing data to test various fission models. With anticipated new radioactive beam facilities, such as FRIB and FAIR, even more exotic nuclei will become available for experiments.

Although important effort has been devoted in the last decades to measure reaction cross sections, experimental data only covers a minute fraction of the whole set of data required for nuclear astrophysics applications. To fill the gaps, theoretical predictions are needed. Many astrophysics applications involve a large number of unstable nuclei and therefore require the use of global approaches. The extrapolation to exotic nuclei or energy ranges far away from experimentally known regions constrains the use of nuclear models to the most reliable ones, even if empirical approaches sometime present a better ability to reproduce experimental data. A subtle compromise between the reliability, accuracy, and applicability of the different theories available has to be found according to the specific application considered.

A continued effort to improve our predictions of the reaction and β-decay rates, including their statistical and systematic uncertainties, for nuclei far away from stability is obviously required. The reliability of our predictions today is still far from being at the level of the requirements in nuclear astrophysics applications. Priority should be given to a better description of the ground-state, fission, and β-decay properties but also nuclear level densities, optical potential, and γ-ray strength functions. A huge amount of work is still needed to make full advantage of the development of state-of-the-art microscopic models in building global universal models that include as much as possible the microscopic character of quantum physics. This effort to improve the microscopic nuclear predictions is concomitant with new development aiming at improving the description of the reaction mechanisms, including the equilibrium, pre-equilibrium, and direct capture processes. This theoretical work requires simultaneously new measurements of structure properties far away from stability but also reaction cross sections on stable targets and any experiments that can provide new insight on the numerous ingredients of the reaction models and their extrapolation far away from stability.

Acknowledgments SG acknowledges financial support from F.R.S.-FNRS (Belgium). This work has been supported by the Fonds de la Recherche Scientifique (F.R.S.-FNRS, Belgium) and the Research Foundation Flanders (FWO, Belgium) under the EOS Project nr O022818F. AK acknowledges financial support from the European Union's Horizon 2020 research and innovation program under Grant Agreement No. 771036 (ERC CoG MAIDEN).

References

1. Y. Ivezić, T. Beers, M. Juric, et al., Ann. Rev. Astron. Astrophys. **50**, 251 (2012)
2. A.B.S. Reddy, D. Lambert, S. Giridhar, Mon. Notices Royal Astron. Soc. **463**, 4366 (2016)
3. B. Barbuy, C. Chiappini, O. Gerhard, Ann. Rev. Astron. Astrophys. **56**, 223 (2018)
4. M. Arnould, S. Goriely, Prog. Part. Nucl. Phys. **112**, 103766 (2020)
5. E. Anders, N. Grevesse, Geochim. Cosmochim. Acta **53**, 197 (1989)
6. M. Asplund, N. Grevesse, A.J. Sauval, P. Scott, Ann. Rev. Astron. Astrophys. **47**, 481 (2009)
7. K. Lodders, in *Principles and Perspectives in Cosmochemistry*, ed. by A. Goswami, B.E. Reddy (Springer Berlin Heidelberg, Berlin, Heidelberg, 2010), pp. 379–417
8. H. Palme, K. Lodders, A. Jones, Planets, Asteroids, Comets Solar Syst. **2**, 15 (2014)
9. S. Goriely, Astron. Astrophys. **342**(3), 881 (1999)
10. M. Arnould, S. Goriely, Phys. Rep. **384**, 1 (2003)
11. M. Arnould, S. Goriely, K. Takahashi, Phys. Rep. **450**, 97 (2007)
12. T. Lee, in *Meteorites and the Early Solar System*, ed. by J. Kerridge, M. Matthews (University of Arizona Press, Tucson, 1988), p. 1063
13. T. Swindle, in *Protostars and Planets III*, ed. by E. Levy, J. Lunine (University of Arizona Press, Tucson, 1993), p. 867
14. E. Anders, E. Zinner, Meteoritics **28**, 490 (1993)
15. E.M. Burbidge, G.R. Burbidge, W.A. Fowler, F. Hoyle, Rev. Mod. Phys. **29**, 547 (1957)
16. A.G.W. Cameron, Pub. Astron. Soc. Pac. **69**, 201 (1957)
17. V. Tatischeff, S. Gabici, Ann. Rev. Nucl. Part. Sci. **68**, 377 (2018)
18. S. Woosley, T. Weaver, Astrophys. J. Suppl. Ser. **101**, 181 (1995)
19. C. Iliadis, *Nuclear Physics of Stars*, 2nd edn. (Wiley-VCH, 2015)
20. M. Limongi, A. Chieffi, Astrophys. J. Suppl. Ser. **237**, 13 (2018)
21. M. Wang, W. Huang, F. Kondev, G. Audi, S. Naimi, Chin. Phys. C **45**(3), 030003 (2021)
22. F. Käppeler, H. Beer, K. Wisshak, Rep. Prog. Phys. **52**, 945 (1989)
23. S. Goriely, N. Mowlavi, Astron. Astrophys. **362**, 599 (2000)
24. A.I. Karakas, J.C. Lattanzio, Pub. Astron. Soc. Pac. **31**, e030 (2014)
25. S. Goriely, L. Siess, Astron. Astrophys. **609**, A29 (2018)
26. G. Meynet, S. Ekström, A. Maeder, Astron. Astrophys. **447**, 623 (2006)
27. U. Frischknecht, R. Hirschi, M. Pignatari, et al., Mon. Notices Royal Astron. Soc. **456**, 1803 (2016)
28. A. Choplin, R. Hirschi, G. Meynet, S. Ekström, C. Chiappini, A. Laird, Astron. Astrophys. **618**, A133 (2018)
29. J. Cowan, C. Sneden, J. Lawler, A. Aprahamian, M. Wiescher, K. Langanke, G. Martínez-Pinedo, F.K. Thielemann, Rev. Mod. Phys. **93**, 015002 (2021)
30. H.T. Janka, *Handbook of Supernovae* (Springer International Pub. AG, 2017)
31. S. Wanajo, B. Müller, H.T. Janka, A. Heger, Astrophys. J **853**, 40 (2018)
32. H.T. Janka, Ann. Rev. Nucl. Part. Sci. **62**, 407 (2012)
33. H.T. Janka, T. Melson, A. Summa, Ann. Rev. Nucl. Part. Sci. **66**, 341 (2016)
34. N. Nishimura, T. Takiwaki, F.K. Thielemann, Astrophys. J **810**, 109 (2015)
35. D. Siegel, J. Barnes, B. Metzger, Nature **569**, 241 (2019)
36. O. Just, I. Kullmann, S. Goriely, A. Bauswein, H.T. Janka, C.E. Collins, Mon. Notices Royal Astron. Soc., submitted (2021). [arXiv]:2109.14617 [astro-ph.HE]

37. C. Winteler, R. Käppeli, A. Perego, A. Arcones, N. Vasset, N. Nishimura, M. Liebendörfer, F.K. Thielemann, Astrophys. J **750**, L22 (2012)
38. L. Roberts, D. Kasen, W. Lee, E. Ramirez-Ruiz, Astrophys. J. Lett. **736**, L21 (2011)
39. S. Goriely, A. Bauswein, H.T. Janka, Astrophys. J. Lett. **738**, L32 (2011)
40. O. Just, A. Bauswein, R. Ardevol Pulpillo, S. Goriely, H.T. Janka, Mon. Notices Royal Astron. Soc. **448**, 541 (2015)
41. A. Bauswein, R. Ardevol Pulpillo, H.T. Janka, S. Goriely, Astrophysical J. Lett. **773**, L9 (2014)
42. S. Wanajo, Y. Sekiguchi, N. Nishimura, K. Kiuchi, K. Kyutoku, M. Shibata, Astrophys. J. Lett. **789**, L39 (2014)
43. B. Abbott, R. Abbott, T. Abbott, et al., Phys. Rev. Lett. **119**, 161101 (2017)
44. D. Watson, C.J. Hansen, J. Selsing, A. Koch, D.B. Malesani, et al., Nature **574**, 497 (2019)
45. S. Goriely, A. Bauswein, O. Just, E. Pllumbi, H.T. Janka, Mon. Notices Royal Astron. Soc. **452**, 3894 (2015)
46. R. Ardevol-Pulpillo, H.T. Janka, O. Just, A. Bauswein, Mon. Notices Royal Astron. Soc. **485**, 4754 (2019)
47. V. Nedora, S. Bernuzzi, D. Radice, B. Daszuta, A. Endrizzi, A. Perego, A. Prakash, M. Safarzadeh, F. Schianchi, D. Logoteta, Astrophys. J **906**, 98 (2021)
48. F. Herwig, M. Pignatari, P.R. Woodward, D.H. Porter, G. Rockefeller, C.L. Fryer, M. Bennett, R. Hirschi, Astrophys. J **727**, 89 (2011)
49. M. Hampel, R.J. Stancliffe, M. Lugaro, B.S. Meyer, Astrophys. J **831**, 171 (2016)
50. O. Clarkson, F. Herwig, M. Pignatari, Mon. Notices Royal Astron. Soc. **474**, L37 (2018)
51. P.A. Denissenkov, F. Herwig, P. Woodward, R. Andrassy, M. Pignatari, S. Jones, Mon. Notices Royal Astron. Soc. **488**(3), 4258 (2019)
52. A. Choplin, L. Siess, S. Goriely, Astron. Astrophys. **648**, A119 (2021)
53. S. Goriely, L. Siess, A. Choplin, Astron. Astrophys. **654**, A129 (2021)
54. M. Rayet, M. Arnould, M. Hashimoto, et al., Astron. Astrophys. **298**, 517 (1995)
55. C. Travaglio, T. Rauscher, A. Heger, M. Pignatari, C. West, Astrophys. J **854**, 18 (2018)
56. K. Nomoto, F.K. Thielemann, K. Yokoi, Astrophys. J **286**, 644 (1984)
57. W. Hillebrandt, J. Niemeyer, Ann. Rev. Astron. Astrophys. **38**, 191 (2000)
58. C. Travaglio, R. Gallino, T. Rauscher, F. Röpke, W. Hillebrandt, Astrophys. J. **799**, 54 (2015)
59. S. Goriely, D. Garcia-Senz, E. Bravo, J. José, Astron. Astrophys. **444**, L1 (2005)
60. C. Fröhlich, G. Martínez-Pinedo, M. Liebendörfer, F.K. Thielemann, E. Bravo, W.R. Hix, K. Langanke, N.T. Zinner, Phys. Rev. Lett. **96**, 142502 (2006)
61. K. Takahashi, K. Yokoi, At. Data Nucl. Data Tables **36**, 375 (1987)
62. C. Iliadis, A. Champagne, A. Chieffi, M. Limongi, Astrophys. J. Suppl. Ser. **193**(1), 16 (2011)
63. R.C. Runkle, A.E. Champagne, J. Engel, Astrophys. J. **556**(2), 970 (2001)
64. S.S. Gupta, B.S. Meyer, Phys. Rev. C **64**, 025805 (2001)
65. G.W. Misch, S.K. Ghorui, P. Banerjee, Y. Sun, M.R. Mumpower, Astrophys. J. Suppl. Ser. **252**(1), 2 (2020)
66. D. Yakovlev, L.R. Gasques, A.V. Afanasjev, et al., Phys. Rev. C **74**, 035803 (2006)
67. A. Potekhin, G. Chabrier, Contrib. Plasma Phys. **53**, 397 (2013)
68. Y. Génolini, D. Maurin, I.V. Moskalenko, M. Unger, Phys. Rev. C **98**, 034611 (2018)
69. S. Goriely, M. Arnould, Astron. Astrophys. **262**, 73 (1992)
70. S. Goriely, Eur. Phys. J. A **51**, 22 (2015)
71. S. Goriely, G. Martínez-Pinedo, Nucl. Phys. A **944**, 158 (2015)
72. J.F. Lemaître, S. Goriely, A. Bauswein, H.T. Janka, Phys. Rev. C **103**, 025806 (2021)
73. M. Aliotta, R. Buompane, M. Couder, A. Couture, R. deBoer, A. Formicola, et al., J. Phys. G: Nucl. Part. Phys. **49**, 010501 (2021)
74. R. Reifarth, C. Lederer, F. Käppeler, J. Phys. G: Nucl. Part. Phys. **41**(5), 053101 (2014)
75. R. Diehl, H. Halloin, K. Kretschmer, G.G. Lichti, V. Schönfelder, A.W. Strong, et al., Nature **439**(7072), 45 (2006)
76. L. Buchmann, M. Hilgemeier, A. Krauss, A. Redder, C. Rolfs, H. Trautvetter, T. Donoghue, Nucl. Phys. A **415**(1), 93 (1984)

77. C. Ruiz, A. Parikh, J. José, L. Buchmann, J.A. Caggiano, A.A. Chen, J.A. Clark, H. Crawford, B. Davids, J.M. D'Auria, C. Davis, C. Deibel, L. Erikson, L. Fogarty, D. Frekers, U. Greife, A. Hussein, D.A. Hutcheon, M. Huyse, C. Jewett, A.M. Laird, R. Lewis, P. Mumby-Croft, A. Olin, D.F. Ottewell, C.V. Ouellet, P. Parker, J. Pearson, G. Ruprecht, M. Trinczek, C. Vockenhuber, C. Wrede, Phys. Rev. Lett. **96**, 252501 (2006)
78. R. Dressler, M. Ayranov, D. Bemmerer, M. Bunka, Y. Dai, et al., J. Phys. G: Nucl. Part. Phys. **39**(10), 105201 (2012)
79. V. Margerin, A. Murphy, T. Davinson, R. Dressler, J. Fallis, et al., Phys. Lett. B **731**, 358 (2014)
80. B.W. Grefenstette, F.A. Harrison, S.E. Boggs, S.P. Reynolds, C.L. Fryer, et al., Nature **506**(7488), 339 (2014)
81. M. Renaud, J. Vink, A. Decourchelle, F. Lebrun, P.R. den Hartog, et al., Astrophys. J. **647**(1), L41 (2006)
82. S.A. Grebenev, A.A. Lutovinov, S.S. Tsygankov, C. Winkler, Nature **490**, 373 (2012)
83. A. Kankainen, P.J. Woods, F. Nunes, C. Langer, H. Schatz, et al., Eur. Phys. J. A **52**(1), 6 (2016)
84. R.B. Vogelaar, L.W. Mitchell, R.W. Kavanagh, A.E. Champagne, P.V. Magnus, M.S. Smith, A.J. Howard, P.D. Parker, H.A. O Brien, Phys. Rev. C **53**, 1945 (1996)
85. A. Kankainen, P.J. Woods, H. Schatz, T. Poxon-Pearson, D.T. Doherty, et al., Phys. Lett. B **769**, 549 (2017)
86. D. Kahl, P.J. Woods, T. Poxon-Pearson, F.M. Nunes, B.A. Brown, H. Schatz, et al., Phys. Lett. B **797**, 134803 (2019)
87. M.B. Bennett, C. Wrede, B.A. Brown, S.N. Liddick, D. Pérez-Loureiro, et al., Phys. Rev. Lett. **116**, 102502 (2016)
88. C. Iliadis, Phys. Rev. C **99**, 065809 (2019)
89. L.S. Brown, G. Gabrielse, Phys. Rev. A **25**, 2423 (1982)
90. L.S. Brown, G. Gabrielse, Rev. Mod. Phys. **58**, 233 (1986)
91. G. Gräff, H. Kalinowsky, J. Traut, Z. Phys. A **297**(1), 35 (1980)
92. M. König, G. Bollen, H.J. Kluge, T. Otto, J. Szerypo, Int. J. Mass Spectrom. Ion Proc. **142**(1–2), 95 (1995)
93. S. George, K. Blaum, F. Herfurth, A. Herlert, M. Kretzschmar, S. Nagy, S. Schwarz, L. Schweikhard, C. Yazidjian, Int. J. Mass Spectrom. **264**(2–3), 110 (2007)
94. M. Kretzschmar, Int. J. Mass Spectrom. **264**(2–3), 122 (2007)
95. M. Vilen, J.M. Kelly, A. Kankainen, M. Brodeur, et al., Phys. Rev. C **101**, 034312 (2020)
96. S. Eliseev, K. Blaum, M. Block, C. Droese, M. Goncharov, et al., Phys. Rev. Lett. **110**, 082501 (2013)
97. S. Eliseev, K. Blaum, M. Block, A. Dörr, C. Droese, et al., Appl. Phys. B **114**(1), 107 (2014)
98. O.T. Kaleja, Ph.D. thesis, Johannes Gutenberg-Universität Mainz (2020)
99. M.B. Comisarow, A.G. Marshall, Chem. Phys. Lett. **25**(2), 282 (1974)
100. T. Eronen, A. Kankainen, J. Äystö, Progr. Part. Nucl. Phys. **91**, 259 (2016)
101. J. Dilling, K. Blaum, M. Brodeur, S. Eliseev, Ann. Rev. Nucl. Part. Sci. **68**, 45 (2018)
102. H. Wollnik, Int. J. Mass Spectrom. **349–350**, 38 (2013)
103. R. Wolf, F. Wienholtz, et al., Int. J. Mass Spectrom. **349–350**, 123 (2013)
104. W.R. Plaß, T. Dickel, C. Scheidenberger, Int. J. Mass Spectrom. **349–350**, 134 (2013)
105. T. Dickel, W. Plaß, et al., Nucl. Instrum. Methods Phys. Res. Sect. A **777**, 172 (2015)
106. M. Reiter, S.A.S. Andrés, J. Bergmann, T. Dickel, J. Dilling, et al., Nucl. Instrum. Methods Phys. Res. Sect. A **1018**, 165823 (2021)
107. V. Manea, J. Karthein, D. Atanasov, M. Bender, K. Blaum, et al., Phys. Rev. Lett. **124**, 092502 (2020)
108. M. Steck, Y.A. Litvinov, Progr. Part. Nucl. Phys. **115**, 103811 (2020)
109. B. Franzke, H. Geissel, G. Münzenberg, Mass Spectrom. Rev. **27**(5), 428 (2008)
110. J. Xia, W. Zhan, B. Wei, Y. Yuan, M. Song, et al., Nucl. Instrum. Methods Phys. Res. A (1), 11 (2002)
111. Y.H. Zhang, Y.A. Litvinov, T. Uesaka, H.S. Xu, Phys. Scripta **91**(7), 073002 (2016)

112. B. Mei, T. Aumann, S. Bishop, K. Blaum, K. Boretzky, et al., Phys. Rev. C **92**, 035803 (2015)
113. J. Glorius, C. Langer, Z. Slavkovská, L. Bott, C. Brandau, et al., Phys. Rev. Lett. **122**, 092701 (2019)
114. Y.H. Zhang, P. Zhang, X.H. Zhou, M. Wang, Y.A. Litvinov, et al., Phys. Rev. C **98**, 014319 (2018)
115. A. Estradé, M. Matoš, H. Schatz, et al., Phys. Rev. Lett. **107**, 172503 (2011)
116. Z. Meisel, S. George, et al., Phys. Rev. Lett. **114**, 022501 (2015)
117. Z. Meisel, S. George, S. Ahn, et al., Phys. Rev. Lett. **115**, 162501 (2015)
118. Z. Meisel, S. George, et al., Phys. Rev. C **101**, 052801 (2020)
119. W. Huang, M. Wang, F. Kondev, G. Audi, S. Naimi, Chin. Phys. C **45**(3), 030002 (2021)
120. F.G. Kondev, M. Wang, W.J. Huang, S. Naimi, G. Audi, Chin. Phys. C **45**(3), 030001 (2021)
121. T. Eronen, et al., Eur. Phys. J. A **48**(4), 46 (2012)
122. M. Vilen, J.M. Kelly, A. Kankainen, M. Brodeur, et al., Phys. Rev. Lett. **120**, 262701 (2018)
123. D. Atanasov, P. Ascher, K. Blaum, R.B. Cakirli, T.E. Cocolios, et al., Phys. Rev. Lett. **115**, 232501 (2015)
124. M.P. Reiter, S. Ayet San Andrés, S. Nikas, J. Lippuner, et al., Phys. Rev. C **101**, 025803 (2020)
125. L. Canete, S. Giraud, A. Kankainen, B. Bastin, F. Nowacki, A. Poves, et al., Phys. Rev. C **101**, 041304 (2020)
126. Z.Y. Xu, S. Nishimura, G. Lorusso, F. Browne, P. Doornenbal, et al., Phys. Rev. Lett. **113**, 032505 (2014)
127. G. Lorusso, S. Nishimura, Z.Y. Xu, A. Jungclaus, Y. Shimizu, et al., Phys. Rev. Lett. **114**, 192501 (2015)
128. R. Caballero-Folch, C. Domingo-Pardo, J. Agramunt, A. Algora, F. Ameil, et al., Phys. Rev. Lett. **117**, 012501 (2016)
129. R. Caballero-Folch, C. Domingo-Pardo, J. Agramunt, A. Algora, F. Ameil, et al., Phys. Rev. C **95**, 064322 (2017)
130. J. Wu, S. Nishimura, G. Lorusso, P. Möller, E. Ideguchi, et al., Phys. Rev. Lett. **118**, 072701 (2017)
131. M. Piersa-Siłkowska, A. Korgul, et al., Phys. Rev. C **104**, 044328 (2021)
132. R.B. Roberts, L.R. Hafstad, R.C. Meyer, P. Wang, Phys. Rev. **55**, 664 (1939)
133. R.E. Azuma, L.C. Carraz, P.G. Hansen, B. Jonson, K.L. Kratz, et al., Phys. Rev. Lett. **43**, 1652 (1979)
134. R. Azuma, T. Björnstad, H. Gustafsson, P. Hansen, B. Jonson, S. Mattsson, G. Nyman, A. Poskanzer, H. Ravn, Phys. Lett. B **96**(1), 31 (1980)
135. J. Dufour, R. Del Moral, F. Hubert, D. Jean, M. Pravikoff, et al., Phys. Lett. B **206**(2), 195 (1988)
136. J. Agramunt, J. Tain, M. Gómez-Hornillos, A. Garcia, F. Albiol, et al., Nucl. Instrum. Methods Phys. Res. A **807**, 69 (2016)
137. I.D. Moore, et al., Nucl. Instrum. Methods Phys. Res. B **317**, 208 (2013)
138. R. Caballero-Folch, I. Dillmann, J. Agramunt, J.L. Taín, A. Algora, et al., Phys. Rev. C **98**, 034310 (2018)
139. A. Tolosa-Delgado, J. Agramunt, J.L. Tain, A. Algora, C. Domingo-Pardo, et al., Nucl. Instrum. Methods Phys. Res. A **925**, 133 (2019)
140. P. Dimitriou, I. Dillmann, B. Singh, V. Piksaikin, K. Rykaczewski, J. Tain, et al., Nucl. Data Sheets **173**, 144 (2021)
141. A. Spyrou, S.N. Liddick, A.C. Larsen, M. Guttormsen, et al., Phys. Rev. Lett. **113**, 232502 (2014)
142. S.N. Liddick, A. Spyrou, B.P. Crider, F. Naqvi, A.C. Larsen, et al., Phys. Rev. Lett. **116**, 242502 (2016)
143. K.L. Jones, A.S. Adekola, D.W. Bardayan, J.C. Blackmon, K.Y. Chae, K.A. Chipps, J.A. Cizewski, L. Erikson, C. Harlin, R. Hatarik, et al., Nature **465**(7297), 454 (2010)
144. A.N. Andreyev, K. Nishio, K.H. Schmidt, Rep. Progr. Phys. **81**(1), 016301 (2017)
145. F.P. Heßberger, Eur. Phys. J. A **53**, 75 (2017)
146. H. Penttilä, D. Gorelov, et al., Eur. Phys. J. A **52**, 104 (2016)

147. V. Rakopoulos, M. Lantz, S. Pomp, A. Solders, A. Al-Adili, et al., Phys. Rev. C **99**, 014617 (2019)
148. J.M. Pearson, Hyperfine Interact. **132**, 59 (2001)
149. D. Lunney, J. Pearson, C. Thibault, Rev. Mod. Phys. **75**(3), 1021 (2003)
150. S. Goriely, N. Chamel, J.M. Pearson, Phys. Rev. C **93**, 034337 (2016)
151. S. Goriely, S. Hilaire, M. Girod, S. Péru, Phys. Rev. Lett. **102**, 242501 (2009)
152. P. Möller, A. Sierk, T. Ichikawa, H. Sagawa, At. Data Nucl. Data Tables **109–110**, 1 (2016)
153. S. Goriely, R. Capote, Phys. Rev. C **89**, 054318 (2014)
154. D. Neudecker, R. Capote, H. Leeb, Nucl. Instrum. Methods Phys. Res. A **723**, 163 (2013)
155. E. Bauge, P. Dossantos-Uzarralde, J. Korean Phys. Soc. **59**, 1218 (2011)
156. M. Bender, P.H. Heenen, P.G. Reinhard, Rev. Mod. Phys. **75**, 121 (2003)
157. M. Stoitsov, J. Dobaczewski, W. Nazarewicz, S. Pittel, D. Dean, Phys. Rev. C **68**, 054312 (2003)
158. N. Wang, M. Liu, X. Wu, J. Meng, Phys. Lett. B **734**, 215 (2014)
159. M. Bender, G. Bertsch, P.H. Heenen, Phys. Rev. C **73**, 034322 (2006)
160. L. Robledo, T. Rodríguez, R. Rodríguez-Guzmán, J. Phys. G: Nucl. Part. Phys. **46**, 013001 (2018)
161. A.F. Fantina, N. Chamel, J.M. Pearson, S. Goriely, Astron. Astrophys. **559**, A128 (2013)
162. J.M. Pearson, N. Chamel, A.Y. Potekhin, A.F. Fantina, C. Ducoin, A.K. Dutta, S. Goriely, Mon. Notices Royal Astron. Soc. **481**(3), 2994 (2018)
163. T. Tachibana, M. Yamada, Y. Yoshida, Prog. Theor. Phys. **84**, 641 (1990)
164. P. Möller, B. Pfeiffer, K.L. Kratz, Phys. Rev. C **67**, 055802 (2003)
165. H. Klapdor, J. Metzinger, T. Oda, At. Data Nuc. Data Tables **31**, 81 (1984)
166. T. Marketin, L. Huther, G. Martinez-Pinedo, Phys. Rev. C **93**, 025805 (2016)
167. I. Borzov, Phys. Part. Nuclei **48**, 885 (2017)
168. J.M. Boillos, P. Sarriguren, Phys. Rev. C **91**, 034311 (2015)
169. M. Martini, S. Péru, S. Goriely, Phys. Rev. C **89**, 044306 (2014)
170. A.P. Severyukhin, V.V. Voronov, I.N. Borzov, N.N. Arsenyev, N.V. Giai, Phys. Rev. C **90**, 044320 (2014)
171. S. Goriely, S. Hilaire, S. Péru, K. Sieja, Phys. Rev. C **98**, 014327 (2018)
172. S. Goriely, P. Dimitriou, M. Wiedeking, et al., Eur. Phys. J. A **55**, 172 (2019)
173. A. Koning, J. Delaroche, Nucl. Phys. A **713**(3–4), 231 (2003)
174. J. Jeukenne, A. Lejeune, C. Mahaux, Phys. Rev. C **16**, 80 (1977)
175. S. Goriely, J.P. Delaroche, Phys. Lett. B **653**, 178 (2007)
176. S. Goriely, S. Hilaire, A.J. Koning, Phys. Rev. C **78**, 064307 (2008)
177. A. Koning, S. Hilaire, S. Goriely, Nucl. Phys. A **810**, 13 (2008)
178. S. Goriely, Eur. Phys. J. A **51**, 172 (2015)
179. Y. Xu, S. Goriely, Phys. Rev. C **86**, 045801 (2012)
180. Y. Xu, S. Goriely, A.J. Koning, S. Hilaire, Phys. Rev. C **90**, 024604 (2014)
181. K. Sieja, S. Goriely, Eur. Phys. J. A **57**, 110 (2021)
182. D. Rochman, S. Goriely, A. Koning, H.Ferroukhi, Phys. Lett. B **764**, 109 (2017)
183. S. Goriely, N. Chame, J. Pearson, Phys. Rev. C **82**, 035804 (2010)
184. J. Kopecky, M. Uhl, Phys. Rev. C **41**, 1941 (1990)

State-of-the-Art Gamma-Ray Spectrometers for In-Beam Measurements

5

Francesco Recchia and Caterina Michelagnoli

Abstract

High-resolution γ-ray spectroscopy is one of the most powerful and sensitive tools to investigate nuclear structure. Significant progress in this field has been achieved through the use of arrays of Compton-suppressed high-purity germanium detectors; however, it is apparent that such devices are not suited to the expected experimental conditions at the planned and under construction radioactive ion beam facilities. Devices with higher efficiency and sensitivity have been developed during the past two decades relying on the possibility to determine the position and the energy deposition of the individual interaction points of a photon within a germanium crystal and on the capability to reconstruct the photon scattering sequence through powerful data analysis algorithms. In these notes a brief introduction to the principles of γ-ray tracking arrays will be given. After a historical overview of the main spectrometers that contributed to the present understanding of the nuclear structure, the principles of advanced γ-ray tracking will be described. A basic technical description of arrays based on this technique will be reported together with some selected results.

F. Recchia (✉)
University and INFN Padova, Padova, Italy
e-mail: francesco.recchia@unipd.it

C. Michelagnoli
Institut Laue-Langevin, Grenoble, France
e-mail: cmichela@ill.eu

© The Author(s), under exclusive license to Springer Nature Switzerland AG 2022
S. M. Lenzi, D. Cortina-Gil (eds.), *The Euroschool on Exotic Beams, Vol. VI*,
Lecture Notes in Physics 1005, https://doi.org/10.1007/978-3-031-10751-1_5

5.1   Introduction

Since germanium detectors became available in the 1960s, they constituted the most used and important tool for γ-ray spectroscopy. In particular, the modern High-Purity Germanium crystals (HPGe) provide an excellent (1–3 keV) energy resolution in the typical energy range for nuclear spectroscopic studies that ranges between approximately 10 keV and 10 MeV. The continuous development of high-resolution γ-ray detector systems has been of vital importance to nuclear structure physics. It has steadily expanded the limits of what can be observed, allowing for the discovery of new phenomena and leading to deeper insights into the nature of the nucleus. Indeed, large arrays of HPGe detectors became in the 1990s the state-of-the-art instruments for γ-ray spectroscopy, having not only a large photo-peak efficiency, but also allowing high selectivity of the reaction channels of interest, through the analysis of multiple γ coincidences. The key concept behind these arrays is to obtain the required selectivity (and efficiency) through the combination of several detectors, each of them with good response function (in other words, with energy resolution and ratio of peak over total, P/T ratio, both as good as possible). It should be strongly remarked that since, in in-beam γ-ray spectroscopy, the photons are emitted by recoiling nuclei, the FWHM of the peak is dominated in most cases by the Doppler broadening due to the finite size of the detector, rather than the intrinsic detector resolution. This means that each element of the array should cover a solid angle as small as possible to keep this broadening within "reasonable" limits. Concerning the P/T ratio, the background generated by partially absorbed photons can be the dominant part of the spectrum for the available size of the HPGe crystals. Such background can be efficiently reduced by surrounding the Ge detectors with the so-called Compton suppression shields, namely, veto detectors that detect photons escaping from the germanium crystal. A major breakthrough in the field of γ-spectroscopy was achieved in the 1990s with the construction of arrays of Compton-shielded large-volume HPGe detectors GASP [1, 2], EUROGAM [3], EUROBALL [4], and GAMMASPHERE [5]. In principle, an array of germanium detectors should have as large as possible photopeak efficiency, in order to limit the time needed to acquire the required statistics to a minimum. Large photopeak efficiency is best obtained by combining several detectors, each of them subtending a small solid angle. In this way, not only the Doppler broadening effects is kept under control but also the probability of multiple hits, i.e., two or more photons entering the same crystal at the same time are minimized. Multiple-hit phenomena can be significant when several photons are emitted simultaneously. The characteristics to be considered to assess the quality of an array of HPGe detectors for high-resolution and high-efficiency in-beam γ-ray spectroscopy can thus be summarized as follows:

- effective energy resolution;
- full-energy (photopeak) efficiency;
- peak-to-total ratio;

- granularity;
- amount of dead materials;
- time resolution;
- counting-rate capabilities.

With the "conventional" Compton suppression techniques, it was not feasible to reach simultaneously total photopeak efficiencies around 50% and P/T ratios around 50%. These two parameters are considered essential features to pursue the physics program at radioactive ion beam facilities [6]. An alternative approach, followed since the 1990s [7], involves the construction of a Ge ball around the target position, with the development of techniques (pulse shape analysis and γ-ray tracking) to "look inside" the germanium crystals and to follow the individual photon scattering.

In these notes the concept of *advanced γ-ray tracking* will be introduced, together with the main subjects connected to this technique. Before going into the details of this detection technique, the composite detector technology, such as clover and cluster, will be briefly discussed. This technology is indeed in use in many facilities and provides still nowadays many advantages.

5.1.1 Clovers and Add-Back Procedure

Following the above discussion, it is clear that in developing an array of HPGe detectors, a careful balance must be found between two contrasting needs. On one hand the detectors should be placed at large distances from the target position in order to subtend small solid angles with each element. As it will be explained in detail in Sect. 5.3, the effective energy resolution strongly depends on the uncertainty on the direction of the γ-ray as a consequence of the Doppler shift of the γ-rays emitted by nuclei recoiling after a nuclear reaction. Thus to obtain a reasonable resolution, it is important that each detector subtend a small angle. On the other hand, in order to maximize the detection efficiency with a given number of detectors, each of them should be placed as close as possible to the target position. The problem can be partially overcome by using composite detectors such as the clover [8] and cluster [9] detectors developed within the EUROBALL collaboration. In these detectors, more crystals (respectively, 4 and 7) share the same cryostat and the same Compton-suppression shield. Efficiency and peak-to-total can be recovered by summing up (adding back) the energy deposited in neighboring detectors.

Clover-based arrays are used in γ-ray spectroscopy in international facilities [10–12]. The working principle of Compton-suppressed clover detectors is schematized in Fig. 5.1. When γ-rays are detected in coincidence by different crystals in a clover, their energies are "added-back" in order to recover the full energy of the original γ ray. This is exemplified in Fig. 5.2 where the spectra corresponding to the energy measured by individual crystals is compared to the spectra corresponding to the energy resulted from the add-back of all the crystals in a clover detector. Because of the electrical coupling within the cryostat of the clover detector, the channels suffer a relevant cross talk that implies that the energy obtained by summing the

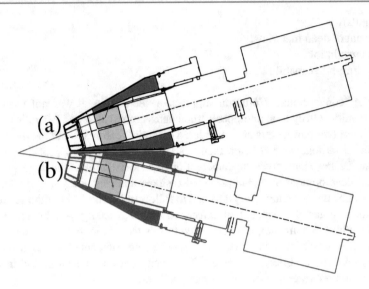

Fig. 5.1 Two clover detectors with the anti-Compton shield. The γ-ray labeled (**a**) scatters in one germanium crystal, and it is absorbed by the anti-Compton shield resulting in a rejected event. The γ-ray labeled (**b**) scatters in one germanium crystal, and it is absorbed by a neighboring germanium resulting in a full-energy count in the add-back spectrum

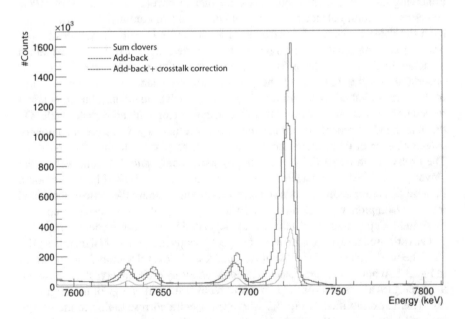

Fig. 5.2 Spectrum obtained with the ^{27}Al(n,γ)^{28}Al reaction with the clover array FIPPS at ILL. In green, spectrum obtained using each single crystal separately; in blue, spectrum obtained using the add-back procedure within each clover detector; in red, add-back spectrum obtained using the add-back procedure and applying a cross talk correction. Taken from [13]

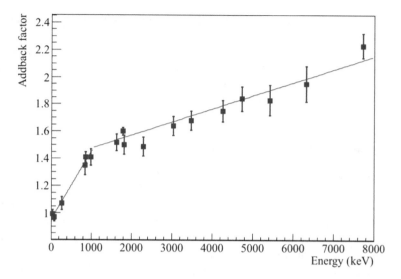

Fig. 5.3 Add-back factor as a function of the energy for the FIPPS clover array at the Institut Laue-Langevin (ILL). See text for more details. Taken from [14]

ones detected by two or more crystals has a deficit that needs to be corrected. The importance of the cross talk is highlighted by the comparison between the add-back spectrum with and without cross talk correction reported in Fig. 5.3.

In Fig. 5.3 the add-back factor as a function of the energy is reported for the FIPPS spectrometer at the Institut Laue-Langevin (ILL), as an example of the importance of add-back for the performance of a clover array. The add-back factor is determined from the ratio of the area of the peak after add-back and the one in the spectrum obtained as sum of all detectors (without add-back). The data in the picture are obtained from (n,γ) reactions on Al and Mn targets. The result of a linear fit is reported. The change of slope at $\approx 1\,$MeV is due to the occurrence of pair production. While the gain in full-energy peak efficiency is evident, especially at high energy, the capabilities of such devices are limited by the following factors:

- Uncertainty in the direction of the γ-ray due to the finite size of the detector: the direction of the γ used for Doppler correction and angular correlations/distributions is assumed to correspond to the one of the crystal with the largest energy deposit. There is no information about where the interaction took place within a crystal. The dimension of a detector and its distance to the target are thus defining how precisely the direction of the γ ray is known;
- In case the γ scatters from one crystal to a neighboring one, given that the sequence in the Compton scattering is only guessed, systematic errors are done in the extraction of observables that require such information (first, second interaction point);

- At high count rate, summing effects can occur (two distinct γ rays are accidentally added-back).
- Eventual nonlinearities in the energy calibrations or cross talk effects (not properly corrected) may affect the resolution in the energy spectrum after add-back. In Fig. 5.2 a comparison of the energy resolution under different conditions is shown. In particular, the effect of cross talk correction is evident.

5.2 Advanced γ-Ray Tracking

5.2.1 General Aspects: The Tracking Concept

The state-of-the-art γ-ray detector arrays aim at overcoming the main limitations of traditional arrays, namely, the "finite" detector size/position resolution and efficiency. γ-Ray tracking arrays, exploiting the position sensitivity obtainable from segmentation of the outer contact of the detector, can achieve a major step forward both in Doppler correction capabilities and in efficiency. The idea is thus to build 4π detector arrays composed entirely of HPGe, as shown in Fig. 5.4, that will be used not simply in a calorimetric mode but instead will allow to reconstruct the path of each γ-ray inside the detector volume. As it will be shown in the following, such a device is highly demanding and challenging in terms of performance and technology. Nowadays two projects exist, one in Europe named AGATA [15], one in the USA, named GRETA [16].

The idea at the basis of the *advanced γ-ray tracking* is to reconstruct the individual γ-ray energies and directions, based on the deposited energy and position

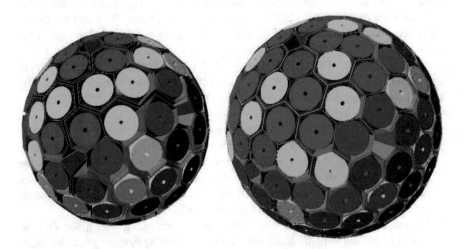

Fig. 5.4 Artistic view of the γ-ray tracking arrays GRETA (120 crystals, left) and the AGATA (180 crystals, right). The colors correspond to different germanium crystal shape tapered in irregular hexagon to guarantee an optimal solid angle coverage

Fig. 5.5 Schematic representation of a photon that undergoes two interactions

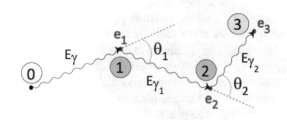

of all the interaction points of an event seen by the detector(s). For each event, the number of photons, their energies, the incident, and scattering directions should be measured, and the events corresponding to incomplete energy release should be discarded. In principle, this could be done by knowing the interaction positions with sufficient precision and the details of the interaction mechanisms. In the energy range between 100 keV and 10 MeV, γ rays interact with Ge mainly via Compton scattering [17]. In the limit of a free electron at rest, this process is analytically described by (5.1):

$$E_{\gamma i} = \frac{E_{\gamma i-1}}{\frac{E_{\gamma i-1}}{m_e c^2}(1 - \cos\theta_i)} \tag{5.1}$$

where $E_{\gamma i}$ is the energy of the γ-ray after the i-th scattering.

Above 1.022 MeV, the γ energy is sufficient for creating an e^+e^- pair, with the positron later annihilating and emitting two 511 keV photons in opposite directions. To clarify the basic concepts behind the technique of γ-ray tracking, let us consider a photon that undergoes two interactions before being absorbed, as shown in Fig. 5.5. Assuming the electrons in the material unbound and at rest, the energy is related to the scattering angle by (5.1). If the source position is known, one can evaluate the Compton formula for each permutation of the interaction points and build a χ^2 function (figure of merit) of the kind:

$$\chi^2 \approx \sum_{n=1}^{N-1} w^n \left(\frac{E_{\gamma'} - E_{\gamma'}^{Pos}}{E_\gamma}\right)_n^2 \tag{5.2}$$

Recalling that the initial photon energy E_γ is the sum of the individual deposited energies, and assuming a given sequence for the photon scattering, the photon energy after the ith scattering is easily computed as:

$$E_{\gamma'} = E_\gamma - \sum_{n=1}^{i} e_n \tag{5.3}$$

$E_{\gamma i}$ can also be computed starting from (5.1). In practice, a figure of merit should be evaluated for all the possible permutations of the interaction points. The event is accepted only if the merit of the best permutation is lower than an empirically defined limit. The discarded events are the "software equivalent" of a Compton suppression. The analysis of all partitions of the measured points for (real) cases of more than one γ per event is normally not feasible (10^{23} partitions for 30 points). The interactions of single photons must be thus grouped by means of clustering techniques [18]. The points generated by a single γ are expected to be localized in a small portion of the total detector volume, compatible with the finite absorption length for the photons and with the forward peaking of the Compton scattering. The interactions belonging to the same track can be clustered on the basis of the angular separation (seen from the source) and their mutual linear distance.

The tracking algorithms can be distinguished in two classes: forward- and back-tracking [19, and references therein]. Using the backtracking method, the interaction points within a given deposited energy interval are considered as last interactions of fully absorbed γ rays. The scattering is then tracked backwards, back to the known position of the source.

The forward tracking method, instead, starts from the identification of a set of clusters, each identified by its energy, sum of the energy depositions in the cluster. The three interaction mechanisms of γ rays in germanium are then tested, and "good clusters" are selected via a χ^2 test. The fundamental effects limiting the performance of the tracking algorithms can be summarized as follows:

- difference between the position of the photon interaction and the one of the energy deposition (electrons drift)
- electrons energy loss via bremsstrahlung
- Rayleigh scattering of low-energy γ rays toward the end of a track (change of direction without change of energy)
- (5.1) is just an approximation; electrons in media are not free nor at rest; the formula should be corrected in order to take the momentum of the electron into account.

All the abovementioned effects are "masked" by the error in the determination of the interaction points. Indeed important uncertainties are introduced by the determination of the interaction points, position resolution, energy threshold, and the presence of dead materials.

The concept of γ-ray tracking can be summarized as represented in Fig. 5.6.

In order to track the γ-ray scattering path, it is necessary to use the detectors in position-sensitive mode. This is described in the next section.

5.2.2 Pulse Shape Analysis

The shape of the signal in a true coaxial detector depends on the distance of the interaction to the central contact [17]. Given the cylindrical symmetry of the system,

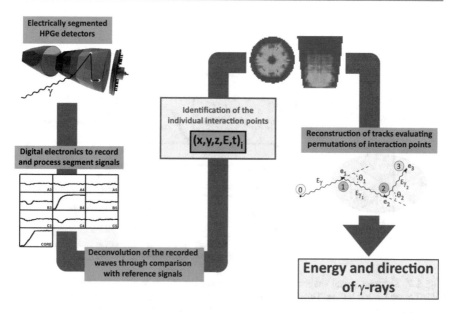

Fig. 5.6 Main steps for the tracking of γ-rays detected by HPGe segmented detectors. The picture summarizes all the different steps detailed in the text

no other information on the position can be extracted, unless this symmetry is broken, for example, by electrically segmenting one of the electrodes. When one of the electrodes is electrically segmented, the motion of the charges within one segment induces signals also in the neighboring electrodes (Fig. 5.7). Contrary to the segment where the interaction takes place (i.e., where there is a net charge release), the total collected charge in the neighboring electrodes is zero. For this reason, the signals induced in the neighboring electrodes are known as *transient signals*. The shapes and amplitudes of net and transient shapes depend also on the angular position of the interaction point, as exemplified in Fig. 5.7. The amplitude of the induced transient signals provides a convenient way to locate the interaction with sub-segment precision, but this is not enough to achieve the required position resolution. In order to reach a better precision, the full shapes of the observed signals should be compared to a set of reference signals, each of them corresponding to interactions taking place in well-defined locations in the crystal. In principle, the basis of reference signals for a given segmented detector should be constructed experimentally, measuring in a semi-automatic way the signals corresponding to specific locations within the crystal, with the possibility to move such locations in any point of the detector. This can be done using collimated γ-ray sources and coincidence detectors to construct what is known as *scanning tables* [20, 21]. However, obtaining scanned bases for all the detectors of an array is extremely time-consuming and in practice not feasible, and as a consequence techniques to

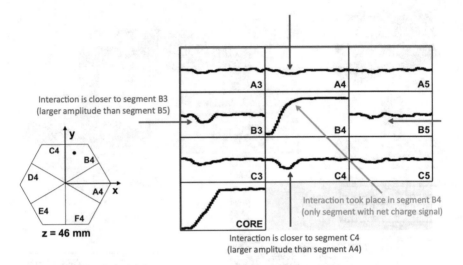

Fig. 5.7 Segment signal shapes following a γ-interaction in one segment at 46 mm depth in an AGATA segmented HPGe. The net charge signal identifies the segment where the interaction took place, while the amplitude of the transient signals suggests its position inside the segment

calculate these reference signals have been developed, with parameters adjusted by comparison with measured pulse shapes in "key-locations" in the crystal [22].

In principle, the knowledge of the electric field (or potential) in the full segmented detector would be required for such calculations. This is quite a complicated problem, that is why, in practice, the Shockley-Ramo theorem is used as a simplification [23, and references therein]. The *weighting field* is calculated for each sensing electrode by solving numerically the Laplace equation with the sensing electrode put at fixed potential while grounding all the others. The Ramo theorem guarantees a direct relation between the charge released in a given position inside the crystal and the weighting potential at that position. For an AGATA detector, which has a 6x6 outer segmentation, the calculation requires the solution of 1 Poisson and 37 Laplace (1 anode and 36 cathodes) equations. The calculated basis is usually considered on a space grid of 1 mm and a time step of 1 ns. Once the reference signal basis is available, the interaction points of the detected γ radiation into the germanium detector are obtained by comparison with the recorded waveforms (an example of detector signal is reported in Fig. 5.7), and this is called pulse shape analysis (PSA). The quality of this procedure strongly determines the performance of the tracking algorithm and depends mainly on two factors, additional to the quality of the signal basis:

- proper handling of the signals to correct for second-order effects due to electronic couplings among the channels
- performance of the algorithm used to make the comparison.

Assuming a "perfectly performing" PSA algorithm, still the experimental waveforms are not at all "ideal," and signal distortions could result in a displacement of the reconstructed interaction point from the actual one. Apart from the electronic noise that has to be minimized by means of proper grounding of the electronics, some other effects can be observed and taken into account/corrected in the data preparation to the PSA.

In any electrically segmented detector, the cross talk phenomenon is present. It produces a shift in the reconstructed energies that is proportional to the segment fold of the event, namely, to the number of segments firing simultaneously. The origin and calculations for the correction of this effect are extensively described in [23, 24]. The cross talk affects the detector energy resolution since the gain varies depending on the pattern of firing segments. The amplitude of a signal from one segment has therefore to be corrected according to a linear combination of the signal amplitudes of the other segments. Experimentally, two kinds of cross talk have been observed: a cross talk which is proportional to the net-charge signal and one which is proportional to its time derivative. Measurements of both kinds of cross talk are reported in [23]. The PSA algorithm compares the measured signals with a reference basis of simulated single interaction signals. This algorithm has to effectively identify the interaction points in the crystal and efficiently process the experimental data in order to obtain the positions and energy depositions from the experimental waveforms in a short time interval. The quality of the PSA algorithm can be tested by comparing the basis signals with experimental data taken in well-defined positions or, in a less direct but practically faster way, by running in-beam experiments with fast moving nuclei and by checking the Doppler correction capabilities.

An algorithm commonly used is the adaptive grid search (AGS) [25]. Basically this algorithm compares the measured signals (net and transient signals of the segments) and calculated ones over a fine grid of points in the crystal. It is suited for searching one or two interaction points per segment. It has been demonstrated that searching one interaction point in a segment is equivalent to consider the energetic barycenter if multiple hits occur, which implies that the signals vary approximately linearly in small distances. The signal comparison is done evaluating the residue R defined as the sum of the squared difference between measured and calculated signals. The algorithm evaluates R over all the points belonging to the real segment, i.e., working as a full grid search. The smallest R identifies the three-dimensional coordinates of the interaction point searched. When two points are searched, the signal in the comparison is a linear combination of signals for two possible points in the real segment, while their amplitudes represent the energy division between the two deposits. This part of data analysis is the bottleneck of the data processing, taking typically up to 95% of the computing resources. To improve the processing time, the PSA algorithm used in AGATA implement the AGS. This is a simple two-step analysis which starts with an evaluation of the residue R on a coarse grid of points and then proceeds with a search on the fine grid in the surroundings of the coarse-grid minimum to refine the position of the interaction point. With a spacing of 2 mm, the full fine grid search implies the evaluation of $\approx$1000 points in a typical AGATA detector segment, resulting in an analysis rate of $\approx$100 events/s in a standard modern CPU. With a coarse grid spacing of 6 mm, the AGS improves the analysis speed by about one order of magnitude, without significant loss of precision. Due to its relative simplicity and its speed of execution, this algorithm is adopted in the standard PSA procedures of the AGATA collaboration and has been used in [26] to obtain a position resolution of 4 mm for γ rays of 1.4 MeV.

5.2.2.1 Radiation Damage by Fast Neutrons

One of the "by-products" of the PSA consists on the correction of the effects produced by the damage of the crystal structure caused by the interaction of fast neutrons in the germanium medium [17]. It is well-known that germanium detectors are sensitive to radiation damage induced by fast neutrons (with energy $E > 1$ keV), which generate charged lattice defects that act as trapping sites for the charge carriers [27]. This problem leads to a worsening of the energy resolution, especially in the form of a tail on the left side of the γ peak in the energy spectrum, as a result of a reduction of the charge collection efficiency (see Fig. 1 in [27] as an example). The main reason for choosing n-type crystals for the detectors used for in-beam γ-ray spectroscopy is that they appear to be less sensitive to neutron damage than p-type ones. This is due to the fact that the defects produced by fast neutrons in the germanium lattice are negatively charged and therefore do not trap the e-carriers which, due to the coaxial configuration of the detectors, dominate the signal formation in the positively biased central electrode, from which the signal is read out. In the case of γ-ray tracking arrays, also the signals induced in the segments of the outer electrode (cathode) are used. They are dominated by the collection of holes, and thus they are affected by charge trapping effects. Starting with undamaged detectors, for which the energy resolution is better in the segments than in the core, a few weeks of operation under beam at medium-high counting rate are sufficient to degrade the energy resolution of the segments to values that would normally require annealing of the crystals. Fortunately, the operation in position-sensitive mode of the germanium crystals offers an interesting way to contrast the trapping effects [27]. In fact, for a given level of neutron damage, the fraction of charge carriers lost in the collection process depends mainly on their travel path, the details of which can be easily obtained from the position of the interaction given by the PSA. The principle is represented in Fig. 5.8 taken from [28]. Ideally, the shape of the pulses is also affected by charge trapping, but for the usual levels of damage, this is not influencing appreciably the pulse shape analysis.

5.2.3 Digital Signal Processing

The choice of Germanium detectors for γ radiation is motivated in first place by their excellent energy resolution conjugated to high detection efficiency. Only super-conducting and bolometer devices can provide better resolution than germanium, but their use for γ radiation is limited by their low efficiency and low count rate capabilities to few specific cases where large production of the decaying isotope

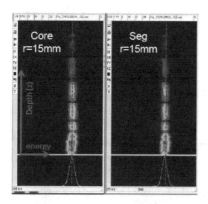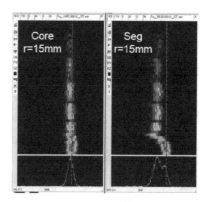

Fig. 5.8 Neutron-damage effects in a HPGe detector operated in position-sensitive mode: depth-energy correlation of the reconstructed interaction points for one of the segmented detectors used in the Legnaro AGATA Demonstrator campaign, before (left) and after (right) 3 months of beam-time operation. Taken from [28]

is available for slow, long-lasting measurements [29]. At present, germanium is the choice for spectroscopy thanks to its resolution that is the most important characteristic of the detectors. The analogue preamplified signal provided by a germanium detector requires a devoted treatment able to preserve the performance in term of intrinsic resolution.

Starting from the late 1980s, the progress in the FADC (*flash analog to digital converter*) technology made possible signal digitization with the needed dynamic range, frequency, and linearity. While a signal bandwidth sufficient to preserve the positional information from pulse shape was estimated to be of the order of 25 MHz [30], in order to obtain an optimal time resolution, a 100 MHz sampling frequency is needed. 100 MHz-FADC with 12 effective bits and sufficient linearity are commercially available nowadays.

Optimal noise filtering is essential to obtain the desired energy resolution. In general, noise filtering requires a change of shape of the detector signal. To ensure full charge collection (in the 100 ns to 500 ns time range), real implementations of preamplifiers transform the current pulse due to charge collection in a voltage signal with a considerably longer exponential shape, with a decay constant of 50 μs. This long decay constant implies that for rates of γ detection in the order of tens of kHz, the waveform for which we need to measure the amplitude in general will seat on the tail of a previous pulse as reported in Fig. 5.9. Each pulse will be superimposed on the residual tail of the former one. The signals need to be elaborated to remove the dependence of the amplitude on the tail, and a possible solution that is commonly used in digital shaping is a waveform formation on a trapezoidal shape. With this processing, the pulses obtain a shape with a much shorter total length but in a way that preserve the proportionality of the amplitude of the signal to the energy deposited in the detector.

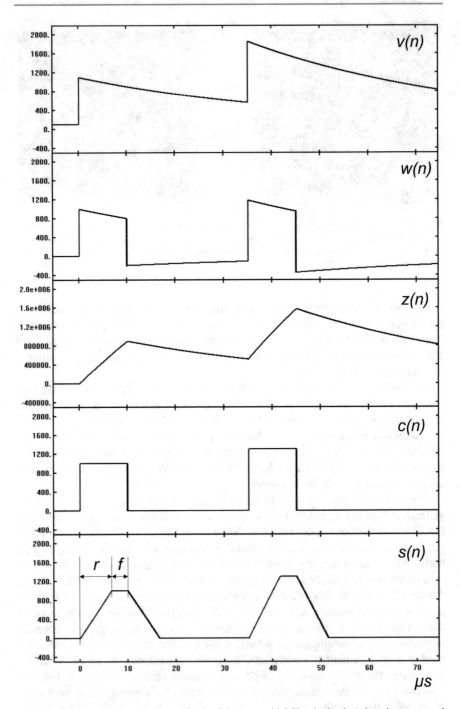

Fig. 5.9 The steps involved in the synthesis of the trapezoidal filter in the time domain as reported in (5.4). Risetime (r) and flattop (f) are defined as depicted in the bottom part of the figure

The first reported implementation of a trapezoidal filter dates back to 1968 by V. Radeka [31]. The problem of optimum signal processing for pulse amplitude spectrometry highlighted the importance of a well-defined weighting function for given noise and rate conditions. The signal and noise can be defined in a conventional way considering a detector signal, for purposes of noise analysis, as an impulse of current. The finite width of the signal is taken later into account. In a simplified scheme, two basic equivalent generators of white noise can be considered in the system. One is a current generator in parallel with the input, representing thermal noise of resistors and dielectrics and the leakage current shot noise. White noise can be represented as a random sequence of impulses, and therefore this noise is for signal processing of the same nature as the signal. The other is the equivalent noise voltage generator in series with the input, representing the overall amplifier noise. This series voltage generator can be converted into a parallel current generator producing equal noise voltage on the input capacitance. The noise-corner time constant is a measure of relative importance of the two noise generators.

Large efficiency germanium arrays are obtained by using large germanium detectors in order to minimize passive material. For example, the crystals of detectors composing the AGATA array have a volume of 370 cm^3. That implies long collection time as well as large detector capacities. In order to obtain a good resolution, it is important to have a constant weighting during charge collection. This implies the use of filters with a weighting function very long with respect to the collection time, i.e., by a long filter time constant [32].

The optimum weighting function for the noise encountered in present detector-FET amplifier systems, and under the constraint of finite measurement time per event, is a cusp [32].

As reported by Jordanov and collaborators [33] with the advent of fast, high-resolution, and high linearity digitizers in the 1990s, the realization of digital filters became practical. To this purpose the trapezoidal filter was implemented to convert a digitized exponential pulse $v(n)$ into a symmetrical trapezoidal pulse $s(n)$:

$$w(n) = v(n) - v(n - r - f)$$
$$z(n) = z(n - 1) + w(n)$$
$$c(n) = w(n) + M \cdot z(n - 1)$$
$$s(n) = s(n - 1) + c(n) - c(n - r)$$

$$(5.4)$$

where r is the length of the risetime and f is the length of the flattop of the trapezoid in digitizer clock cycle units; see Fig. 5.9. The parameter M depends only on the decay time constant τ of the exponential pulse, and it is given in clock cycle units:

$$M = 1 - \exp\left(-\frac{1}{\tau}\right).$$

$$(5.5)$$

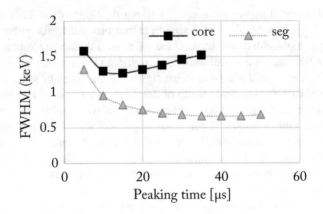

Fig. 5.10 Resolutions obtained for different peaking time of the shaper applied to the central contact (core) and to a segment (seg) of an AGATA detector, in the low energy limit

This effect of this kind of shaping is represented in Fig. 5.9 where examples of input and output signals are depicted. It is remarkable for the implementation in real-time processing electronics that the operations reported in (5.4) are all simple sums or subtractions, with the only exception of the multiplication by the constant M, (5.5). These kind of operations are cheap in terms of computing resources and can be implemented in FPGA.

The trapezoidal filter that optimizes the resolution has to be tuned as a function of the noise spectrum. The largest the capacitance of the detection element, the larger will be the integration of the low-frequency noise. As a consequence, in segmented germanium detectors, the compromise to be adopted to obtain the best possible resolution, in the low-count rate conditions, will be different for the filter used on the signals of the segments (low capacity) with respect to the one used for the central contact. In Fig. 5.10 the resolution of a segment and the resolution of the central contact of an AGATA detector are compared as a function of the peaking time (risetime + flat-top of the trapezoid).

The response function to a trapezoidal filter and the weight that is given to signal and noise can be analyzed in the frequency domain. While filtering in time domain implies performing a convolution of experimental noise with impulse response of the shaper, filtering in the frequency domain can be represented by the multiplication of the Fourier transform of experimental noise with the Fourier transform of the impulse response of the shaper.

In Fig. 5.11 a noise spectrum is reported as measured from an AGATA detector. The effect of trapezoidal filtering is visible in the same figure. The filtered spectrum was obtained multiplying the noise spectrum by the Fourier transform of the impulse response function of the trapezoidal filter.

Further filtering of the noise is obtainable using a time variant baseline restorer that consists in a long timescale running average over the input signal. This is needed to get rid of excess low-frequency noise, and it is demonstrated to be beneficial in

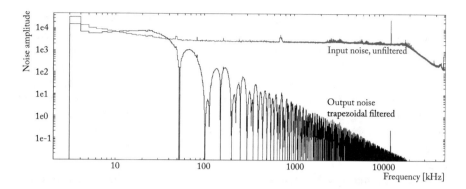

Fig. 5.11 Example of trapezoidal filtering on the noise of an AGATA detector, represented in the frequency domain. The fast Fourier transform of the input noise is depicted in blue. The output noise after the trapezoidal filtering is depicted in black

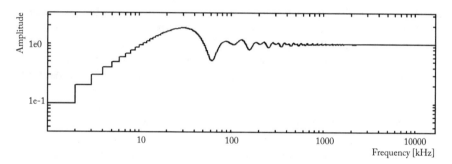

Fig. 5.12 Effect of baseline restorer filter in the frequency domain. For a running average baseline restorer of 10 μs length, the frequencies below 10 kHz result suppressed

actual systems. In the low count rate limit, the effect of the noise suppression by the baseline restorer filter is represented in Fig. 5.12 in the frequency domain.

It should be remarked that real implementations of the amplification network commonly result in a non-ideal shaping of the signal. Stray inductive and capacity couplings often result in distortions of the preamplifier response function (e.g., over-shoots) that differs from a simple exponential function. Non-exponential response function impacts the outcome of the trapezoidal filtering and calls for compensating filters like anti-overshoot in the output of the preamplifier or more complex filtering based on multi-pole deconvolution.

5.2.4 Count Rate Capabilities

The performance of a γ-ray spectrometer based on digital electronics has a strong dependence on the count rate. Each individual detector of a tracking array covers a larger solid angle than previous generation of Compton-suppressed array. Moreover, until the 4π array will not be completed, it is convenient to use the array in a closer configuration with respect to the nominal position (at 23 cm distance from the target in the case of AGATA). This allows to achieve high efficiencies at the cost of a larger count rate in each detection element, concentrating all the germanium detectors in the small fraction of solid angle around the target. In order to get the very best energy resolutions, large-volume germanium detectors are operated at low counting rates and relatively large shaping times. Therefore, we have to carefully consider the efficiency and energy-resolution losses that can arise at counting rates of 50 kHz or higher.

In order to study this effect, a measurement of the efficiency has been performed using a simplified version of the "two-sources-method" [17]. Six different measurement configurations of the two sources, ^{60}Co and ^{137}Cs, have been used. The ^{60}Co source was kept in the same position for all the runs in order to measure the resolution and the efficiency, while the ^{137}Cs source was moved to vary the total counting rate of the crystal between 13 kHz (corresponding to the ^{60}Co source alone) and 200 kHz. For every position of the sources (i.e., count rate), a scan of the possible values of the shaping time and baseline restorer width was performed. Resolution was measured for different values of the trapezoidal-filter risetime in the range between 1.25 and 10 μs and of the baseline-restorer width in the range between 0.64 and 20.48 μs. The flat top of the trapezoid was kept at 1 μs for all measurements.

Resolution strongly depends on the count rate as can be seen in Fig. 5.13 where the resolution is reported as a function of the shaping time and the baseline restorer length. Using this graphs, it is possible to optimize the response of the detectors for the actual count rate of each experiment.

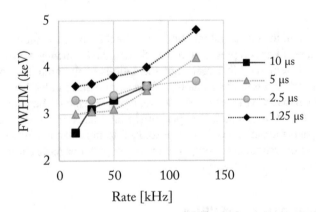

Fig. 5.13 Resolution at energy 1332.5 keV versus count rate for several trapezoidal-filter risetimes from 1.25 to 10 μs. The baseline restorer lengths was optimized for each case, resulting in 20 μs for all rates but the largest one, where the best resolution is found for a baseline restorer of 10 μs

5.3 Doppler Correction Capabilities

It is of foremost interest to have the best possible Doppler corrections capabilities. For this reason in-beam experiment performance of a γ-ray tracking array like AGATA depends critically on the precision achieved in locating the individual photon interaction points.

An experiment using the first prototype of AGATA triple cluster was performed, in order to compare the performance of different pulse shape analysis algorithms under realistic experimental conditions.

The basic idea is that, when the γ rays are emitted in-flight by a recoiling nucleus, the width of peaks in the Doppler-corrected spectra will depend on three factors, namely, the intrinsic detector energy resolution, the error on the velocity vector of the emitting nucleus, and the uncertainty on the photon direction. The last factor depends on the position resolution of the PSA algorithm used. If the other causes of Doppler broadening are known, the position resolution of the detector can be inferred from the observed energy resolution. This is not an easy task because all the direct and indirect sources of Doppler broadening have to be tracked down and, when not negligible, accurately quantified.

The broadening of the peaks has been predicted using a Monte Carlo simulation, but, as a first approximation, it can also be calculated using the propagation of errors and some schematic assumptions. The importance of this approach is to make the results intuitive. The Doppler-shift formula is the following:

$$E_\gamma^{cm} = E_\gamma \frac{1 - \beta \cos \theta}{\sqrt{1 - \beta^2}} \tag{5.6}$$

where E_γ^{cm} is the intrinsic energy of the γ-ray, E_γ is the energy of the photon in the laboratory (in other words the energy seen by the detector), β is the velocity of the emitting nucleus, and θ is the angle between the direction of the recoiling nucleus and the direction of the photon in the laboratory.

Each of the parameters entering the formula contributes to the final uncertainty. For instance, the θ angle is determined experimentally from the position of the first interaction of the photon and the target position. Thus, an error in the position is translated into an error in the direction of the γ-ray, giving an imperfect Doppler correction. Quantitatively, the contribution of each parameter to the final position resolution is evaluated through the propagation of errors on E_γ^{cm}, giving:

$$\left(\Delta E_\gamma^{cm} \right)^2 = \left(\frac{\partial E_\gamma^{cm}}{\partial \theta} \right)^2 (\Delta\theta)^2 +$$

$$+ \left(\frac{\partial E_\gamma^{cm}}{\partial \beta} \right)^2 (\Delta\beta)^2 +$$

$$+ \left(\frac{\partial E_\gamma^{cm}}{\partial E_\gamma} \right)^2 (\Delta E_\gamma)^2. \tag{5.7}$$

In this calculation, the different broadening sources are considered as statistically independent contributions, neglecting for simplicity any correlation between them. In (5.7), $\Delta\beta$ and $\Delta\theta$ are, respectively, the uncertainty on the velocity module and on the direction of the nucleus emitting the radiation. Even if the recoil velocity vector can be measured on an event-by-event basis, $\Delta\beta$ and $\Delta\theta$ will be generally non-zero.

The term ΔE_γ in (5:7) describes the contribution of the intrinsic energy resolution of the detector.

The partial derivatives are:

$$\frac{\partial E_\gamma^{cm}}{\partial\theta} = E_\gamma \frac{\beta\sin\theta}{\sqrt{1-\beta^2}}$$

$$\frac{\partial E_\gamma^{cm}}{\partial\beta} = E_\gamma \frac{\beta-\cos\theta}{\left(1-\beta^2\right)^{3/2}}$$

$$\frac{\partial E_\gamma^{cm}}{\partial E_\gamma} = \frac{1-\beta\cos\theta}{\sqrt{1-\beta^2}} \tag{5.8}$$

The angular error is propagated to the error in the determination of the intrinsic energy of the γ ray by the coefficient given in the first raw of (5.8). As an example, the contributions of the three sources of Doppler broadening are sketched in Fig. 5.14, for the case of photons of 1 MeV emitted from a nucleus in motion with $\beta = 0.2\%$ and detected with an uncertainty $\Delta\theta = 1°$ on its direction.

Besides the finite resolution on the position of the interaction, in an indirect measurement, the other relevant sources of error in the determination of the intrinsic energy of the photon are the energy resolution of the detector and the uncertainty on the velocity vector of the emitting recoil. To simplify the experimental setup, it is possible to measure indirectly the velocity vector of the emitting nucleus through a kinematic calculation based on the measurement of the velocity vector of the other reaction products. As a consequence, some other contributions have to be considered to explain the experimental broadening, namely:

- angular and energy dispersion of beam due to the accelerator and the transport;
- straggling of the beam and reaction products inside the target.

The position resolution provided by the PSA algorithm can be deduced from the quality of the Doppler correction, provided a proper reaction is selected to maximize the Doppler broadening originating from the position resolution and to minimize the other sources. A reaction that was used to investigate this was the ^{48}Ti(d, p) in inverse kinematics [26].

The results of this procedure are shown in Fig. 5.15, together with the spectra obtained by Doppler correcting at detector or segment level. The improvement in quality of the spectra is apparent. For the 1382 keV peak of ^{49}Ti, FWHM=4.8 keV is obtained following the PSA algorithm, which should be compared to 14 keV and 35 keV at segment and detector level, respectively.

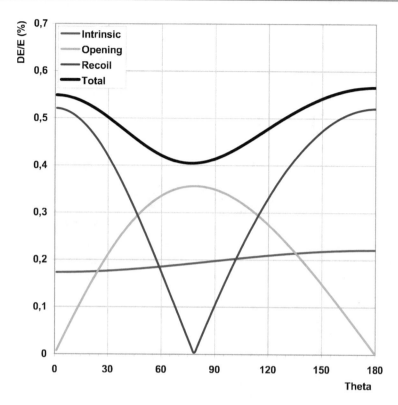

Fig. 5.14 The contributions of the different Doppler broadening sources as a function of the azimuthal angle of the detector with respect to the direction of the recoil emitting the radiation. A photon energy of 1 MeV is assumed, with a typical energy resolution for a germanium detector, producing the "Intrinsic" contribution (in red); a source velocity of $\beta = 20.0\%$ with an error of 0.5%, giving the "Recoil" contribution (in blue); an uncertainty $\Delta\theta = 1°$ in the source direction, obtaining the "Opening" contribution (in green)

The resulting position resolution is extracted quantitatively by comparing the experimental peak width to the simulated value using the curves shown in Fig. 5.15. In this case the simulated data from the three individual crystals were summed up, obtaining the curve plotted in Fig. 5.16. The observed FWHM of 4.8 keV of the peak at 1382 keV corresponds to a position resolution of 3.8 mm at this energy.

The improvement in Doppler correction capabilities have been exploited, for example, in beam-induced fission experiments at GANIL, using the VAMOS magnetic spectrometer for fission fragment identification and velocity reconstruction. In Fig. 5.17 the comparison of the Doppler corrected spectrum for ^{98}Zr obtained with a standard HPGe array (EXOGAM) and AGATA is shown. Already from such preliminary analysis of the commissioning data, a gain in energy resolution is evident.

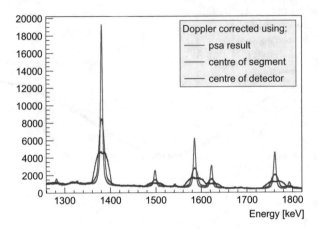

Fig. 5.15 Doppler-corrected spectra for the full cluster, deducing the direction of the photon, respectively, from the center of the detector, center of the segment, and the PSA information. Taken from [34]

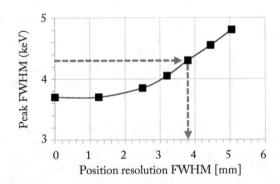

Fig. 5.16 Width of the simulated 1382 keV peak as a function of the position smearing for the full triple cluster. Individual crystal energy resolution have been considered. The horizontal arrow indicates the experimental width. Taken from [34]

5.4 Lifetime Measurements with Doppler Techniques

The lifetime of a nuclear state is the observable that is measured in γ-ray spectroscopy experiments in order to derive the reduced transition probabilities, crucial for the comparison with nuclear theoretical models. The Doppler-shift techniques are powerful tools for lifetimes in the 10^{-14}–10^{10} s interval. When the nuclear level of interest is populated in a heavy-ion reaction, the radioactive decay will happen from the moving reaction product. The finite time it takes to travel a given distance can be considered as a "unit of measurement" of the decay lifetime of the populated level. When the radiation with energy E_γ (center of mass energy) is emitted by a nucleus moving with a velocity $\beta = v/c$, the γ-ray energy observed

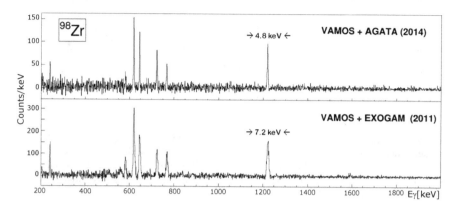

Fig. 5.17 Doppler correction capabilities in the AGATA+VAMOS campaign at GANIL. The Doppler-corrected spectra corresponding to ^{98}Zr (selected in VAMOS) obtained with AGATA (upper panel) and EXOGAM (lower panel) revel the improvement due to the AGATA position resolution

at an angle θ between the recoil and the γ-ray emission directions follows the Doppler relation. Depending on the order of magnitude of the lifetime and on the kind of nuclear reaction involved, different variants of the technique have been developed. All methods take great advantage of the use of HPGe detectors, whose excellent energy resolution allows for the observation of fine details in the γ energy spectrum. In particular, a big improvement in the accuracy of the determination of the lifetimes by means of these techniques was achieved with the use of arrays of HPGe detectors, placed symmetrically all around the target position. The possibility of a simultaneous measurement at many angles is important for discovering the presence of contaminants as well as for increasing the accuracy since the lifetime is obtained in a number of independent evaluations. When the lifetime of the level of interest is comparable to the slowing down time of the emitting nucleus in a given material, the Doppler-shift attenuation method (DSAM) is used. This technique is widely adopted for the measurement of sub-ps lifetimes, corresponding to the typical slowing down times of heavy ions in solid mediums, which are $\approx 10^{-13}$–10^{-12} s for kinetic energies below or about 10 MeV per nucleon. Thanks to the angular resolution of a tracking array as AGATA, this technique can be pushed to shorter lifetimes. In case of lifetimes of few fs, the shape of the peak in the γ spectrum is mainly determined by the kinematics, since most of the nuclear states decay as soon as they are populated. The first experiment performed with the AGATA array to measure $\approx$ fs lifetimes aimed to the determination of the lifetime of an excited state in ^{15}O, of interest for nuclear astrophysics. The details of the experiment are reported in [35]. In Fig. 5.18 the effect on the observed peak position of different lifetimes is shown in the "AGATA-like" case and in a standard coaxial detector. As evident from the picture, the extremely good angular resolution allows for distinguishing among the different lifetimes. Such angular sensitivity for γ rays de-exciting short-lived levels is depicted in Fig. 5.19, where the observation angle is

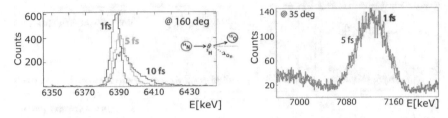

Fig. 5.18 Examples of sensitivity to fs lifetimes. 6.79 MeV γ-rays emitted by ^{15}O nuclei produced in the simplify direct process reported in the figure, with the produced 15O recoiling at a fixed direction with respect to the beam axis (35 deg). The sensitivity to $\approx$ fs lifetimes obtained in the spectra of AGATA (left) is lost when using a traditional HPGe detector, as one from the GASP array (right). Taken from [35]

sorted as a function of the detected energy. The broadening of the γ-ray peaks due to short lifetimes is evident as compared to the γ rays emitted at rest. Such line shapes can be analyzed by means of Geant4 simulations in order to extract the lifetime. In order to establish the very lower limit of sensitivity of the measurement, Geant4 simulations of the whole production, slowing down, gamma emission, and detection process can be used. For the specific experiment considered here as an example, one obtains the χ^2 curves reported in Fig. 5.20. Those curves are useful to establish the minimum value of lifetime that one can reach with a given setup. In the considered case, one sees the appearance of a minimum in the χ^2 curve for lifetimes of about 0.7 fs.

5.5 Linear Polarization Measurements

The "almost continuous" angular distribution achievable with tracking arrays can be exploited to gain a new degree of sensitivity in γ polarization measurements, necessary for the determination of the magnetic or electric character of γ transitions and, thus, of the parity of nuclear states. Clover detectors (see Sect. 5.1) are usually used for such measurements [36], by evaluating the asymmetry A_S of the Compton scattering of γ rays corresponding to a given transition inside the clover detector. This asymmetry is typically evaluated with respect to the reaction plane, and it is given by:

$$A_S = \frac{N_\parallel - N_\perp}{N_\parallel + N_\perp} \tag{5.9}$$

where $N_\parallel$ and $N_\perp$ are the number of Compton scattered γ rays detected, respectively, parallel and perpendicular to the reaction plane.

Using the tracking capabilities of arrays as AGATA, the azimuthal Compton scattering angle ϕ can be evaluated with a 1 degree precision, as shown in Fig. 5.18. The linear polarization can be thus determined, instead of using (5.9), by considering that the average Compton scattering cross section depends on ϕ via

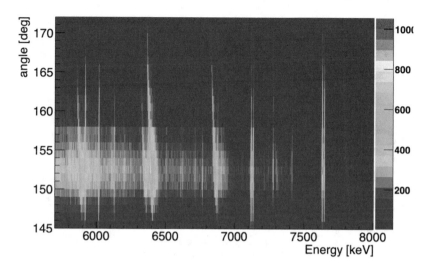

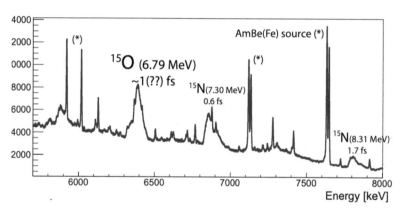

Fig. 5.19 Angular position of the first interaction point, with respect to the beam direction, sorted as a function of the energy of the reconstructed γ ray. The narrow straight lines correspond to the emission at rest from the AmBe(Fe) source, while the broad "tilted" lines are gammas emitted while the excited nucleus is moving in the stopper. In the projection on the energy axis (lower panel), the broad and composite structure are assigned to ^{15}O and ^{15}N, produced in the reaction. Taken from [35]

the relation [37]:

$$N(\phi) = a + b \, \cos(2\phi) \tag{5.10}$$

The analyzing power on the AGATA triple clusters for γ linear polarization has been determined in [37] using (5.10). The measured $N(\phi)$, normalized to a isotropic distribution obtained with ^{137}Cs calibration source, is reported in Fig. 5.21 for the $2^+ \rightarrow 0^+$ transitions in ^{106,108}Pd populated in Coulomb excitation experiments. The impact of position resolution and other systematic effects is discussed in [37].

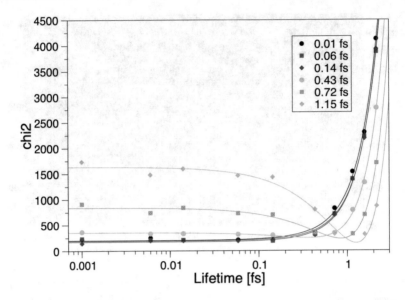

Fig. 5.20 Sensitivity of the simulated line shape to changes of the lifetime in the fs region, for the experiment considered in the text. A minimum in the χ^2 curve starts being evident for lifetimes around 0.7fs. Taken from [35]

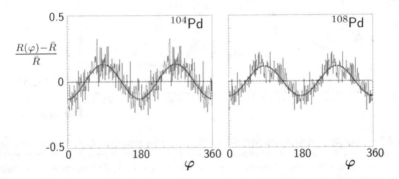

Fig. 5.21 Asymmetry distribution for the $2^+ \rightarrow 0^+$ transitions in ^{106,108}Pd populated in Coulomb excitation experiments. The γ polarization is obtained by fitting the experimental data sorted with a binning of 1 degree, to (5.10). Taken from [37]

5.6   Conclusions

In this lecture notes, the limits of previous generation Compton-suppressed γ-detector arrays have been discussed, motivating the development and the construction of the state-of-the-art γ-tracking arrays. The basics of the techniques in use for the operation of such detectors have been presented. The tracking arrays present improved detection sensitivity, thanks to enhanced efficiency and Doppler correction capabilities, extremely important for nuclear structure studies using low-intensity radioactive beams. Moreover, the position sensitivity of the tracking array can be exploited, not only with the aim of Doppler correction but also for gaining a new degree of sensitivity in lifetime measurements and polarization studies. In order to highlight those new capabilities, a limited selection of results has been discussed.

References

1. D. Bazzacco, Proc. Confer. Phys. Large γ-Ray Detector Arrays **32**, 376 (1992)
2. C. Rossi Alvarez, Nucl. Phys. News **3**, 10–13 (1993)
3. C.W. Beausang et al., Nucl. Instrum. Methods A **313**, 37–49 (1992)
4. J. Simpson, Zeitschrift für Physik A Hadrons Nuclei **358**(2), 358 (1997)
5. I.-Y. Lee, Nucl. Phys. A, **520**, c641–c655 (1990)
6. SPES Technical Design Report, INFN-LNL-223 (2008)
7. Th. Kroell, D. Bazzacco, Nucl. Instrum. Methods A **463**, 227–249 (2001)
8. G. Duchêne et al., Nucl. Instrum. Methods A **432**, 90–110 (1999)
9. M. Wilhelm et al., Nucl. Instrum. Methods A **381**, 462–465 (1996)
10. U. Rizwan et al., Nucl. Instrum. Methods A **820**, 126–131 (2016)
11. F. Azaiez, Nucl. Phys. A **654**, 1003c–1008c (1999)
12. C. Michelagnoli et al., EPJ Web Conf. **193**, 04009 (2018)
13. G. Colombi, Master Thesis, University of Milan and ILL (2020)
14. D. Reygadas, PhD Thesis, University of Grenoble and ILL (2021)
15. S. Akkoyun et al., Nucl. Instrum. Methods A **668**, 26–58 (2012)
16. S. Paschalis et al., Nucl. Instrum. Methods A **709**, 44–55 (2013)
17. G.F. Knoll, *Radiation Detection and Measurement,* 4th edn. (Wiley, New York, 2010)
18. G.J. Schmid et al., Nucl. Instrum. Methods A **430**, 69–83 (1999)
19. A. Lopez-Martens et al., Nucl. Instrum. Methods A **533**, 454–466 (2004)
20. L. Nelson, et al., Nucl. Instrum. Methods A **573**, 153–156 (2007)
21. T.M. Hoa, A. Korichi, *5th AGATA Week* (Orsay, 2007).
22. B. Bruyneel et al., Eur. Phys. J. A **52**, 70 (2016)
23. B. Bruyneel, PhD Thesis, University of Cologne (2006)
24. B. Bruyneel, P. Reiter, G. Pascovici, Nucl. Instrum. Methods A **569**, 774–789 (2006)
25. R. Venturelli, D. Bazzacco. LNL Annual Report (2005)
26. F. Recchia et al., Nucl. Instrum. Methods A **604**, 555–562 (2009)
27. B. Bruyneel et al., Eur. Phys. J. A **49**, 1–9 (2013)
28. D. Bazzacco, *4th Summer School in Radiation Detection and Measurements*, Munich (2011)
29. A. Yamaguchi et al., Phys. Rev. Lett. **123**, 222501 (2019)
30. F.C.L. Crespi et al., Nucl. Instrum. Methods A **593**, 440–447 (2008)
31. V. Radeka, IEEE Trans. Nucl. Sci. **15**, 455–470 (1968)
32. V. Radeka, Nucl. Instrum. Methods **99**, 525–539 (1972)
33. V.T. Jordanov et al., Nucl. Instrum. Methods A **353**, 261–264 (1994)
34. F. Recchia, PhD Thesis, University of Padova (2008)
35. C. Michelagnoli, PhD Thesis, University of Padova (2013)
36. L. Garcia-Rafi et al., Nucl. Instrum. Mathods A **391**, 461 (1997)
37. P.G. Bizzeti et al., Eur. Phys. J. A **49**, 51 (2015)

Nuclear Structure Studies with Active Targets

6

Riccardo Raabe

Abstract

The use of gaseous detectors in nuclear structure studies presents several challenges and interesting opportunities. In the last twenty years, the challenges have been addressed with the development of various active targets, designed to perform measurements with very weak radioactive ion beams. In this paper we review the characteristics of these instruments and how they can be used to great effect in a wide range of physics cases.

6.1 Introduction

Gaseous detectors are largely used in experimental physics. They exploit the ionisation created by charged particles when they traverse any matter; in gases, the charged particles leave behind long "tracks" of electron and ion pairs. The ionisation electrons can be collected by an electric field and transported to a region where they are amplified, typically by a wire at high potential, creating an electric signal. Different configurations of the electric field and amplification region allow to use the signals either for counting the number of particles traversing the gas volume, for timing purposes, or, in more complex cases, to record the full three-dimensional information about the trajectory of the particle.

The latter is the case of a time projection chamber (TPC) [1], where electrons drift to a segmented detection plane which provides a two-dimensional projection of the trajectory. The third dimension is reconstructed from the difference in time between the arrival of the electrons on the plane.

R. Raabe (✉)
KU Leuven, Instituut voor Kern- en Stralingsfysica, Leuven, Belgium
e-mail: riccardo.raabe@kuleuven.be

© The Author(s), under exclusive license to Springer Nature Switzerland AG 2022
S. M. Lenzi, D. Cortina-Gil (eds.), *The Euroschool on Exotic Beams, Vol. VI*,
Lecture Notes in Physics 1005, https://doi.org/10.1007/978-3-031-10751-1_6

In high-energy nuclear and particle physics, very large TPCs containing gas at low pressure are often placed around the vertex of complex reactions, where hundreds of particles are created at the same time. Gaseous detectors are "slow": while particles fly out in a time scale of nanoseconds, it takes several tens of microseconds for the ionisation electrons to drift to the detection plane and form an image of the trajectories. The reactions of interest, however, are rare for this kind of physics, and TPCs can be used without mixing tracks from different events. The particles produced are mostly light (sub-nuclear) and very fast; the energy that they deposit in the gas and thus the number of electrons created is small (but sufficient) and does not depend much on the kind of particle [2]: we speak about *minimum ionising* particles. To identify the particles, large magnetic fields are usually employed to bend the trajectories according to the mass-to-charge ratio of the particles (the geometry is such that the magnetic fields do not affect the drift of the electrons).

In low-energy nuclear physics (also referred to as *nuclear structure* physics), things can be very different, and several challenges arise for the use of TPCs.

The charged particles emitted in the reactions can be light ions (p, d, t, ^{3}He, α particles) or heavy ions throughout the whole chart of nuclei. Since the energy loss per unit length is roughly proportional to the mass, the charge square and inversely proportional to the energy of the particle [2], we may expect signals that differ by more than three orders of magnitude: it is very difficult to realise a detection system with such a large *dynamic range*. Then, the total event rate can be high, and the events of interest need to be selected by employing an appropriate trigger. Sometimes the radiated particles do not exit the TPC, and a prompt event selection must be based on the pattern of the detected trajectories in the TPC. Finally, we would like to use TPCs as *active targets*, by studying the reactions that take place on the nuclei of the detection gas. The choice of the gas, and often its pressure, is thus dictated by the physics case that we want to study. For example, to study resonant reactions with α particles, we will fill our active target with (mostly) He gas which, however, is not ideal for the collection of the signals from the ionisation electrons.

On top of those issues, we intend to use these instruments with radioactive ion beams, by performing reaction studies in *inverse kinematics*: a heavy projectile impinging on a light target nucleus. In general, this brings in additional complications, such as *kinematical compression* and a worsening in energy resolution due to the energy loss of the beam in the target (see, e.g. a discussion of these effects in [3]). As we will see, however, active targets can overcome some of those issues; also, they are very versatile instruments and have specific advantages that make them very attractive tools for nuclear structure studies.

In Sect. 6.2 of this chapter, we illustrate the working principles of active targets and have a brief overview of existing instruments. In Sect. 6.3 we discuss how they can be used in a number of different physics cases.

6.2   Working Principles

As mentioned, an active target is a time projection chamber, where the gas is both the detector and the target. The working concept is illustrated in Fig. 6.1. The electric field can be oriented perpendicular to the beam with the segmented plane on one of the sides, as in the figure, or parallel to it, with the segmented plane perpendicular to the beam and placed either on the beam entrance side or on the opposite side. The electrons are amplified using different technologies and cause a signal on each *pad* of the segmented plane, either by induction or by direct collection. Using the TPC scheme, the two-dimensional projection of the track is complemented by the timing information to reconstruct the third dimension.

From the detection of the tracks, the reaction vertex can be directly reconstructed. This is one of the main advantages of active targets: since the energy loss of the beam in the gas is known, the energy available at the reaction vertex can be calculated with much better accuracy than in the case of a solid target foil, where usually the energy at the mid-point is assumed. This allows for the use of a much larger amount of target material, which compensates for the weak intensity of radioactive ion beams, without degradation in energy resolution when reconstructing the kinematics of the reaction process.

The energy deposited by the particles in the gas can be estimated from the charges collected on the pad plane. If the particles are stopped in the gas volume, a more accurate value of their total energy can be obtained from the *range*, the length of the track. If the charged particles leave the volume, they can be intercepted by *auxiliary detectors*. In either case, information on the specific energy loss per unit length

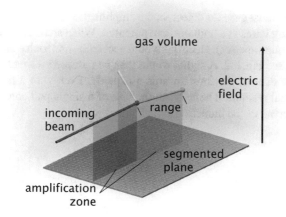

Fig. 6.1 Working principle of an active target. The incoming beam enters the gas volume, where a reaction may take place. The ionisation electrons, produced by all the charged particles traversing the gas, drift in an electric field towards an amplification zone. The signals are read out on a segmented plane

can be used to identify the particle, especially when combined with the kinematic reconstruction.

The *geometric efficiency* of an active target covers potentially the full solid angle. In practice, however, limitations arise when particle tracks are too short to be reliably identified or when the energy deposited per unit length exceeds the detection limits.

The choice of the detection gas is dictated by the physics case of interest. The pressure is usually a compromise between the need of a large amount of target material and the need of tracks with a reasonable length for the events of interest; this has to be combined with the best working conditions for the electron ionisation, drift, and amplification. For the latter, one can use *gas mixtures* to achieve better results, taking into account that nuclei of different species can lead to unwanted background processes. In practice, pressures between about 50 mbar and 1 bar are usually employed.

The main limitation in the use of these detectors is related to the long time needed for the electrons to drift to the detection plane, of the order of (tens of) microseconds. This constrains the rate of events of interest that can be recorded and often requires a "smart" trigger to select them among many other background events. For these characteristics, active targets are well-suited to study rare events or events induced by very low-intensity radioactive ion beams. In this sense, in many cases they are complementary to measurements using thin solid targets.

We will now discuss some aspects in more detail. For a complete overview of the technical details and performances, we refer to the work of Y. Ayyad and collaborators in [4].

6.2.1 Amplification Technology

Until recently, in most gaseous detectors the amplification of the electron signal was achieved by using wires at a high positive potential. When arriving in the vicinity of the wire (a few micrometres), electrons would accelerate and ionise other atoms in a fast *avalanche*; see Fig. 6.2. A signal would then be induced on a pad plane, mainly due to the slow drift of the ions towards a cathode. Even with a limited number of wires, a spatial resolution better than the size of a single pad could be obtained thanks to the spread of the induction signal.

Fig. 6.2 Electron amplification obtained with wires. The signal on the pads is generated by induction

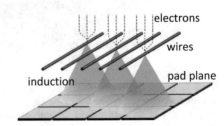

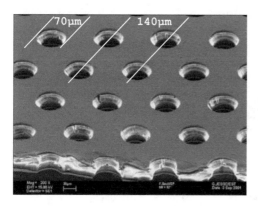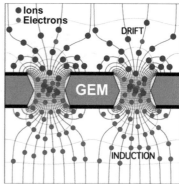

Fig. 6.3 Left: electron microscope picture of a GEM device. Right: sketch of the working principle of GEMs. Figures reproduced from [7] under a Creative Commons Attribution 4.0 International License

In the last 20 years, the amplification technology, driven by its large use in high-energy physics, has evolved towards micro-pattern gaseous detectors (MPGDs) [5]. These are planar structures that use various geometries to generate a strong electric field. The amplification gap is typically in the order of $\sim$100 μm, compared to the mm size of the wire-based structures. In the existing active targets, mainly gas electron multipliers and micro-mesh gaseous structures are used.

Gas electron multipliers (GEMs) [6] consist of a $\sim$50 μm Kapton foil covered with copper on both sides and perforated by bi-conical channels. Figure 6.3 shows an electron microscope picture of the device and its working principle. A potential difference of a few hundred volts is applied between the two surfaces; the electrons are driven by the electric field lines through the holes, where the field is sufficiently strong to cause electron multiplication. GEMs are decoupled from the readout structure and thus very versatile: several GEMs can be combined to achieve a stronger multiplication. In active targets, a segmented anode collects the electrons to generate the signals (see Fig. 6.3).

An example of a micro-mesh gaseous structures (Micromegas) device [8] is shown in Fig. 6.4. A thin metallic micromesh separates the drift space from the amplification gap, which is about 100 μm. The electron multiplication is obtained with a strong electric field across the gap. Different technologies can be used to obtain a uniform gap.

With these devices, gains of a factor 10^3 and up to 10^5 can be reached with excellent intrinsic resolution [9, 10]. Both GEMs and Micromegas have potentially a high rate capability; in addition, their characteristics allows to build detection devices that can sustain sparks somewhat better than a wire system. The latter is an important condition, because often, in active targets, the settings of the gas and its pressure are at the limits of the working range, to accommodate the requirements of the physics case.

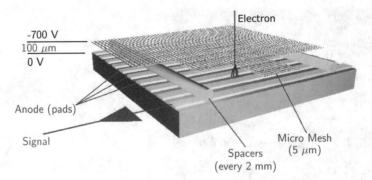

Fig. 6.4 Arrangement of a Micromegas electron amplification device

6.2.2 Configurations and Auxiliary Detectors

As we have mentioned, the energy loss in gas of charged particles (ions) emitted in nuclear reactions may differ considerably. Even for one specific particle, the inverse kinematics of reactions with radioactive ion beams dictates that their energy depends very strongly upon the emission angle. This reflects in very different lengths of the tracks of the emitted particles, which cannot all be stopped in the active volume.

Two strategies have been used to improve on this aspect. A magnetic field can be used to bend the trajectories of the charged particles. This way the energy and the identification of the particles can be obtained from the curvature of their trajectories, even for tracks which are not completely stopped in the active gas volume. Usually, the magnetic field is oriented parallel to the electric drift field, to avoid deviating the trajectories of the ionisation electrons from the electric field lines. The magnetic field may be generated by a dipole and oriented perpendicularly to the beam axis: it is the configuration of a spectrometer, and we find an example with the SπRIT TPC/active target [11]. A solenoid, on the other hand, can give a magnetic field oriented parallel to the beam axis. The active target is then built with a cylindrical symmetry, and the ionisation electrons are collected on a pad plane perpendicular to the beam. This is the case of AT-TPC [12] and SpecMAT [13] active targets. This configuration has specific advantages for some physics cases, when the detection of the beam tracks is not crucial, as we will see in Sect. 6.3.

Alternatively, auxiliary charged particle detectors (e.g. silicon detectors) can be placed around the active volume to intercept the escaped particles. This solution can be applied if the active volume is not too large; depending on the physics case of interest, however, one can choose to cover only part of the total solid angle. The measurement of the energy loss in the gas, combined with the residual energy deposited in the auxiliary detectors, provides an effective way of identifying the particle using the ΔE-E method. This configuration is used, for example, in ACTAR TPC [14] and its demonstrator [15].

Detection of the γ-rays emitted in a reaction or decay can also provide valuable information. Active targets are optimised for use with low-intensity radioactive ion beams, so it is essential to have an efficient system to detect the γ radiation. High-Z scintillation detectors developed in the last 15–20 years, such as $LaBr_3(Ce)$ and $CeBr_3$ [16], combine that intrinsic efficiency with a good resolution. The modular PARIS array [17] of $LaBr_3(Ce)$-$Na(Tl)$ phoswich detectors has been considered for use in combination with ACTAR TPC; the SpecMAT active target is designed with an array of $CeBr_3$ detectors placed around the active volume. For the latter, the use in strong magnetic fields was successfully tested [18].

6.2.3 Electronics and Trigger

The design of active targets for nuclear structure studies was initially limited by the maximum number of electronic channels available. The Maya detector [19], developed at GANIL, already represented a significant progress, with more than a thousand pads read out by the Gassiplex chips [20], which had been developed for high-energy physics. With its wire amplification technology, however, Maya only allowed the readout of two particle tracks per event. For the new generation of active targets, that in Europe started with the ACTAR TPC project, the development of a new, dedicated electronic readout system was immediately identified as a central goal. The system, named General Electronics for TPCs (GET) [21], was realised by a collaboration between the laboratories of CEA Saclay, CENBG Bordeaux, GANIL Caen, and the Michigan State University.

The design (Fig. 6.5) is modular and strongly integrated. A front-end card (AsAd) can process signals from 256 channels through 4 AGET chips and a 4-channel

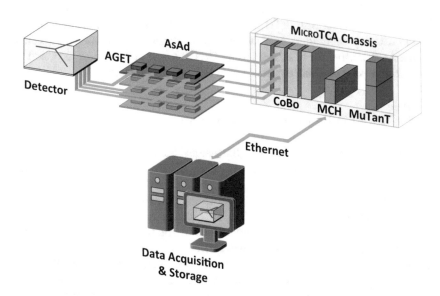

Fig. 6.5 Scheme of the GET electronics; see text for the details. Figure reproduced from [21] with permission from Elsevier

analog-to-digital converter (ADC). Each AGET features, for each of its 64 channels, a charge-sensitive pre-amplifier, a shaper, a discriminator, and a 512-cell circular switch capacitor array that stores the charge information collected from the pad. The frequency of the readout can be adjusted between 1 MHz and 100 MHz, so that the circular array corresponds to a time interval between $\approx$5 μs and $\approx$500 μs. When a trigger is generated, the information of the 512 cells is passed over to the 12-bit ADC, processed and sent to a Concentration Board (CoBo), that can receive data from 4 AsAd cards (1024 channels). Up to 10 CoBo boards can be placed in a micro-TCA chassis, controlled by a Multiplicity Trigger and Time (MuTanT) module, and up to 3 chassis can be combined for a total number of 30,720 channels. All channels are aligned through a time stamp distributed by the MuTanT module. Event by event one can thus record the "history" of all the pads of the active target in the time interval, preceding the trigger, set by the choice of the readout frequency.

The goal was to have a system capable of a throughput of 1000 events per second. It is easy to calculate that this would correspond to a data rate of $\approx$200 Gb per second, if all the cells of all the channels were to be read out. Obviously, the system offers the possibility of suppressing all the data below a threshold (zero-suppression), thus avoiding reading out all the pads without a signal. The implementation of a smart trigger system is also crucial. In GET, a trigger can be generated from an external signal (e.g. a beam detector or an auxiliary detector); from the active target itself, if the number of pads with a signal exceeds a selected threshold; and from a high-level user-defined hit pattern of the pads.

GET is now used in many active targets and TPCs worldwide. Its modular design allows customisation and updates of the different stages. Beside the readout of the pads from a TPC, the system can be adapted for the readout of different sorts of detectors. This is, for example, the case of the auxiliary $CeBr_3$ scintillators in SpecMAT [18] or the Si-CsI(Tl) telescopes in FARCOS [22]. Adaptation boards and/or custom preamplifiers were developed for those applications with excellent results. An extensive list of devices using GET is given in Table 1 of [21], and the same work reports of a successful use with HPGe γ-ray detectors.

6.2.4 Track Finding and Reconstruction

The possibility of recording the three-dimensional tracks of the particles involved in a reaction is a great advantage of active targets. To extract the relevant information, however, a complex analysis of the data is necessary. Particle tracks have first to be identified and then properly fitted to extract vertex position, emission angles, lengths, and energy deposition profiles. This clearly requires the development of automated procedures. The job is complicated by the fact that the tracks are often not "ideal", due to the intrinsic issues related to gaseous detectors (uniformity of response, dynamic range), combined with the electro-mechanic challenges of collecting signals from an extremely dense pad plane, and the inevitable presence of background/noise signals. Again, we benefit from many methods that have been

developed in high-energy physics. An extensive discussion is found in [4]; here we give a short overview.
Track finding algorithms that have been used with active targets include:

- The random sample consensus method (RANSAC) [23] finds a model dependence in data with outliers, by sampling a data set a given number of times and excluding the outliers. In a track reconstruction, one chooses parameters such as the expected track shape (e.g. linear), the width of the track (as determined by the spread of ionisation electrons), and the number of sampling attempts. The method is fast and robust especially for the identification of points belonging to a rectilinear track.
- The Hough transform [24] was developed to analyse photographs of tracks in bubble chambers, and it is still widely used to find shapes in any type of image. The principle is to describe the looked-for shape in parametric form and then map the image in the parameter space by applying the transformation to each data point. Usually, for straight lines, the Hesse normal form is used for the transformation [25]. In the parameter space, the best value of the parameters can be selected by constructing histograms for the calculated value of each point. The transformation can be adapted to better suit specific cases according to the detector geometry and track shapes, including curved ones. Examples are given in [26, 27].
- The Riemann tracking [28] is more adapted for the reconstruction of helical trajectories. A circular arc is mapped on the Riemann sphere, therefore transforming the nonlinear problem of circle fitting into a direct and fast linear fitting of a plane. This method is in principle well-suited for active target like AT-TPC and SpecMAT, which are meant to be used inside a solenoid; the large energy loss of ions in the gas, however, means that the curvature radius is not constant, complicating the problem significantly.
- Machine learning methods using neural networks could become the best option to identify complicated patterns as the spiral tracks of ions in magnetic fields. The first encouraging attempts in this directions have been made in the analysis of events in the AT-TPC [29, 30].

The fitting of the identified tracks aims at the determination of the track parameters or at their improvement, if some values were available after the initial identification step. The characteristics of the active target (geometry, presence of a magnetic field, gas pressure) are crucial in guiding the choice of the parameter space to be investigated.

Methods for the fit of rectilinear tracks are presented in [31] for the active target Maya [19]. The hyperbolic secant squared (SECHS) method was developed to account for the hexagonal shape of the pads in Maya. The centroid of the charge collected along each row of pads (as selected in the track-finding step) is determined, and the track is fitted from the position of the centroids. A minimum of three pads per row is required. The method gives slightly different results for the three different directions of the rows of pads, and usually the orientation giving the largest number of maxima (i.e. the "most perpendicular" to the track) is chosen. For tracks for which the charge deposited is small, the orthogonal distance regression method is used instead. The method finds the line that minimises the sum of the distances from the centres of the pads belonging to the track, weighted by the charge of the pad. The last step of the fit, for tracks that stop in the gas volume, is to fit their length, to determine the energy of the particle. If possible, the energy loss along the track is fitted using a known parameterisation, for example, from the energy-loss tables of SRIM [32]. The end of the track is then identified as the point where the energy loss drops to a given fraction of the maximum (Bragg peak). If the track is too short and a fit is not possible, the stopping point is calculated from the slope of the charge profile of the last part of the track.

The complex tracks of ions in a magnetic field have their standard example with the AT-TPC and have been so far mostly analysed using Monte Carlo procedures [12]. A large number of candidate tracks are generated in a simulation that account for all the effects of the detector, including the response of the electronics, by randomising six parameters: the three coordinates of the reaction vertex, the two angles of the track, and the energy of the particle. The parameters of the simulated track that best reproduce the hit pattern, the energy loss profile, and the vertex position of a real track are then assigned to the real track. The energy of the particle, in this case, is derived from the (decreasing) curvature of the track. The procedure is quite expensive computationally; a preliminary good knowledge of the parameters is therefore necessary. This may be especially crucial for the energy-loss functions, which are known with sizeable uncertainties of the order of 10% or higher, in particular at low energies.

By employing the technologies and methods listed in the sections above, active targets could ideally achieve excellent resolution in energy and angle. If the settings of the instrument are tuned to optimise a specific kind of events, values of the order of a degree for the angle and few tens of keV for the energy seem to be within reach [4]. The actual values, however, may vary significantly depending on the geometry of the detector, type of particles, length of the tracks, specific energy deposition, and of course the reaction process to be studied. In Sect. 6.3 we will consider various physics cases and present the result of actual measurement, which will give us a better idea of the performance of active targets.

6.2.5   Active Targets for Nuclear Structure Studies

We present a short list of detectors, which we will also encounter in Sect. 6.3 when discussing some physics cases. For a more exhaustive list, we refer to the work of S. Beceiro-Novo et al. in [33].

The IKAR detector at GSI Darmstadt [34] can be considered the first instrument where detection gas and reaction target were the same. IKAR was designed to detect light particles scattered in low-momentum transfer collisions. The large vessel was filled with hydrogen at a pressure of 10 bar and divided into six ionisation chambers

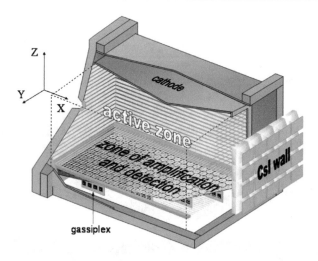

Fig. 6.6 Scheme of the Maya detector. The beam enters from the left, along the x axis. See text for the details. Figure reproduced from [19] with permission from Elsevier

by a succession of anode and cathode plates, on which the ionisation charges created by the scattered protons were collected.

The Maya active target [19] had a configuration with the electric drift field perpendicular to the beam direction and a volume of $28 \times 26 \times 20$ cm^3, see Fig. 6.6. The amplification was achieved with wire technology, and the charges were collected on a plane of 32×32 hexagonal pads. When a trigger was issued by the wires, the signals were kept in the memory of the Gassiplex multiplexed chips [20] until they were sent to the data acquisition. In the forward direction, an array of Si and CsI detectors allowed to identify and collect the energy of charged particles scattered outside the active volume. This general-purpose detector was used in several successful measurements of transfer reactions, inelastic scattering, and resonant reactions (see Sect. 6.3). The limitations of Maya concerned mostly the dynamic range of the Gassiplex, the low event rate, and detection of maximum two scattered tracks due to the wire geometry. It was to overcome these problems that the GET electronics described above (Sect. 6.2.3) was designed and adopted by various active targets.

ACTAR TPC [14] is similar to Maya in concept, with an active volume of $295 \times 295 \times 255$ mm^3 and a Micromegas detection plane divided in $16,384$ 2×2 mm^2 pads. The complex pad plane required a dedicated development, for which two electro-mechanic solutions were explored [14, 35]. The uniformity of the drift field (perpendicular to the beam direction, as in Maya) is realised with a double-wire cage, allowing high-energetic charged particles to escape the active volume and be detected in solid-state detectors placed around the field cage. Figure 6.7 shows the detector scheme, with the cathode on the bottom of the chamber and the segmented plane on the top. An array of Si detectors is shown in the forward direction, but

Fig. 6.7 Top, CAD drawing of the ACTAR TPC detector. Bottom, picture of the detector chamber; the front-end part of the GET electronics is on the top of the chamber. Figure on top reproduced from [14] with permission from Elsevier

room is available to place other detectors on the four sides of the active volume. The first measurements with ACTAR TPC have confirmed the projected performances [14, 36, 37]. A smaller-scale "demonstrator" version of ACTAR TPC [15] was realised prior to the full-size detector, with an active volume of $64 \times 128 \times 170 \, mm^3$. Besides working as a test bench for various solutions used in ACTAR TPC, the demonstrator can be conveniently used when the aim is to visualise the vertex of the reaction, while particles can be stopped in solid-state detectors surrounding the active volume with a good spatial efficiency.

Other detectors using the same arrangement, with the drift field perpendicular to the beam direction, are CAT-S and CAT-M [38], the active targets of the Center for Nuclear Study of the University of Tokyo. These instruments utilise Thick-GEMs [39, 40] to reduce the amplification gain along the beam track, thus effectively increasing the dynamic range. The CATS are intended to be used for inelastic scattering measurements (Sect. 6.3.4); like in ACTAR TPC, the drift volume is delimited by a field cage and can be surrounded by solid-state detectors.

We also mention, in this category, the MAIKo active target [41] of the Research Center for Nuclear Physics (RCNP) of Osaka University. MAIKo is meant to be used with He gas to study cluster structures in light exotic nuclei, in measurements where the low energy of multiple particles must be measured with a very good resolution. The amplification technology employs micro-pixel chambers [42]: anode pixels, 400 μm apart, are surrounded by cathode strips to provide an average spatial resolution of about 300 μm with a total 256+256 electronic channels.

Differently from the above, the following active targets have the drift field collinear with the beam direction. The AT-TPC of the National Superconducting Cyclotron Laboratory at Michigan State University [12] (Fig. 6.8) is actually often placed slightly tilted with respect to the beam, to spread the charges created by the beam on multiple pads and to separate piling-up beam tracks. The cylindrical chamber is 1 m long and has a radius of 29.2 cm; it is intended to be placed in a strong, uniform magnetic field ($B \simeq 2$ T), collinear with the beam axis. The magnetic field bends the trajectory of charged particles emitted in reactions, helping in their identification and energy measurement (see Sect. 6.2.2). The anode uses Micromegas for the amplification; it is organised in two concentric detection areas, an inner one divided in 6144 small triangular pads (with 0.5 cm height) and an outer

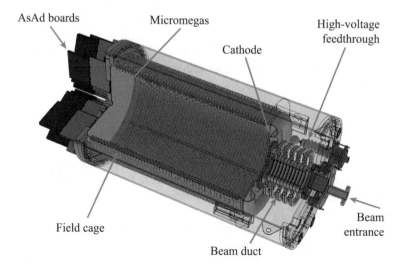

Fig. 6.8 Schematic view (cutaway) of the AT-TPC. The main components are indicated. Figure reproduced from [12] with permission from Elsevier

one with 4096 larger pads (with 1 cm height). As mentioned in Sect. 6.2.4, dedicated tools were developed for the analysis of the tracks that are generated in the AT-TPC. Reference [43] describes some physics cases that will exploit the advantages of the detector. For the AT-TPC as well, a half-size prototype version (pAT-TPC) was developed [44] and used in a few measurements without the external magnetic field.

SpecMAT (Spectroscopy of exotic nuclei in a Magnetic Active Target) [13], developed at the KU Leuven, also has a cylindrical active volume (Fig. 6.9), with the anode (pad plane) placed downstream with respect to the beam direction. The current version of the pad plane is a Micromegas detector divided into 2916 triangular pads (with 4.6 mm height). The size of SpecMAT is smaller than the AT-TPC: the field cage is 323.5 mm long, with a diameter of 220 mm. The volume is surrounded by an array of 45 $48 \times 48 \times 48$ mm^3 CeBr$_3$ scintillators for γ-ray detection, with a measured resolution of 4.4% at 661.7 keV and a geometric efficiency around 25% with respect to the centre of the field cage. A hole at the centre of the pad plane allows evacuation of the radioactive beam, which is essential to reduce the background in the γ-ray array. The whole system is designed to be placed in the ISOLDE Solenoidal Spectrometer [45], in a uniform magnetic field up to 3 T, again exploiting the cyclotron motion of the charged particles to measure their energy. The main physics goal for SpecMAT is the measurement of direct reactions [46].

6.3   Physics Cases

Active targets can be built for a specific purpose, optimised for the measurement of a particular type of reaction process. Or, they can be constructed as more generic instruments that can be adapted to various cases of interest by changing parameters such as the target gas, its pressure, possibly modifying the geometry or adding auxiliary detectors. In the sections that follow we will see examples of both, in a selected list of experiments performed so far. For a systematic discussion of applications of active targets to nuclear structure physics, we refer to the work of D. Bazin et al. in [47].

6.3.1   Decay Through Charged Particle Emission

Exotic decays like two-proton radioactivity or β-delayed charged particle emission are sensitive probes to study nuclear structure at the limits of nuclear stability. Their probability, or that of the reverse capture process, may also be relevant for nuclear astrophysics processes.

In such studies, very rare parent nuclei (down to a few per day) are implanted in a detector as a radioactive ion beam, along with much more abundant contaminants, where they undergo decay. In the most challenging cases, the half-life may be of the order of milliseconds or shorter. Another concern is the low energy of the emitted charged particles. Dedicated fast electronics has been developed to overcome these

Fig. 6.9 The SpecMAT active target. Top, a cutaway view; bottom, the assembled detector. Three rings of $CeBr_3$ scintillation detectors surround the active volume. Figures reproduced from [46] with permission from KU Leuven, Faculty of Science

issues when implanting the nuclei in solid-state detectors: as the particles cannot escape, the method provides accurate measurements of the half-life of the decay, the total emitted energy, and the branching ratio. If multiple particles are emitted, however, one would like to access information about the energy sharing and relative emission angle to study the correlations. This is possible if the implantation is made in a gas volume. In this case the detector is used as a TPC, not as an active target; however, we mention it here to illustrate the versatility of these devices.

The first challenging studies of this kind were performed at GANIL (France) with the CENBG TPC [48], which allowed the first direct observation of the two protons emitted in the decay of ^{45}Fe ($t_{1/2} = 2.45$ ms) [49] and ^{54}Zn ($t_{1/2} = 1.59$ ms) [50]. With ACTAR TPC, the aim was to access much faster decays; this was achieved in the study of the proton-emission decay from the 10^+ isomeric state in ^{54}Ni to two states in ^{53}Co [37], which takes place with a very short half-life $t_{1/2} = 155$ ns.

The measurement took place at GANIL, where the ions of interest were identified among all those implanted in the ACTAR TPC by using the energy loss vs. time of flight method in the LISE3 spectrometer. When the implantation of one parent nucleus took place, the system was triggered to record tracks for the following 10 μs. About 0.4% of those nuclei were in the 10^+ isomeric state, and about half of these decayed via proton emission. The pressure of the gas (an Ar-CF$_4$ mixture) was set to have tracks of a few centimetres for the two groups of emitted protons, which allowed to measure their energy (1.2 MeV and 2.5 MeV) with about 10% precision.

An event of interest is shown in Fig. 6.10. The left panel shows the charge collected on the pad plane: the proton track is clearly separated from that of the implanted ion. The track are both visible despite the very different energy loss, because ACTAR TPC allows increasing the dynamic range by setting a different amplification on different zones of the pad plane: a polarisation of the pads reduces the amplification voltage across the Micromegas gap. The right panel shows a three-dimensional view of the event: the start of the proton track appears separated from the end of the ion track, because of the time difference between the two moments, which reflects in a different arrival time of the ionisation electrons on the pad plane. From the distance between the two points and the drift velocity of the electrons (fitted from the data), the decay time of each event can be calculated, and the half-life can be fitted. In this case, the about 3000 identified events led to a precision of about 4 ns [37].

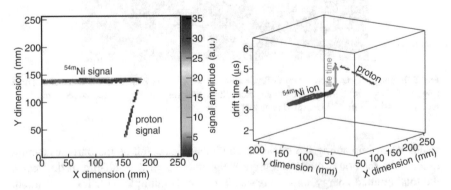

Fig. 6.10 Left, charges collected on the pad plane of ACTAR TPC after implantation of a ^{54m}Ni decaying via proton emission; Right, three-dimensional view of the same event. Figure reproduced from [37] under a Creative Commons Attribution 4.0 International License

Other TPCs were built specifically for the study of low-energy charged particles emitted in decays. Astrobox [51] is a cylindrical gas detector, with a pad plane with only five elements equipped with a Micromegas electron amplifier. It allowed measuring β-delayed protons from the decay of ^{23}Al down to an energy of about 100 keV, with an excellent resolution of 15 keV. The optical TPC (OTPC) of the University of Warsaw [52] uses a different technology: the UV rays generated by the electrons in the amplification stage are transformed into visible light by a wavelength shifter, forming an image which is recorded on a CCD camera. The third dimension of the tracks is provided by the drift time profile of the electrons, measured in a photomultiplier placed next to the CCD camera. The relatively simple system was very successful in measuring exotic decays like two- and three-proton emission [53].

6.3.2 Fusion and Reaction Cross Sections

The study of the reaction processes involving exotic nuclei has been a topic of interest since the production of the first in-flight fragmented radioactive ion beams in the mid-1980s. It was the measurement of unexpectedly large interaction cross sections with beams of ^{6}He and ^{11}Li [54] that led to the discovery of their halo structures [55]. The effect of halos, cluster structures, and the weak binding of exotic nuclei on the reaction mechanism has been the subject of an intense theoretical and experimental effort [56, 57]. The problem is particularly interesting at energies around and below the Coulomb barrier, where the interplay of various reaction channels can lead to a modification of the tunnelling probability and thus the fusion cross section.

Experimentally, the beam particle is a light (exotic) nucleus, impinging on a heavier target: the reaction takes place in direct kinematics. Still, the measurements are extremely challenging, since they have to be repeated at different energies, with weak beam intensities and cross sections that drop exponentially at energies below the Coulomb barrier. The different reaction channels (fusion, breakup, transfer) are identified through their signatures; this requires the detection of most of the outgoing radiation, energies, and angular distributions.

Active targets offer several advantages for this kind of measurements. With an appropriate choice of the gas pressure, the beam particles can lose a significant energy as they progress through the active volume: each position along the beam trajectory corresponds then to a well-defined centre-of-mass energy. By selecting appropriate values for the energy of the incoming beam, the energy region around the Coulomb barrier can be covered in a single measurement. The energy loss profiles along the tracks (dE/dx) can be crucial for the identification of charged particles and reaction processes. For example, in a complete fusion event, one expects the beam track to show a large, sudden increase of the dE/dx near the end, with little or no deviation in the direction (small deviations could be due to the recoil from evaporated particles). Transfer and breakup events could be separated from fusion by the kinematics of the detected charged particles.

An example is the ^{10}Be + ^{40}Ar measurement [58] performed with the prototype AT-TPC at the TwinSol facility of the University of Notre Dame [59]. The pressure was set to stop the beam particles right before the detection plane (perpendicular to the beam in the pAT-TPC), and the trigger selected only events for which the tracks were shorter than that, indicating that the beam had interact with the nuclei of the gas. Figure 6.11 shows, in the top panel, the dE/dx of a ^{10}Be + ^{40}Ar fusion event with, possibly, evaporation of neutrons but no charged particles. The peak near the end of the track reveals the much higher charge of the compound nucleus. In [58] other tracks are presented, where different processes are identified. In the bottom panel in Fig. 6.11, the fusion cross section is shown, as derived from the number of identified events as function of the position in the target translated into centre-of-mass energy. The events below 14 MeV are mostly due to the fusion of ^{10}Be on the carbon nuclei in the P10 gas mixture (90% Ar and 10% methane); the energy loss resolution was not sufficient to separate them from the fusion of ^{40}Ar. The calculations with the PACE4 evaporation code [60] reproduce very well the behaviour, though an overall scaling was necessary because of issues encountered in the normalisation of the experimental data. Remarkably, the method allowed to measure the fusion cross section with a beam intensity as low as 100 particles per second.

Despite its potential, this use of active targets has not been explored much to date. One limitation is represented by the choice of the target nuclei, which have to be components of a gas; and, if the gas is not pure, like in the example above, separation of background events can be problematic.

6.3.3 Resonant Reactions

Resonant reactions are a powerful tool for the study of unbound states. The state is formed for a short time from the combination of the nucleons in the projectile and the target (thus only states above the threshold for breakup into projectile plus target are explored). It then decays, either through the same channel in a resonant *elastic* scattering or possibly through different channels. The reaction probability and the decay pattern carry the information about the structure of the state. For example, cluster structures [61] can be investigated with reactions involving α particles; resonant proton scattering, on the other hand, can be used to explore proton single-particle widths or to access states in very proton-rich nuclei. Extension to neutron-rich nuclei is possible by using isospin symmetry, through the population of isobaric analog states (IAS) [62, 63]. Additionally, many resonances in light and medium-mass nuclei are important for the cross section of reactions which are relevant for nucleosynthesis processes and energy generation in stars [64].

Experimentally, the particles emitted in the process are measured at the backward centre-of-mass angles, where the resonant process dominates, while the direct ones are suppressed. When radioactive nuclei are involved, the measurement is performed in inverse kinematics, and those angles correspond to the forward direction in the reference frame of the laboratory. Similar to the physics case discussed in Sect. 6.3.2

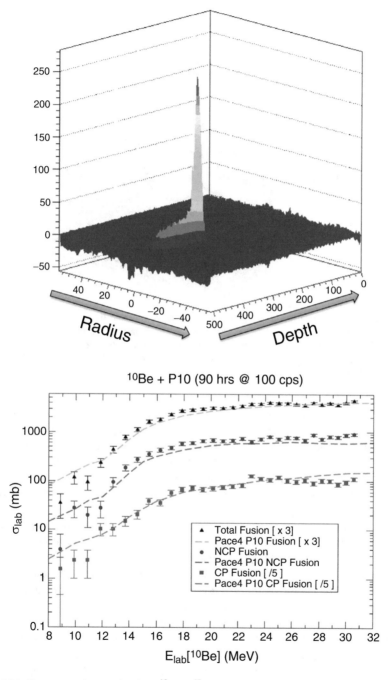

Fig. 6.11 Top, energy loss profile for a ^{10}Be + ^{40}Ar fusion event measured in the prototype AT-TPC. Notice the sudden increase near the end of the track. Bottom, data (points) and calculations (dashed lines) of the total fusion cross section and fusion followed by charged particle emission (CP) or not (NCP), for ^{10}Be on the nuclei of the P10 detection gas. The cross sections are scaled for clarity as indicated in the figure. Figure reproduced from [58] with permission from Elsevier

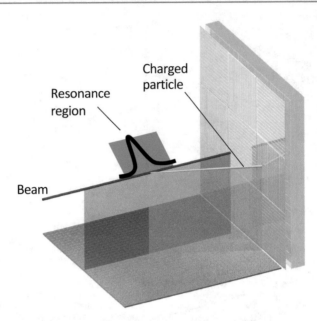

Fig. 6.12 Illustration of the arrangement for the study of resonant reactions in an active target. The incoming beam loses energy as it traverses the target gas: each position corresponds to a well-defined centre-of-mass energy of the compound nucleus. At the right energy the resonance is excited, and it decays by, for example, emitting a charged particle which can be detected in auxiliary detectors positioned in the forward direction

above, in active targets, the energy region of interest is explored via the energy loss of the beam as it progresses through the gas volume. The situation is illustrated in Fig. 6.12. If the gas pressure is set to stop the beam before the end of the active volume, then the energy range in the compound nucleus down to the breakup threshold is explored. Alternatively, one can choose the initial beam energy and pressure in order to explore in detail a particular energy region in the compound nucleus.

The technique of studying resonances in a thick target in inverse kinematics (TTIK) has been in fact developed with the first radioactive ion beams, both for solid [65] and gas targets [66,67]. With a passive target, when the beam-like particle is stopped in the target, the only information available is that of the emitted light particle; the kinematics of the reaction can be reconstructed either under some assumptions (two-body process, no excitation of the beam-like residue) or with additional information such as the time-of-flight of the particle with respect to the incoming beam [67]. With an active target, redundant information is available that helps constraining the exact process: the centre-of-mass energy of the reaction (from the vertex position), the energy of the beam-like residue (from its range), and a handle on the particle identification through the energy loss profiles. Because of the different energy loss, the charged particle emitted in the decay has a longer range than the beam-like particle and is captured in solid-state detectors in the forward hemisphere; see Fig. 6.12.

A number of measurements have been performed with active targets. Resonant scattering on α particles was measured in the pAT-TPC, using ^{6}He [68] and ^{10}Be [69] beams to look for α particle chains in ^{10}Be and ^{14}C, respectively. Proton scattering was used to populate resonances in ^{9}C [70] and in the unbound ^{10}N [71] nuclei using the TexAT active target [72]. In a measurement of the ^{12}C+p resonant elastic scattering at ISOLDE/CERN with the Maya detector, a resolution of $\approx$50 keV (FWHM) in the excitation energy of ^{13}N was obtained [73].

Even better results were obtained in the ^{18}O+p commissioning measurement of ACTAR TPC [14], shown in Fig. 6.13. The top panels show the energy of the scattered light particle, detected in forward-placed silicon detectors, plotted against the range of the beam-like particle. They show how inelastic scattering can be

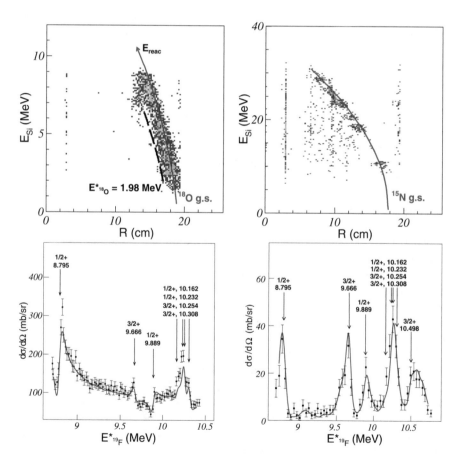

Fig. 6.13 Measurement of the ^{18}O+p resonant scattering in ACTAR TPC. Top, energy of the scattered light particle plotted against the range of the beam-like particle for protons (left) and α particles (right). The plots allow to disentangle elastic and inelastic scattering. Bottom, excitation function of ^{19}F from the (p,p) (left) and (p,α) (right) data at $\theta_{c.m.} = (160\pm50)°$. Figure reproduced from [14] with permission of Elsevier

separated: for example, if the residual ^{18}O nucleus is left in the $E^* = 1.98$ MeV
excited state, its range is shorter than in the case of elastic scattering. The bottom
panels show the excitation function of ^{19}F from the (p,p) and (p,α) channels. The
red curve is the result of an R-matrix calculation [74] where the known parameters
of the resonance states were fixed and the only free parameter was the resolution of
the detector. Resolutions of 38 keV for (p,p) and 54 keV for (p,α) were obtained.
Clearly, active targets will be in the future the instrument of choice to perform this
kind of studies with radioactive ion beams.

6.3.4 Inelastic Scattering to Excite Giant Resonances

A very specific class of reactions, for which active targets have opened new
possibilities, is that of low-momentum transfer scattering in inverse kinematics. In
those processes a light target nucleus is scattered at low energies and laboratory
angles close to 90 degrees: the energy loss in a solid target would strongly limit the
measurement to some high threshold. Likewise, the momentum change of the beam-
like particle would be too small to be measured, for example, in a spectrometer at
forward angles. Currently, only gaseous targets allow these measurements, either as
active targets or inside a storage ring.

The IKAR active target [34] was the first to exploit this feature, to measure proton
elastic scattering from a number of light exotic He, Li, and Be nuclei and derive their
matter density distribution [34, 75–78].

Next, this feature was applied to the study of giant resonances, excited through
inelastic scattering.

Giant resonances (GRs) are collective, high-frequency excitations of nuclei [79]
which exhaust the main part of the corresponding strength, indicating that a large
number of nucleons participate in the process. They can be characterised by changes
in spin and/or isospin quantum numbers and by their multipolarity. The well-known
giant dipole resonance was observed for the first time in ^{238}U in 1947 [80]; it is an
isovector excitation, where neutrons and protons oscillate with opposite phases. The
isoscalar modes, in which all nuclear matter oscillates together in a compression
motion, were first observed only 30 years later [81, 82].

GRs are very important for modern nuclear physics, because their correct
prediction is a robust test for mean-field approaches based on energy density
functionals. These models [83] have a very broad applicability, to describe finite
nuclei throughout the whole chart, but also to predict the properties of infinite
nuclear matter through the parameters of its equation of state (EoS). The latter,
in turn, is a fundamental ingredient in the description of celestial bodies in
extreme conditions, such as core-collapse supernovae, neutron stars, and neutron-
star mergers [84].

One important parameter of the EoS is the incompressibility of nuclear matter
K_∞, which can be extrapolated from the isoscalar compression modes of nuclei
[85] (monopole ISGMR, dipole ISGDR, and quadrupole ISGQR). Uncertainties on

K_∞ can be reduced by measuring the energy of the ISGMR for different values of isospin [86], i.e. across a chain of isotopes or isobars. To extend these chains, it is necessary to measure the ISGMR in unstable isotopes.

Experimentally, the isoscalar modes are excited with inelastic scattering on deuterons or α particles. The cross section is largest at small centre-of-mass angles, and it is necessary to have a very good angular resolution at small angles to disentangles the contribution of the different multipoles and identify the mode of the GR. In direct kinematics, spectrometers are used to measure the momentum of the scattered deuterons or α particles on the stable targets. In inverse kinematics, the situation is very challenging, because the recoil light particles have a very low energy in the angular range of interest. The situation is illustrated in Fig. 6.14: if a solid target (a foil) is used, the light particles are stopped in the target material, preventing detection below a certain threshold, which increases rapidly with the emission angle. With a gas target the threshold is dictated by the length of the tracks, which for a given energy can be increased by reducing the pressure. Close to the beam direction, however, it becomes difficult to separate the (very short) tracks from the beam track. Still, a large region of low centre-of-mass angles becomes accessible, which is essential for the multipole analysis of the resonances.

These measurements are also possible in storage rings: this was demonstrated at the ESR ring in GSI (Darmstadt), where a He gas-jet target was used to measure the

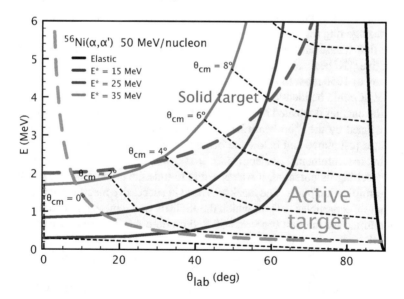

Fig. 6.14 Kinematic plot (energy vs. laboratory angles) for α particles emitted in the inelastic scattering from an incident ^{56}Ni beam at 50 MeV/nucleon. The black (elastic) and blue curves are for scattering at different excitation energies in ^{56}Ni. Also indicated are the corresponding centre-of-mass angles of the scattering. The red and orange dashed lines indicate the thresholds for detection of the α particles using, respectively, a solid foil and an active target. The yellow area indicates the gain by using an active target

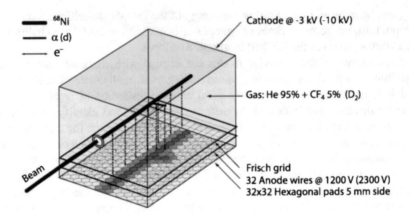

Fig. 6.15 Arrangement of the Maya active target for inelastic scattering measurements. The pressure is chosen to stop the light particles in the gas, for the energy and angular range of interest. The tracks need to be long enough to identify the reaction vertex and evaluate their length with sufficient precision. Figure reprinted from [92] with permission from the American Physical Society

ISGMR in ^{58}Ni in inverse kinematics [87]. The inelastically scattered α particles were detected in solid-state telescopes, of which the first stage, double-sided silicon detectors, were designed to function as active barriers to preserve the high vacuum of the storage ring [88].

The first measurement with an active target was performed at GANIL with a radioactive ^{56}Ni beam at 50 MeV/nucleon, using Maya filled with a deuterium gas at a pressure of 1050 mbar [89]; see the arrangement in Fig. 6.15. Both the ISGMR and the ISGQR could be identified, and their energy was measured. Maya was tuned to detect the charges deposited by the α particles, which means that the much stronger signal caused by the ^{56}Ni beam had to be limited. This was achieved by placing two plates just above and below the beam to collect the ionisation charges. In the following measurements, an electrostatic masks was developed for the same purpose [90]: by setting its potential, it was possible to tune the amount of charges drifting to the amplification plane. The mask was used in successive measurements: again of ^{56}Ni, but on α particles [91], in which the ISGMR was remeasured and the ISGDR was identified for the first time, and of ^{68}Ni on both deuterons and α particles [92, 93]. In the latter, next to the observation of the ISGMR and ISGQR, indications were found of the dipole strength and of a low-energy ("soft") part of the monopole strength. Some of the results of this measurement are shown in Fig. 6.16.

The measurement on ^{68}Ni has been recently repeated with the ACTAR TPC active target; at the same time, ^{58}Ni has been measured to benchmark the performances of ACTAR TPC. The data are still being analysed, but preliminary results indicate an improvement in the resolution of a factor 2 to 3 thanks to the finer spatial resolution of the pad plane. With ACTAR TPC, the use of a mask was not necessary: through the GET electronics, an offset voltage was put on the central

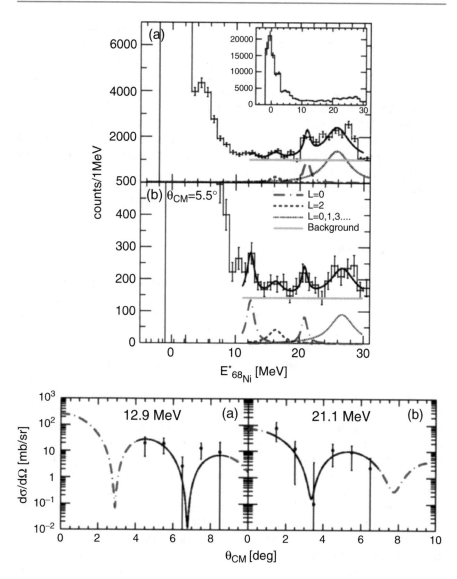

Fig. 6.16 Top panel: excitation energy in ^{68}Ni from (α,α') scattering, for (**a**) all angles and (**b**) at 5.5°. The lines are Lorentzian fits for the different multipoles and background as indicated. Bottom panel: angular distributions for the events in the peaks at (**a**) 12.9 MeV and (**b**) 21.1 MeV, fitted with the result of DWBA calculations using microscopic RPA predictions with isoscalar $L = 0$ multipolarity. Figure reprinted from [93] with permission from the American Physical Society

pads, along the beam track, effectively reducing the amplification voltage gap across the Micromegas and thus attenuating the signal from the beam particles.

Similar measurements have been performed with the active targets (CAT) of the Center for Nuclear Study in Tokyo. CAT-S [38] was used for the inelastic scattering of ^{132}Sn on deuterons at RIKEN [94], while the larger CAT-M was employed at HIMAC (Chiba, Japan) for the measurement of ^{136}Xe on protons and ^{132}Xe on deuterons [95]. With such high-Z beams, the problem of avoiding signals from the beam track is made even more difficult by the production of highly energetic δ electrons. Those electrons travel for few centimetres through the gas, effectively blinding a large region around the beam path. Recently, the issue was dealt with by creating a magnetic field along the beam path to confine the electrons. The first tests at HIMAC seem to indicate that such a solution is very effective.

6.3.5 Transfer Reactions

Transfer reactions are particularly useful as a spectroscopic tool due to their selectivity, to populate states with pronounced single-particle or cluster characters. The application of such reactions to exotic nuclei is a major tool to explore the evolution of the shell structure with isospin. The measurements provide the energy of the states, indication of their spin and parity, and, through comparison of the cross section with a model, spectroscopic factors for given configurations. Measurements on unstable nuclei are performed in inverse kinematics, usually on a thin foil containing protons or deuterons, surrounded by charged particles and, in some cases, γ-ray detectors. This arrangement present two well-known problems. The kinematical compression reduces the energy difference between the detected light particles by a factor 2 to 4, with respect to the difference in excitation energies of the beam-like particle [3]; hence, the resolution of the populated states is worsened by the same factor. Secondly, the energy loss of the heavy beam particle traversing the target means that the energy at the vertex position is not well determined: the same process could produce light particles at slightly different energies, again worsening the resolution. Because of the weak intensity of the radioactive beams, a compromise with the target thickness has to be made. The resolution in excitation energy, resulting from these effects, is often of a few hundreds of keV (see, e.g. [96]). A γ-ray array can help significantly [97], but it also has limitations in efficiency.

The kinematical compression problem can be solved by placing the target in a solenoid and measuring the distance travelled by the light charged particles on the beam axis instead of their angles. This configuration is adopted by the HELIOS spectrometer in Argonne [98] and recently by the ISOLDE Solenoidal Spectrometer (ISS) in CERN [45].

Active targets can solve the second problem, because the energy at the reaction point can be directly derived from the position of the vertex in the gas, through the known energy loss of the beam. This way, a much larger thickness can be used than with a solid foil without a loss in resolution. As an example, if a solid CD_2 target is used, its thickness must be typically less than ≈ 500 $\mu g/cm^2$ to achieve a sufficient resolution in the particle spectra; this corresponds to 125 $\mu g/cm^2$ deuterium. A 25-cm-long active target filled with deuterium gas at 1 bar pressure, on the other hand, contains 4 mg/cm^2, 30 times more than the foil; and a resolution on the vertex position of ≈ 1 mm corresponds to a very thin slice of 16 $\mu g/cm^2$. With these properties, active targets should allow performing transfer-reaction measurements with beam intensities of at least one order of magnitude less than target foils. The actual resolution in excitation energy will ultimately depend on the precision in the determination of the energy of the light particles, either through their range or the energy deposited in auxiliary detectors.

To date there are no examples yet of transfer reactions performed with active targets, with the exception of the $^{12}C(^8He,^7H)^{13}N$ proton transfer at GANIL [99] and the $^{11}Li(p,t)^9Li$ two-neutron transfer at TRIUMF [100], both with Maya; in the latter a resolution in Q-value of about 400 keV was reached. For a typical (d,p) reaction, simulations such as ACTARSim [101] for ACTAR TPC indicate that 100 keV resolution in Q-value should be achievable [102]. The complementary detection of γ-rays, such as foreseen in SpecMAT (see Sect. 6.2.5), will help improving on this value [46].

Transfer and inelastic reactions were also used in the Maya active target to populate excited states in fissile nuclei, with the purpose of studying the fission process [103]. A weak ^{238}U beam was accelerated at an energy of 6.5 MeV/nucleon at GANIL and sent into Maya filled with isobutane (C_4H_{10}) at 50 mbar. The logic of the detection is shown in Fig. 6.17: the detection of the scattered light particle (from elastic, inelastic, and transfer reactions) in the forward wall allowed to separate the reaction channels and calculate the excitation energy in the beam-like particle that decayed by fission, with the fragments detected in the active volume. The technique allows to measure the fission barrier distribution in the populated nuclei, the distribution of fragments, and the cross sections as a function of beam energy and is of extreme interest for use with heavy radioactive ion beams. The results of such measurements can be used to test models predicting fission probabilities, for example, in nuclei at the end of the r-process nucleosynthesis [105].

6.4   Conclusion

Active targets are instruments designed to detect primarily rare events, with high efficiency and good resolution properties. They are versatile, and specific applications have led to a rich variety of technological solutions and clever detection techniques. For example, the limitations in dynamic range have been dealt with, on one side, by using various methods to reduce or eliminate the ionisation signal from the beam particles and, on the other side, by employing either magnetic fields or auxiliary solid-state detectors to increase the energy and types of particles detected in the reaction. Smart trigger techniques can select the events of interest among a large background, induced by beam contaminants and/or other processes.

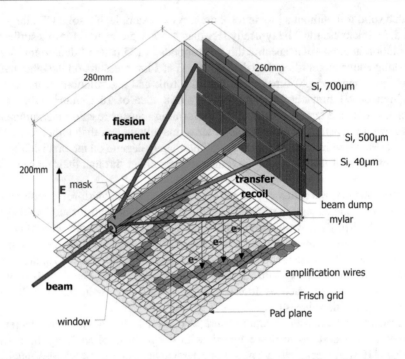

Fig. 6.17 Scheme of the detection of a reaction-induced fission event in Maya. The forward wall of detectors was upgraded with respect to the original one, and a new version of the mask to collect the charges created by the beam particles was used [104]. Figure reprinted from [103] with permission from Elsevier

In some cases new domains have become accessible, such as the study of correlations between particles emitted in very fast decays or low-momentum transfer reactions to populate compression modes in unstable nuclei. In others, like resonant reactions, active targets can provide additional experimental information to disentangle competing processes.

Still, further developments are needed. Among the areas of concern, there is the improvement of the performances of gas detectors when using pure gases of interest for nuclear structure studies, such as H_2, D_2, ^{4}He, and ^{3}He. The general behaviour of active targets with very heavy and high-energy beam particles needs to be better investigated, to verify the limits due to saturation of the signals on the pad plane and control the space-charge effects that can instantly modify the electric drift field and produce distorted tracks. The analysis of complex event patterns will also need improvements, to reduce the time that is necessary at present to extract all the information from the raw data and eventually build reliable online analysis algorithms. A coordination between the work of the different research groups on this topic is already present, but it can certainly be reinforced.

With the development of new radioactive beam facilities (FRIB, FAIR, SPIRAL2, SPES) and the upgrade of existing ones (ISOLDE), active targets are going to play an important role: their characteristics make them particularly well-suited for use with the weakest radioactive ion beams, thus pushing the boundaries of our knowledge of nuclear structure.

References

1. J.A. MacDonald (ed.), *The Time Projection Chamber* (AIP Conf Proc. Vol. 108, New York, 1984). https://doi.org/10.1063/1.34317
2. P.A. Zyla, et al., *Review of Particle Physics* (Prog. Theor. Exp. Phys., p. 083C01, 2020), chap. PassageofParticlesThroughMatter. https://doi.org/10.1093/ptep/ptaa104
3. R. Raabe, *Nucleon-Transfer Reactions with Radioactive Ion Beams* (IOS Press, 2019), p. 209. Proceedings of the International School of Physics "Enrico Fermi". https://doi.org/10.3254/978-1-61499-957-7-209
4. Y. Ayyad, D. Bazin, S. Beceiro-Novo, M. Cortesi, W. Mittig, Eur. Phys. J. A **54**, 181 (2018). https://doi.org/10.1140/epja/i2018-12557-7
5. S.D. Pinto, in *IEEE Nuclear Science Symposuim Medical Imaging Conference* (2010), pp. 802–807. https://doi.org/10.1109/NSSMIC.2010.5873870
6. F. Sauli, Nucl. Instrum. Methods Phys. Res. A **386**, 531 (1997). https://doi.org/10.1016/S0168-9002(96)01172-2
7. F. Sauli, Nucl. Instrum. Methods Phys. Res. A **805**, 2 (2016). https://doi.org/10.1016/j.nima.2015.07.060
8. Y. Giomataris, P. Rebourgeard, J. Robert, G. Charpak, Nucl. Instrum. Methods Phys. Res. A **376**, 29 (1996). https://doi.org/10.1016/0168-9002(96)00175-1
9. P. Bhattacharya, S. Bhattacharya, N. Majumdar, S. Mukhopadhyay, S. Sarkar, P. Colas, D. Attie, Nucl. Instrum. Methods Phys. Res. A **732**, 208 (2013). https://doi.org/10.1016/j.nima.2013.07.086. Vienna Conference on Instrumentation 2013
10. M. Cortesi, J. Yurkon, W. Mittig, D. Bazin, S. Beceiro-Novo, A. Stolz, J. Instrum. **10**, P09020 (2015). https://doi.org/10.1088/1748-0221/2015/09/p09020
11. R. Shane, A.B. McIntosh, T. Isobe, W.G. Lynch, H. Baba, J. Barney, Z. Chajecki, M. Chartier, J. Estee, M. Famiano, B. Hong, K. Ieki, G. Jhang, R. Lemmon, F. Lu, T. Murakami, N. Nakatsuka, M. Nishimura, R. Olsen, W. Powell, H. Sakurai, A. Taketani, S. Tangwancharoen, M.B. Tsang, T. Usukura, R. Wang, S.J. Yennello, J. Yurkon, Nucl. Instrum. Methods Phys. Res. A **784**, 513 (2015). https://doi.org/10.1016/j.nima.2015.01.026
12. J. Bradt, D. Bazin, F. Abu-Nimeh, T. Ahn, Y. Ayyad, S. BeceiroăNovo, L. Carpenter, M. Cortesi, M.P. Kuchera, W.G. Lynch, W. Mittig, S. Rost, N. Watwood, J. Yurkon, Nucl. Instrum. Methods Phys. Res. A **875**, 65 (2017). https://doi.org/10.1016/j.nima.2017.09.013
13. O. Poleshchuk, R. Raabe, S. Ceruti, A. Ceulemans, H. De Witte, T. Marchi, A. Mentana, J. Refsgaard, J. Yang, Nucl. Instrum. Methods Phys. Res. A **1015**, 165765 (2021). https://doi.org/10.1016/j.nima.2021.165765
14. B. Mauss, P. Morfouace, T. Roger, J. Pancin, G.F. Grinyer, J. Giovinazzo, V. Alcindor, H. Álvarez-Pol, A. Arokiaraj, M. Babo, B. Bastin, C. Borcea, M. Caamaño, S. Ceruti, B. Fernández-Domínguez, E. Foulon-Moret, P. Gangnant, S. Giraud, A. Laffoley, G. Mantovani, T. Marchi, B. Monteagudo, J. Pibernat, O. Poleshchuk, R. Raabe, J. Refsgaard, A. Revel, F. Saillant, M. Stanoiu, G. Wittwer, J. Yang, Nucl. Instrum. Methods Phys. Res. A **940**, 498 (2019). https://doi.org/10.1016/j.nima.2019.06.067
15. T. Roger, J. Pancin, G.F. Grinyer, B. Mauss, A.T. Laffoley, P. Rosier, H. Alvarez-Pol, M. Babo, B. Blank, M. CaamaÑo, S. Ceruti, J. Daemen, S. Damoy, B. Duclos, B. Fernández-Domínguez, F. Flavigny, J. Giovinazzo, T. Goigoux, J.L. Henares, P. Konczykowski, T. Marchi, G. Lebertre, N. Lecesne, L. Legeard, C. Maugeais, G. Minier, B. Osmond, J.L. Pedroza, J. Pibernat, O. Poleshchuk, E.C. Pollacco, R. Raabe, B. Raine, F. Renzi, F. Saillant, P. Sénécal, P. Sizun, D. Suzuki, J.A. Swartz, C. Wouters, G. Wittwer, Yang, Nucl. Instrum. Methods Phys. Res. A **895**, 126 (2018). https://doi.org/10.1016/j.nima.2018.04.003

16. E.V.D. van Loef, P. Dorenbos, C.W.E. van Eijk, K. Krämer, H.U. Güdel, Appl. Phys. Lett. **79**, 1573 (2001). https://doi.org/10.1063/1.1385342
17. A. Maj, The PARIS Collaboration, Acta Phys. Pol. B **40**, 565 (2009)
18. O. Poleshchuk, J.A. Swartz, A. Arokiaraj, S. Ceruti, H. De Witte, G.F. Grinyer, A.T. Laffoley, T. Marchi, R. Raabe, M. Renaud, J. Yang, Nucl. Instrum. Methods Phys. Res. A **987**, 164863 (2021). https://doi.org/10.1016/j.nima.2020.164863
19. C.E. Demonchy, W. Mittig, H. Savajols, P. Roussel-Chomaz, M. Chartier, B. Jurado, L. Giot, D. Cortina-Gil, M. Caamaño, G. Ter-Arkopian, A. Fomichev, A. Rodin, M.S. Golovkov, S. Stepantsov, A. Gillibert, E. Pollacco, A. Obertelli, H. Wang, Nucl. Instrum. Methods Phys. Res. A **573**, 145 (2007). https://doi.org/10.1016/j.nima.2006.11.025
20. J.C. Santiard, W. Beusch, S. Buytaert, C.C. Enz, E.H.M. Heijne, P. Jarron, F. Krummenacher, K. Marent, F. Piuz. Gasplex: a low-noise analog signal processor for readout of gaseous detectors. CERN-ECP-94-17 (1994)
21. E.C. Pollacco, G.F. Grinyer, F. Abu-Nimeh, T. Ahn, S. Anvar, A. Arokiaraj, Y. Ayyad, H. Baba, M. Babo, P. Baron, D. Bazin, S. Beceiro-Novo, C. Belkhiria, M. Blaizot, B. Blank, J. Bradt, G. Cardella, L. Carpenter, S. Ceruti, E. De Filippo, E. Delagnes, S. De Luca, H. De Witte, F. Druillole, B. Duclos, F. Favela, A. Fritsch, J. Giovinazzo, C. Gueye, T. Isobe, P. Hellmuth, C. Huss, B. Lachacinski, A.T. Laffoley, G. Lebertre, L. Legeard, W.G. Lynch, T. Marchi, L. Martina, C. Maugeais, W. Mittig, L. Nalpas, E.V. Pagano, J. Pancin, O. Poleshchuk, J.L. Pedroza, J. Pibernat, S. Primault, R. Raabe, B. Raine, A. Rebii, M. Renaud, T. Roger, P. Roussel-Chomaz, P. Russotto, G. Saccà, F. Saillant, P. Sizun, D. Suzuki, J.A. Swartz, A. Tizon, N. Usher, G. Wittwer, J.C. Yang, Nucl. Instrum. Methods Phys. Res. A **887**, 81 (2018). https://doi.org/10.1016/j.nima.2018.01.020
22. E. Pagano, L. Acosta, L. Auditore, C. Boiano, G. Cardella, A. Castoldi, M. D'Andrea, D. Dell'aquila, E. De Filippo, S. De Luca, F. Fichera, L. Francalanza, N. Giudice, B. Gnoffo, A. Grimaldi, C. Guazzoni, G. Lanzalone, I. Lombardo, T. Minniti, S. Norella, A. Pagano, M. Papa, S. Pirrone, G. Politi, F. Porto, L. Quattrocchi, F. Rizzo, P. Russotto, G. Saccá, A. Trifirò, M. Trimarchi, G. Verde, M. Vigilante, EPJ Web Confer. **117**, 10008 (2016). https://doi.org/10.1051/epjconf/201611710008
23. M.A. Fischler, R.C. Bolles, Commun. ACM **24**, 381 (1981). https://doi.org/10.1145/358669.358692
24. P.V.C. Hough, Conf. Proc. C **590914**, 554 (1959)
25. R.O. Duda, P.E. Hart, Commun. ACM **15**, 11 (1972). https://doi.org/10.1145/361237.361242
26. C. Santamaria, et al., JPS Conf. Proc. **6**, 030130 (2015). https://doi.org/10.7566/JPSCP.6.030130
27. T. Furuno, T. Kawabata, M. Murata, A track finding algorithm for maiko tpc using hough transform method, Technical report, Research Center for Nuclear Physics (RCNP), Osaka University (2014)
28. A. Strandlie, J. Wroldsen, R. Frühwirth, B. Lillekjendlie, Comput. Phys. Commun. **131**, 95 (2000). https://doi.org/10.1016/S0010-4655(00)00086-2
29. M. Kuchera, R. Ramanujan, J. Taylor, R. Strauss, D. Bazin, J. Bradt, R. Chen, Nucl. Instrum. Methods Phys. Res. A **940**, 156 (2019). https://doi.org/10.1016/j.nima.2019.05.097
30. R. Solli, D. Bazin, M. Hjorth-Jensen, M. Kuchera, R. Strauss, Nucl. Instrum. Methods Phys. Res. A **1010**, 165461 (2021). https://doi.org/10.1016/j.nima.2021.165461
31. T. Roger, M. Caamaño, C. Demonchy, W. Mittig, H. Savajols, I. Tanihata, Nucl. Instrum. Methods Phys. Res. A **638**(1), 134 (2011). https://doi.org/10.1016/j.nima.2011.02.061
32. J.F. Ziegler, J.P. Biersack, U. Littmark, *The Stopping and Range of Ions in Solids* (Pergamon Press, New York, 1985). https://www.srim.org/
33. S. Beceiro-Novo, T. Ahn, D. Bazin, W. Mittig, Prog. Part. Nucl. Phys. **84**, 124 (2015). https://doi.org/10.1016/j.ppnp.2015.06.003
34. S.R. Neumaier, G.D. Alkhazov, M.N. Andronenko, A.V. Dobrovolsky, P. Egelhof, G.E. Gavrilov, H. Geissel, H. Irnich, A.V. Khanzadeev, G.A. Korolev, A.A. Lobodenko, G. Munzenberg, M. Mutterer, W.S.D.M. Seliverstov, T. Suzuki, N.A. Timofeev, A.A. Vorobyov, V.I. Yatsoura, Nucl. Phys. A **712**, 247 (2002). https://doi.org/10.1016/S0375-9474(02)01274-5

35. J. Giovinazzo, J. Pibernat, T. Goigoux, R. de Oliveira, G.F. Grinyer, C. Huss, B. Mauss, J. Pancin, J.L. Pedroza, A. Rebii, T. Roger, P. Rosier, F. Saillant, G. Wittwer, Nucl. Instrum. Methods Phys. Res. A **892**, 114 (2018). https://doi.org/10.1016/j.nima.2018.03.007

36. J. Giovinazzo, J. Pancin, J. Pibernat, T. Roger, Nucl. Instrum. Methods Phys. Res. A **953**, 163184 (2020). https://doi.org/10.1016/j.nima.2019.163184

37. J. Giovinazzo, T. Roger, B. Blank, D. Rudolph, B.A. Brown, H. Alvarez-Pol, A.A. Raj, P. Ascher, M. Caamaño-Fresco, L. Caceres, D.M. Cox, B. Fernández-Domínguez, J. Lois-Fuentes, M. Gerbaux, S. Gréy, G.F. Grinyer, O. Kamalou, B. Mauss, A. Mentana, J. Pancin, J. Pibernat, J. Piot, O. Sorlin, C. Stodel, J.C. Thomas, M. Versteegen, Nat. Commun. **12**, 4805 (2021). https://doi.org/10.1038/s41467-021-24920-0

38. S. Ota, H. Tokieda, C.S. Lee, Y.N. Watanabe, J. Radioanal. Nucl. Chem. **305**, 907 (2015). https://doi.org/10.1007/s10967-015-4130-5

39. C. Shalem, R. Chechik, A. Breskin, K. Michaeli, Nucl. Instrum. Methods Phys. Res. A **558**, 475 (2006). https://doi.org/10.1016/j.nima.2005.12.241

40. C.K. Shalem, R. Chechik, A. Breskin, K. Michaeli, N. Ben-Haim, Nucl. Instrum. Methods Phys. Res. A **558**, 468 (2006). https://doi.org/10.1016/j.nima.2005.12.219

41. T. Furuno, T. Kawabata, S. Adachi, Y. Ayyad, T. Baba, T. Hashimoto, Y. Ishii, S. Kabuki, H. Kubo, Y. Matsuda, Y. Matsuoka, T. Mizumoto, M. Murata, H.J. Ong, T. Sawano, A. Takada, J. Tanaka, I. Tanihata, T. Tanimori, M. Tsumura, H.D. Watanabe, J. Phys. Confer. Ser. **569**, 012042 (2014). https://doi.org/10.1088/1742-6596/569/1/012042

42. A. Ochi, T. Nagayoshi, S. Koishi, T. Tanimori, T. Nagae, M. Nakamura, Nucl. Instrum. Methods Phys. Res. A **471**, 264 (2001). https://doi.org/10.1016/S0168-9002(01)00996-2

43. Y. Ayyad, N. Abgrall, T. Ahn, H. Álvarez-Pol, D. Bazin, S. Beceiro-Novo, L. Carpenter, R.J. Cooper, M. Cortesi, A.O. Macchiavelli, W. Mittig, B. Olaizola, J.S. Randhawa, C. Santamaria, N. Watwood, J.C. Zamora, R.G.T. Zegers, Nucl. Instrum. Methods Phys. Res. A **954**, 161341 (2020). https://doi.org/10.1016/j.nima.2018.10.019

44. D. Suzuki, M. Ford, D. Bazin, W. Mittig, W.G. Lynch, T. Ahn, S. Aune, E. Galyaev, A. Fritsch, J. Gilbert, F. Montes, A. Shore, J. Yurkon, J.J. Kolata, J. Browne, A. Howard, A.L. Roberts, X.D. Tang, Nucl. Instrum. Methods Phys. Res. A **691**, 39 (2012). https://doi.org/10.1016/j.nima.2012.06.050

45. T.L. Tang, B.P. Kay, C.R. Hoffman, J.P. Schiffer, D.K. Sharp, L.P. Gaffney, S.J. Freeman, M.R. Mumpower, A. Arokiaraj, E.F. Baader, P.A. Butler, W.N. Catford, G. de Angelis, F. Flavigny, M.D. Gott, E.T. Gregor, J. Konki, M. Labiche, I.H. Lazarus, P.T. MacGregor, I. Martel, R.D. Page, Z. Podolyák, O. Poleshchuk, R. Raabe, F. Recchia, J.F. Smith, S.V. Szwec, J. Yang, Phys. Rev. Lett. **124**, 062502 (2020). https://doi.org/10.1103/PhysRevLett.124.062502

46. O. Poleshchuk, SpecMAT, the active target for transfer reaction studies and gamma-ray spectroscopy in a strong magnetic field. Ph.D. thesis, Katholieke Universiteit Leuven (2021)

47. D. Bazin, T. Ahn, Y. Ayyad, S. Beceiro-Novo, A. Macchiavelli, W. Mittig, J. Randhawa, Prog. Part. Nucl. Phys. **114**, 103790 (2020). https://doi.org/10.1016/j.ppnp.2020.103790

48. B. Blank, M. Ploszajcak, Rep. Prog. Phys. **71**, 046361 (2008). https://doi.org/10.1088/0034-4885/71/4/046361

49. J. Giovinazzo, B. Blank, C. Borcea, G. Canchel, J.C. Dalouzy, C.E. Demonchy, F. de Oliveira Santos, C. Dossat, S. Grevy, L. Hay, J. Huikari, S. Leblanc, I. Matea, J.L. Pedroza, L. Perrot, J. Pibernat, L. Serani, C. Stodel, J.C. Thomas, Phys. Rev. Lett. **99**, 102501 (2007). https://doi.org/10.1103/PhysRevLett.99.102501

50. P. Ascher, L. Audirac, N. Adimi, B. Blank, C. Borcea, B.A. Brown, I. Companis, F. Delalee, C.E. Demonchy, F. de Oliveira Santos, J. Giovinazzo, S. Grevy, L.V. Grigorenko, T. Kurtukian-Nieto, S. Leblanc, J.L. Pedroza, L. Perrot, J. Pibernat, L. Serani, P.C. Srivastava, J.C. Thomas, Phys. Rev. Lett. **107**, 102502 (2011). https://doi.org/10.1103/PhysRevLett.107.102502

51. E. Pollacco, L. Trache, E. Simmons, A. Spiridon, M. McCleskey, B.T. Roeder, A. Saastamoinen, R.E. Tribble, G. Pascovici, M. Kebbiri, J.P. Mols, M. Raillot, Nucl. Instrum. Methods Phys. Res. A **723**, 102 (2013). https://doi.org/10.1016/j.nima.2013.04.084

52. K. Miernik, W. Dominik, H. Czyrkowski, R. Dąbrowski, A. Fomitchev, M. Golovkov, Z. Janas, W. Kuśmierz, M. Pfützner, A. Rodin, S. Stepantsov, R. Slepniev, G. Ter-Akopian, R. Wolski, Nucl. Instrum. Methods Phys. Res. A **581**, 194 (2007). https://doi.org/10.1016/j.nima.2007.07.076

53. M. Pfützner, M. Karny, L.V. Grigorenko, K. Riisager, Rev. Mod. Phys. **84**, 567 (2012). https://doi.org/10.1103/RevModPhys.84.567

54. I. Tanihata, H. Hamagaki, O. Hashimoto, Y. Shida, N. Yoshikawa, K. Sugimoto, O. Yamakawa, T. Kobayashi, N. Takahashi, Phys. Rev. Lett. **55**, 2676 (1985). https://doi.org/10.1103/PhysRevLett.55.2676

55. P.G. Hansen, B. Jonson, Europhys. Lett. **4**, 409 (1987)

56. N. Keeley, R. Raabe, N. Alamanos, J.L. Sida, Prog. Part. Nucl. Phys. **59**, 579 (2007)

57. L.F. Canto, P.R.S. Gomes, R. Donangelo, J. Lubian, M.S. Hussein, Physics Reports **596**, 1 (2015). https://doi.org/10.1016/j.physrep.2015.08.001

58. J.J. Kolata, A.M. Howard, W. Mittig, T. Ahn, D. Bazin, F.D. Becchetti, S. Beceiro-Novo, Z. Chajecki, M. Febbrarro, A. Fritsch, W.G. Lynch, A. Roberts, A. Shore, R.O. Torres-Isea, Nucl. Instrum. Methods Phys. Res. A **830**, 82 (2016). https://doi.org/10.1016/j.nima.2016.05.036

59. T. Ahn, D.W. Bardayan, D. Bazin, S. Beceiro Novo, F.D. Becchetti, J. Bradt, M. Brodeur, L. Carpenter, Z. Chajecki, M. Cortesi, A. Fritsch, M.R. Hall, O. Hall, L. Jensen, J.J. Kolata, W. Lynch, W. Mittig, P. O'Malley, D. Suzuki, Nucl. Instrum. Methods Phys. Res. A **376**, 321 (2016). https://doi.org/10.1016/j.nimb.2015.12.042

60. O.B. Tarasov, D. Bazin, Nucl. Instrum. Methods Phys. Res. B **266**, 4657 (2008). https://doi.org/10.1016/j.nimb.2008.05.110

61. M. Freer, H. Horiuchi, Y. Kanada-En'yo, D. Lee, U.G. Meißner, Rev. Mod. Phys. **90**, 035004 (2018). https://doi.org/10.1103/RevModPhys.90.035004

62. V.Z. Goldberg, in *Exotic Nuclei and Atomic Masses (ENAM 98)*, vol. 455 (AIP Conference Proceedings, 1998), p. 319. https://doi.org/10.1063/1.57227

63. G.V. Rogachev, V.Z. Goldberg, J.J. Kolata, G. Chubarian, D. Aleksandrov, A. Fomichev, M.S. Golovkov, Y.T. Oganessian, A. Rodin, B. Skorodumov, R.S. Slepnev, G. Ter-Akopian, W.H. Trzaska, R. Wolski, Phys. Rev. C **67**, 041603(R) (2003). https://doi.org/10.1103/PhysRevC.67.041603

64. M. Wiescher, T. Ahn, *Nuclear Particle Correlations and Cluster Physics* (World Scientific, 2017), chap. Clusters in Astrophysics, p. 203. https://doi.org/10.1142/9789813209350_0008

65. R. Coszach, T. Delbar, W. Galster, P. Leleux, I. Licot, E. Liénard, P. Lipnik, C. Michotte, A. Ninane, J. Vervier, C.R. Bain, T. Davinson, R.D. Page, A.C. Shotter, P.J. Woods, D. Baye, F. Binon, P. Descouvemont, P. Duhamel, J. Vanhorenbeeck, M. Vincke, P. Decrock, M. Huyse, P. Van Duppen, G. Vancraeynest, F.C. Barker, Phys. Rev. C **50**, 1695 (1994). https://doi.org/10.1103/PhysRevC.50.1695

66. K. Artemov, O.P. Belyanin, A.L. Vetoshkin, R. Wolski, M.S. Golovkov, V.Z. Goldberg, M. Madeja, V.V. Pankratov, I.N. Serikov, V.A. Timofeev, V.N. Shadrin, Sov. J. Nucl. Phys. **52**, 406 (1990)

67. K. Markenroth, L. Axelsson, S. Baxter, M.J.G. Borge, C. Donzaud, S. Fayans, H.O.U. Fynbo, V.Z. Goldberg, S. Grevy, D. Guillemaud-Mueller, B. Jonson, K.M. Kallman, S. Leenhardt, M. Lewitowicz, T. Lonnroth, P. Manngard, I. Martel, A.C. Mueller, I. Mukha, T. Nilsson, G. Nyman, N.A. Orr, K. Riisager, G.V. Rogachev, M.G. Saint-Laurent, I.N. Serikov, N.B. Shulgina, O. Sorlin, M. Steiner, O. Tengblad, M. Thoennessen, E. Tryggestad, W.H. Trzaska, F. Wenander, J.S. Winfield, R. Wolski, Phys. Rev. C **62**, 034308 (2000). https://doi.org/10.1103/PhysRevC.62.034308

68. D. Suzuki, A. Shore, W. Mittig, J.J. Kolata, D. Bazin, M. Ford, T. Ahn, F.D. Becchetti, S. Beceiro Novo, D. Ben Ali, B. Bucher, J. Browne, X. Fang, M. Febbraro, A. Fritsch, E. Galyaev, A.M. Howard, N. Keeley, W.G. Lynch, M. Ojaruega, A.L. Roberts, X.D. Tang, Phys. Rev. C **87**, 054301 (2013). https://doi.org/10.1103/PhysRevC.87.054301

69. A. Fritsch, S. Beceiro-Novo, D. Suzuki, W. Mittig, J.J. Kolata, T. Ahn, D. Bazin, F.D. Becchetti, B. Bucher, Z. Chajecki, X. Fang, M. Febbraro, A.M. Howard, Y. Kanada-En'yo, W.G. Lynch, A.J. Mitchell, M. Ojaruega, A.M. Rogers, A. Shore, T. Suhara, X.D. Tang, R. Torres-Isea, H. Wang, Phys. Rev. C **93**, 014321 (2016). https://doi.org/10.1103/PhysRevC. 93.014321

70. J. Hooker, G.V. Rogachev, E. Koshchiy, S. Ahn, M. Barbui, V.Z. Goldberg, C. Hunt, H. Jayatissa, E.C. Pollacco, B.T. Roeder, A. Saastamoinen, S. Upadhyayula, Phys. Rev. C **100**, 054618 (2019). https://doi.org/10.1103/PhysRevC.100.054618

71. J. Hooker, G.V. Rogachev, V.Z. Goldberg, E. Koshchiy, B.T. Roeder, H. Jayatissa, C. Hunt, C. Magana, S. Upadhyayula, E. Uberseder, A. Saastamoinen, Phys. Lett. B **769**, 62 (2017). https://doi.org/10.1016/j.physletb.2017.03.025

72. J. Bishop, G.V. Rogachev, E. Aboud, S. Ahn, M. Assunção, M. Barbui, A. Bosh, V. Guimarães, J. Hooker, C. Hunt, D. Jayatissa, E. Koshchiy, S. Lukyanov, R. O'Dwyer, Y. Penionzhkevich, E. Pollacco, C. Pruitt, B.T. Roeder, A. Saastamoinen, L. Sobotka, E. Uberseder, S. Upadhyayula, J. Zamora, J. Phys. Confer. Ser. **1308**, 012006 (2019). https:// doi.org/10.1088/1742-6596/1308/1/012006

73. S. Sambi, R. Raabe, M.J.G. Borge, M. Caamano, S. Damoy, B. Fernandez-Dominguez, F. Flavigny, H. Fynbo, J. Gibelin, G.F. Grinyer, A. Heinz, B. Jonson, M. Khodery, T. Nilsson, R. Orlandi, J. Pancin, D. Perez-Loureiro, G. Randisi, G. Ribeiro, T. Roger, D. Suzuki, O. Tengblad, R. Thies, U. Datta, Eur. Phys. J. A **51** (2015). https://doi.org/10.1140/epja/ i2015-15025-0

74. A.M. Lane, R.G. Thomas, Rev. Mod. Phys. **30**, 257 (1958). https://doi.org/10.1103/ RevModPhys.30.257

75. G.D. Alkhazov, A.V. Dobrovolsky, P. Egelhof, H. Geissel, H. Irnich, A.V. Khanzadeev, G.A. Korolev, A.A. Lobodenko, G. Munzenberg, M. Mutterer, S.R. Neumaier, W. Schwab, D.M. Seliverstov, T. Suzuki, A.A. Vorobyov, Nucl. Phys. A **712**, 269 (2002). https://doi.org/10. 1016/S0375-9474(02)01273-3

76. O.A. Kiselev, F. Aksouh, A. Bleile, O.V. Bochkarev, L.V. Chulkov, D. Cortina-Gil, A.V. Dobrovolsky, P. Egelhof, H. Geissel, M. Hellström, N.B. Isaev, B.G. Komkov, M. Mátos, F.V. Moroz, G. Münzenberg, M. Mutterer, V.A. Mylnikov, S.R. Neumaier, V.N. Pribora, D.M. Seliverstov, L.O. Sergueev, A. Shrivastava, K. Sümmerer, H. Weick, M. Winkler, V.I. Yatsoura, in *The 4th International Conference on Exotic Nuclei and Atomic Masses*, ed. by C.J. Gross, W. Nazarewicz, K.P. Rykaczewski (Springer, Berlin, Heidelberg, 2005), p. 215. https://doi.org/10.1007/3-540-37642-9_62

77. A.V. Dobrovolsky, G.D. Alkhazov, M.N. Andronenko, A. Bauchet, P. Egelhof, S. Fritz, H. Geissel, C. Gross, A.V. Khanzadeev, G.A. Korolev, G. Kraus, A.A. Lobodenko, G. Münzenberg, M. Mutterer, S.R. Neumaier, T. Schäfer, C. Scheidenberger, D.M. Seliverstov, N.A. Timofeev, A.A. Vorobyov, V.I. Yatsoura, Nucl. Phys. A **766**, 1 (2006). https:// doi.org/10.1016/j.nuclphysa.2005.11.016

78. S. Ilieva, F. Aksouh, G.D. Alkhazov, L. Chulkov, A.V. Dobrovolsky, P. Egelhof, H. Geissel, M. Gorska, A. Inglessi, R. Kanungo, A.V. Khanzadeev, O.A. Kiselev, G.A. Korolev, X.C. Le, Y.A. Litvinov, C. Nociforo, D.M. Seliverstov, L.O. Sergeev, H. Simon, V.A. Volkov, A.A. Vorobyov, H. Weick, V.I. Yatsoura, A.A. Zhdanov, Nucl. Phys. A **875**, 8 (2012). https://doi. org/10.1016/j.nuclphysa.2011.11.010

79. M.N. Harakeh, A. van der Woude, *Giant Resonances: Fundamental High Frequency Modes of Nuclear Excitation* (Oxford University Press, Oxford, 2001)

80. G.C. Baldwin, G.S. Klaiber, Phys. Rev. **71**, 3 (1947). https://doi.org/10.1103/PhysRev.71.3

81. M.N. Harakeh, K. van der Borg, T. Ishimatsu, H.P. Morsch, A. van der Woude, F.E. Bertrand, Phys. Rev. Lett. **38**, 676 (1977). https://doi.org/10.1103/PhysRevLett.38.676

82. D.H. Youngblood, C.M. Rozsa, J.M. Moss, D.R. Brown, J.D. Bronson, Phys. Rev. Lett. **39**, 1188 (1977). https://doi.org/10.1103/PhysRevLett.39.1188

83. M. Bender, P.H. Heenen, P.G. Reinhard, Rev. Mod. Phys. **75**, 121 (2003). https://doi.org/10. 1103/RevModPhys.75.121

84. J. Lattimer, Ann. Rev. Nucl. Part. Sci. **71**, 433 (2021). https://doi.org/10.1146/annurev-nucl-102419-124827
85. U. Garg, G. Colò, Prog. Part. Nucl. Phys. **101**, 55 (2018). https://doi.org/10.1016/j.ppnp.2018.03.001
86. E. Khan, J. Margueron, I. Vidaña, Phys. Rev. Lett. **109**, 092501 (2012). https://doi.org/10.1103/PhysRevLett.109.092501
87. J.C. Zamora, T. Aumann, S. Bagchi, S. Bönig, M. Csatlós, I. Dillmann, C. Dimopoulou, P. Egelhof, V. Eremin, T. Furuno, H. Geissel, R. Gernhäuser, M.N. Harakeh, A.L. Hartig, S. Ilieva, N. Kalantar-Nayestanaki, O. Kiselev, H. Kollmus, C. Kozhuharov, A. Krasznahorkay, T. Kröll, M. Kuilman, S. Litvinov, Y.A. Litvinov, M. Mahjour-Shafiei, M. Mutterer, D. Nagae, M.A. Najafi, C. Nociforo, F. Nolden, U. Popp, C. Rigollet, S. Roy, C. Scheidenberger, M. von Schmid, M. Steck, B. Streicher, L. Stuhl, M. Thürauf, T. Uesaka, H. Weick, J.S. Winfield, D. Winters, P.J. Woods, T. Yamaguchi, K. Yue, J. Zenihiro, Phys. Lett. B **763**, 16 (2016). https://doi.org/10.1016/j.physletb.2016.10.015
88. H. Moeini, S. Ilieva, F. Aksouh, K. Boretzky, A. Chatillon, A. Corsi, P. Egelhof, H. Emling, G. Ickert, J. Jourdan, N. Kalantar Nayestanaki, D. Kiselev, O. Kiselev, C. Kozhuharov, T. Le Bleis, X.C. Le, Y.A. Litvinov, K. Mahata, J.P. Meier, F. Nolden, S. Paschalis, U. Popp, H. Simon, M. Steck, T. Stöhlker, H. Weick, D. Werthmüller, A. Zalite, Nucl. Instrum. Methods Phys. Res. A **634**, 77 (2011). https://doi.org/10.1016/j.nima.2011.01.036
89. C. Monrozeau, E. Khan, Y. Blumenfeld, C.E. Demonchy, W. Mittig, P. Roussel-Chomaz, D. Beaumel, M. Caamaño, D. Cortina-Gil, J.P. Ebran, N. Frascaria, U. Garg, M. Gelin, A. Gillibert, D. Gupta, N. Keeley, F. Marechal, A. Obertelli, J.A. Scarpaci, Phys. Rev. Lett. **100**, 042501 (2008). https://doi.org/10.1103/PhysRevLett.100.042501
90. J. Pancin, J. Gibelin, M. Goth, P. Gangnant, J.F. Libin, R. Raabe, T. Roger, P. Roussel-Chomaz, J. Instrum. **7**, P01006 (2012). https://doi.org/10.1088/1748-0221/7/01/P01006
91. S. Bagchi, J. Gibelin, M.N. Harakeh, N. Kalantar-Nayestanaki, N.L. Achouri, H. Akimune, B. Bastin, K. Boretzky, H. Bouzomita, M. Caamaño, L. Càceres, S. Damoy, F. Delaunay, B. Fernández-Domínguez, M. Fujiwara, U. Garg, G.F. Grinyer, O. Kamalou, E. Khan, A. Krasznahorkay, G. Lhoutellier, J.F. Libin, S. Lukyanov, K. Mazurek, M.A. Najafi, J. Pancin, Y. Penionzhkevich, L. Perrot, R. Raabe, C. Rigollet, T. Roger, S. Sambi, H. Savajols, M. Senoville, C. Stodel, L. Suen, J.C. Thomas, M. Vandebrouck, J. Van de Walle, Phys. Lett. B **751**, 371 (2015). https://doi.org/10.1016/j.physletb.2015.10.060
92. M. Vandebrouck, J. Gibelin, E. Khan, N.L. Achouri, H. Baba, D. Beaumel, Y. Blumenfeld, M. Caamaño, L. Càceres, G. Colò, F. Delaunay, B. Fernandez-Dominguez, U. Garg, G.F. Grinyer, M.N. Harakeh, N. Kalantar-Nayestanaki, N. Keeley, W. Mittig, J. Pancin, R. Raabe, T. Roger, P. Roussel-Chomaz, H. Savajols, O. Sorlin, C. Stodel, D. Suzuki, J.C. Thomas, Phys. Rev. C **92**, 024316 (2015). https://doi.org/10.1103/PhysRevC.92.024316
93. M. Vandebrouck, J. Gibelin, E. Khan, N.L. Achouri, H. Baba, D. Beaumel, Y. Blumenfeld, M. Caamaño, L. Càceres, G. Colò, F. Delaunay, B. Fernandez-Dominguez, U. Garg, G.F. Grinyer, M.N. Harakeh, N. Kalantar-Nayestanaki, N. Keeley, W. Mittig, J. Pancin, R. Raabe, T. Roger, P. Roussel-Chomaz, H. Savajols, O. Sorlin, C. Stodel, D. Suzuki, J.C. Thomas, Phys. Rev. Lett. **113**, 032504 (2014). https://doi.org/10.1103/PhysRevLett.113.032504
94. S. Ota, H. Tokieda, C. Iwamoto, M. Dozono, U. Garg, S. Hayakawa, K. Kawata, N. Kitamura, M. Kobayashi, S. Masuoka, S. Michimasa, A. Obertelli, D. Suzuki, R. Yokoyama, J. Zenihiro, H. Baba, O. Beliuskina, P. Chebotaryov, P. Egelhof, T. Harada, M.N. Harakeh, K. Howard, N. Imai, M. Itoh, Y. Kiyokawa, C.S. Lee, Y. Maeda, Y. Matsuda, E. Milman, V. Panin, H. Sakaguchi, P. Schrock, S. Shimoura, L. Stuhl, M. Takaki, K. Taniue, S. Terashima, T. Uesaka, Y.N. Watanabe, K. Wimmer, K. Yako, Y. Yamaguchi, Z. Yang, , K. Yoneda, CNS Annual Report 2016 **96**, 19 (2018)
95. C. Iwamoto, S. Ota, S. Hayakawa, K. Kawata, Y. Mizoi, R. Kojima, H. Tokieda, N. Zhang, O. Beliuskina, M. Dozono, T. Harada, N. Imai, T. Isobe, N. Kitamura, S. Masuoka, S. Michimasa, T. Murakami, D. Suzuki, R. Tsunoda, J. Zenihiro, E. Takada, CNS Annual Report 2019 **99**, 17 (2021)

96. K.L. Jones, A.S. Adekola, D.W. Bardayan, J.C. Blackmon, K.Y. Chae, K.A. Chipps, J.A. Cizewski, L. Erikson, C. Harlin, R. Hatarik, R. Kapler, R.L. Kozub, J.F. Liang, R. Livesay, Z. Ma, B.H. Moazen, C.D. Nesaraja, F.M. Nunes, S.D. Pain, N.P. Patterson, D. Shapira, J.F. Shriner, M.S. Smith, T.P. Swan, J.S. Thomas, Nature (London) **465**, 454 (2010). https://doi.org/10.1038/nature09048

97. J. Diriken, N. Patronis, A.N. Andreyev, S. Antalic, V. Bildstein, A. Blazhev, I.G. Darby, H. De Witte, J. Eberth, J. Elseviers, V.N. Fedosseev, F. Flavigny, C. Fransen, G. Georgiev, R. Gernhauser, H. Hess, M. Huyse, J. Jolie, T. Kroell, R. Kruecken, R. Lutter, B.A. Marsh, T. Mertzimekis, D. Muecher, F. Nowacki, R. Orlandi, A. Pakou, R. Raabe, G. Randisi, P. Reiter, T. Roger, M. Seidlitz, M. Seliverstov, K. Sieja, C. Sotty, H. Tornqvist, J. Van De Walle, P. Van Duppen, D. Voulot, N. Warr, F. Wenander, K. Wimmer, Phys. Lett. B **736**, 533 (2014). https://doi.org/10.1016/j.physletb.2014.08.004

98. A.H. Wuosmaa, J.P. Schiffer, B.B. Back, C.J. Lister, K.E. Rehm, Nucl. Instrum. Methods Phys. Res. A **580**, 1290 (2007). https://doi.org/10.1016/j.nima.2007.07.029

99. M. Caamaño, D. Cortina-Gil, W. Mittig, H. Savajols, M. Chartier, C.E. Demonchy, B. Fernandez, M.B. Gomez Hornillos, A. Gillibert, B. Jurado, O. Kiselev, R. Lemmon, A. Obertelli, F. Rejmund, M. Rejmund, P. Roussel-Chomaz, R. Wolski, Phys. Rev. Lett. **99**, 062502 (2007). https://doi.org/10.1103/PhysRevLett.99.062502

100. T. Roger, H. Savajols, I. Tanihata, W. Mittig, M. Alcorta, D. Bandyopadhyay, R. Bieri, L. Buchmann, M. Caamaño, B. Davids, N. Galinski, A. Gallant, D. Howell, R. Kanungo, W. Mills, S. Mythili, M. Notani, R. Openshaw, E. Padilla-Rodal, P. Roussel-Chomaz, G. Ruprecht, G. Savard, G. Sheffer, A.C. Shotter, M. Trinczek, P. Walden, Phys. Rev. C **79**, 031603(R) (2009). https://doi.org/10.1103/PhysRevC.79.031603

101. P. Konczykowski, B. Fernández-Dominguez, H. Alvarez-Pol, M. Caamaño, G.F. Grinyer, A.T. Laffoley, B. Mauss, J. Pancin, D. Pérez-Loureiro, T. Roger, Nucl. Instrum. Methods Phys. Res. A **927**, 125 (2019). https://doi.org/10.1016/j.nima.2019.02.013

102. D. Perez-Loureiro. Private communication

103. C. Rodríguez-Tajes, F. Farget, L. Acosta, H. Alvarez-Pol, M. Babo, F. Boulay, M. Caamaño, S. Damoy, B. Fernàndez-Domìnguez, D. Galaviz, G. Grinyer, J. Grinyer, M.N. Harakeh, P. Konczykowski, I. Martel, J. Pancin, G. Randisi, F. Renzi, T. Roger, A.M. Sànchez-Benìtez, P. Teubig, M. Vandebrouck, Nucl. Phys. A **958**, 246 (2017). https://doi.org/10.1016/j.nuclphysa.2016.12.003

104. C. Rodríguez-Tajes, J. Pancin, S. Damoy, T. Roger, M. Babo, M. Caamaño, F. Farget, G. Grinyer, B. Jacquot, D. Pérez-Loureiro, D. Ramos, D. Suzuki, Nucl. Instrum. Methods Phys. Res. A **768**, 179 (2014). https://doi.org/10.1016/j.nima.2014.08.046

105. S. Goriely, Eur. Phys. J. A **51**, 22 (2015). https://doi.org/10.1140/epja/i2015-15022-3

Gamma Ray Emission Imaging in the Medical and Nuclear Safeguards Fields

7

Peter Dendooven and Tatiana A. Bubba

Abstract

Gamma rays emitted from within an object can reveal information about that object in a non-destructive way, i.e. without physically opening the object and looking inside. This makes gamma ray emission imaging very useful in widely varying applications. In these notes, we highlight its application to the medical field, where we discuss molecular imaging in nuclear medicine and in vivo dose delivery verification in particle beam radiotherapy, and nuclear safeguards field, where imaging of spent nuclear fuel assemblies is part of monitoring the non-proliferation of nuclear weapons. The purpose and basic principles of gamma ray emission imaging are discussed as the foundation to look in more detail into the essential instrument design considerations and the iterative image reconstruction procedures. These notes are not intended to be a comprehensive review; their purpose is to introduce gamma ray emission imaging to those that are new to this technique. The examples of implementation that are presented were thus chosen in order to introduce the reader to a fairly wide range of applications and practical implementations.

P. Dendooven (✉)
Particle Therapy Research Center (PARTREC), Department of Radiation Oncology, University Medical Center Groningen, Groningen, The Netherlands

Helsinki Institute of Physics, University of Helsinki, Helsinki, Finland
e-mail: p.g.dendooven@umcg.nl; peter.dendooven@helsinki.fi

T. A. Bubba
Department of Mathematical Sciences, University of Bath, Bath, UK
e-mail: tab73@bath.ac.uk

© The Author(s), under exclusive license to Springer Nature Switzerland AG 2022
S. M. Lenzi, D. Cortina-Gil (eds.), *The Euroschool on Exotic Beams, Vol. VI*,
Lecture Notes in Physics 1005, https://doi.org/10.1007/978-3-031-10751-1_7

7.1   Introduction

Being able to look inside an object without physically opening it is very appealing
in many situations. For instance, being able to diagnose a patient without surgical
intervention has clear advantages, in terms of patient risk and comfort and potentially also cost. Some objects, such as the spent nuclear fuel assemblies discussed in
these notes, are so radioactive that getting close to open them and have a look inside
is a major and costly undertaking, requiring hot cells and remote handling. The
penetrating nature of high-energy photons allows them to escape from objects and
be measured by a suitable camera outside the object. Imaging these gamma rays,
i.e. determining their place of origin inside the object, reveals information about
the object. In some applications, the gamma rays need to be "introduced" into the
object for the purpose of imaging; in other applications the gamma rays result from
the processes which the object undergoes. Looking at the applications discussed in
these notes, imaging in nuclear medicine belongs to the former category, whereas
in vivo dose delivery verification in particle beam radiotherapy and the imaging of
spent nuclear fuel assemblies belongs to the latter category.

We discuss the imaging of gamma rays, being defined as photons emitted during
a transition between nuclear energy levels and positron annihilation photons. For
the sake of brevity, we adopt a common definition with "gamma ray" referring to
high-energy photons no matter what their origin.

These notes are meant as an introduction to the technique of gamma ray emission
imaging and its use in the medical and nuclear safeguards fields. They are not
intended to be a comprehensive review. References have largely been chosen based
on their relevance to this purpose. These notes contain three major sections. First,
the fields considered and the purpose of gamma ray imaging in them are explained.
Next, the basic principles of gamma ray emission imaging are discussed as the
foundation to look in more detail into the essential instrument design considerations
and the iterative image reconstruction procedure. Finally, examples of the practical
implementation in each field are discussed in some detail.

When embarking on the development of gamma ray emission imaging for a
certain application, remember that in the end, it is not about the camera, nor about
the image; it is about what the image can do for you.

7.2   Applications of Gamma Ray Emission Imaging

We introduce the fields of application of gamma ray emission imaging being considered here: nuclear medicine, particle beam radiotherapy and nuclear safeguards.
Then, we explain the purpose of gamma ray emission imaging in these fields.
Gamma ray imaging is very much at the origin and the basis of the field of
nuclear medicine. In particle beam radiotherapy, gamma ray emission imaging is
not a widely established technique, with various practical implementations under
investigation and development. In nuclear safeguards, gamma ray emission imaging

was recently introduced and greatly improves the sensitivity of detecting the undeclared diversion of spent nuclear fuel material.

7.2.1 Nuclear Medicine

7.2.1.1 The Dual Use of Radioactive Isotopes

Nuclear medicine is the field that makes use of radioactive isotopes, more specifically of the radiation emitted in radioactive decay, for two purposes: therapy and diagnostics. The therapeutic use is based on the ionizing power of the radiation, most often with the purpose of killing tumour cells, and is thus a branch of radiotherapy. Radioactive isotopes are introduced into the patient in the form of a tracer (see Sect. 7.2.1.2) or sealed radioactive sources, resulting in an internal irradiation, so-called brachytherapy. Radioactive isotopes are used for diagnostic purposes by imaging the radiation emitted in their radioactive decay. Some isotopes have dual and simultaneous use (both therapy and diagnostics), or a therapeutic and diagnostic isotope can be used simultaneously, resulting in what is called a theranostics procedure.

7.2.1.2 The Tracer Principle

Nuclear medicine (with the exception of brachytherapy using sealed radioactive sources) is based on the tracer principle, discovered by George de Hevesy in the early 1900s, a discovery awarded with the 1943 Nobel Prize in Chemistry [1]. The tracer principle states that the chemical behaviour of a molecule containing a radioactive isotope of a certain chemical element is identical to that of the molecule containing a stable isotope of that same element. This reflects the fact that chemical behaviour, and thus also biological, physiological and medical behaviour, depends on the chemical element to which the isotope belongs, not on the specific isotope. In other words, chemistry is essentially determined by the atomic electrons, not by the atomic nucleus. A molecule containing a radioactive isotope is said to be labelled with that isotope.

The detection of radiation emitted in the radioactive decay allows to track the transport and distribution of the labelled molecule, hence the name "tracer". The tracer principle is used in nuclear medicine to study the behaviour in the human body of a molecule with a biologically/physiologically relevant function by labelling this molecule with a radioactive isotope and imaging its distribution via the radiation emitted in radioactive decay. Such a labelled molecule is called a radiopharmaceutical or radiotracer. The imaging procedure thus starts with the production of a radioactive isotope and the synthesis of a radiotracer. The radiotracer is injected into the patient, after which the radiotracer predominantly travels to locations where a certain biological/physiological function is active. Imaging the radiation emitted thus enables to image a certain function; imaging based on the tracer principle is therefore called functional or molecular imaging. Molecular imaging is mostly used to diagnose diseases, but also used in fundamental research into the functioning of the human body.

7.2.2 Particle Beam Radiotherapy

7.2.2.1 The Rationale of Particle Beam Radiotherapy

Radiotherapy, the therapeutic use of ionizing radiation, started within a year of the discovery of X-rays by Wilhelm Röntgen in 1895 [2]. During the First World War, Marie Sklodowska-Curie used ionizing radiation to sterilize wounds. Until shortly after the second world war, radiotherapy was performed using X-rays (photon energies up to the order of 100 keV), with some attempts to use neutrons. Around that time, MeV photons became available by the development of linear electron accelerators (generating high-energy bremsstrahlung photons) and ^{60}Co irradiation sources.

In 1946, Robert Wilson published a seminal paper pointing out the advantage of radiotherapy using energetic proton beams instead of photons [3]. The difference stems from the fundamentally different way in which photons and protons interact with matter. Photons interact essentially via all-or-nothing processes; if a photon interacts with matter, it disappears. We thus consider that Compton scattered photons (having a lower energy than the original photon) no longer belong to the beam of photons. Protons mainly interact with electrons in matter via the electromagnetic interaction. The average energy loss in the interaction with one electron is very small; hence a proton gradually loses energy until it comes to a standstill. When travelling in matter, a photon beam does not stop, but gradually weakens (i.e. its intensity decreases), whereas a proton beam stops at a certain point. The location at which the proton beam stops can be controlled by selecting the proton energy. An additional benefit of protons is that their stopping power (the energy loss per distance travelled in matter) increases as their energy decreases. This leads to a peak in radiation dose (energy deposited per kg of matter) at the end of the proton range (the distance at which the proton stops), the Bragg peak. Figure 7.1 illustrates the importance of the fundamentally different behaviour of photons and protons for external beam radiotherapy. Photons exhibit a skin-sparing effect (the dose is lower at the entrance in the human body), reaching a maximum dose at a shallow depth after which the dose decreases gradually. The proton dose is fairly constant over a sizeable fraction of its range and increases sharply before the proton stops, resulting in the Bragg peak. The region in the body to be treated, most often a tumour, is typically larger than the width of the Bragg peak. A uniform dose over a large volume can be achieved by a spread-out Bragg peak: adding a set of proton beams of different energy, and with well-chosen intensities, results in a flat dose distribution over an extended region, with a sharp fall off at the distal edge (here, distal means furthest in the direction of the beam). This however leads to a smaller ratio of the dose in the target area to the dose before the target area. The advantage of protons compared to photons is clear: because a proton beam deposits less dose before the target area and no dose beyond it, a proton irradiation enables to deposit the same dose in, e.g. a tumour with less dose deposited in healthy tissue and organs. This is the main rationale of proton beam radiotherapy: the reduction of side effects of the treatment and a better quality of life for the patient.

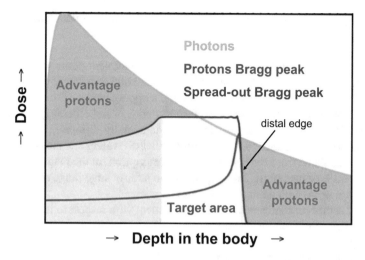

Fig. 7.1 Illustration of the dose (energy deposited per kg) versus depth in the human body of high-energy photons and protons. Modified from [4] with permission from J.A. Langendijk

The same arguments made above obviously apply to heavier ion beams. Carbon ions have been widely used for a long time, helium ions are experiencing a revival, and oxygen beams are being considered. The Bragg peak for heavier ions is sharper, resulting in a larger dose ratio between the Bragg peak and the entrance region. Carbon and heavier ions undergo a fair amount of fragmentation while being stopped, resulting in some dose beyond the Bragg peak, where the fragmentation products stop.

7.2.2.2 The Need of In Vivo Range Verification

Section 7.2.2.1 explains why there is a global, overall, reduction in dose to healthy tissue when using particles instead of photons, leading to a benefit for the patient. However, concerning dose to healthy tissue in particle therapy, we also need to consider the edges of the high-dose region, this region being most often the tumour to be killed. The Bragg peak enables a very precise deposition of a high dose in a particular region and thus a very accurate irradiation. However, if the dose is not delivered according to plan, the consequences can be serious: a maximum dose in healthy tissue, causing severe damage and side effects, and/or part of a tumour not being irradiated, resulting in the patient not being adequately treated. Such scenarios are avoided by very careful, i.e. conservative, treatment planning, ensuring that potential deviations between an actually deposited dose distribution and the planned dose distribution do not lead to an unacceptable treatment outcome. Presently, robust treatment planning is used to ensure a safe dose delivery [5, 6]. The main causes of uncertainty in the dose delivery to be considered during treatment planning are the following:

- Particle range calculation: for the purpose of treatment planning, a computed tomography (CT) image[1] of the patient is taken and translated, voxel-by-voxel, into the relative stopping power (RSP), the stopping power of the tissue relative to that of water (water is the reference material in radiotherapy). Because the interaction of X-rays and particles is fundamentally different, this translation cannot be perfect; there is an uncertainty in the RSP values and thus in the particle range calculated from the RSP. A typical prescription for the range uncertainty is 1 mm + 3% of the range. This, e.g. results in a fairly large range uncertainty of 4 mm for a beam with a range of 100 mm.
- Patient setup: presently, positioning of a patient with respect to the proton beam has an error that is at most about 3 mm.
- Changes in patient anatomy: treatment planning is typically done one to a few weeks before the treatment starts. During this time, as well as during the 5 to 8 weeks of daily irradiations, the patient anatomy will change. This will change the location of the Bragg peak inside the patient. More and more, image-guided radiotherapy is being introduced in order to see such changes in the patient and, if needed, modify the treatment plan accordingly.

In order to detect deviations in dose deposition between what is actually delivered and what is planned, the dose distribution that is actually delivered to an individual patient during each single irradiation has to be determined: this means that there is a need for in vivo dose delivery verification. In practice, if the distal edge of the dose distribution (see Fig. 7.1) is verified to be in the correct place, one has good confidence that the full dose distribution is delivered according to plan. Also, some methods for in vivo dose delivery verification have the best precision at the distal edge. Because of these two reasons, in practice, in vivo *dose* delivery verification actually means in vivo *range* verification, the term we use in these notes.

7.2.2.3 In Vivo Range Verification by Gamma Emission Imaging

As the particles stop in the patient, the particles themselves cannot be measured for in vivo range verification; one is forced to use secondary radiation which is generated by the particles in the patient. Various types of secondary radiation can be used:

- Prompt gamma rays: emitted within less than about 1 ns in nuclear interactions between the beam particles and the atomic nuclei in tissue.

---

[1] In CT imaging, an X-ray source and an X-ray detector rotate around the patient, and a 3D X-ray image of the patient is obtained by measuring X-ray attenuation by different tissues inside the body.

- Positron annihilation photons: positron emitting radioactive nuclides are produced in nuclear interactions between the beam particles and the atomic nuclei
  in tissue. The positron emitted during radioactive decay stops in the patient and
  then annihilates with an electron, with two back-to-back 511 keV photons being
  emitted. The signal is thus on average delayed by the radioactive lifetime.
- Light charged particles, mostly protons: these are emitted by carbon and heavier
  ion beams and can have sufficient energy to leave the patient.
- Acoustic waves: the high dose delivered in the Bragg peak during a short beam
  pulse causes a spike in temperature, resulting in an acoustic wave which can
  be picked up by an acoustic transducer. This resembles traditional ultrasound
  imaging.

For more details on in vivo range verification, we refer to the review papers
of Knopf and Lomax [7] and Parodi and Polf [8]. Examples of emission imaging
gamma rays for in vivo range verification are discussed in Sect. 7.4.2. We finish this
section by illustrating in Fig. 7.2 the relation between the creation of gamma rays
and dose deposition.

7.2.3 Nuclear Safeguards

The Treaty on the Non-Proliferation of Nuclear Weapons (NPT) entered into force
on 5 March 1970 [11]. The aims of the NPT are the prevention of the spread of
nuclear weapons and weapons technology, furthering the goal of disarmament and
the promotion of the peaceful uses of nuclear energy. On 11 May 1995, the NTP
was extended indefinitely. There are 191 State parties to the treaty, including five
nuclear-weapon States, defined as those that manufactured and exploded a nuclear
weapon before 1 January 1967. These nuclear weapon States are committed not
to assist, encourage or induce in any way a non-nuclear weapon State party to
acquire nuclear weapons. The non-nuclear weapon States that are party to the NTP
committed themselves not to acquire nuclear weapons.

The International Atomic Energy Agency (IAEA) is not a party to the NTP.
However, the NTP establishes a safeguards system under the responsibility of the
IAEA, entrusting it with key verification responsibilities. "(Nuclear) Safeguards"
refers to the system that has been put in place to ensure that non-nuclear weapon
State parties to the NTP honour their international legal obligations. The objective
of IAEA Safeguards is thus "to deter the spread of nuclear weapons by the early
detection of the misuse of nuclear material or technology. This provides credible
assurances that States are honouring their legal obligations that nuclear material
is being used only for peaceful purposes" [12]. Safeguards uses a set of technical
measures to verify, independently from the NTP parties, nuclear materials and
activities [13]. Safeguards activities of the IAEA are enabled via a mandatory
comprehensive safeguards agreement between the IAEA and each non-nuclear
weapon State party to the NTP. At the end of 2020, 10 out of 186 non-nuclear

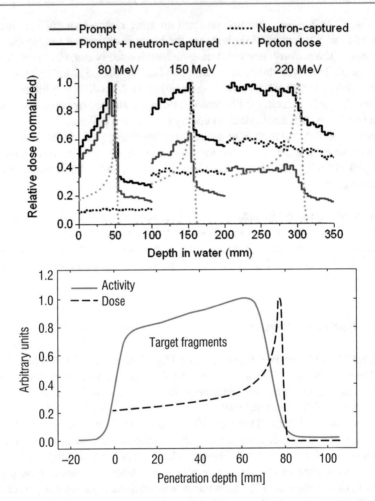

Fig. 7.2 Top: Gamma ray distributions vs. depth when stopping 80, 150 and 220 MeV proton beams in water. Reprinted from [9] with permission from John Wiley and Sons. The distributions were measured by moving a gamma ray detector behind a linear array of slit collimators located perpendicularly to the proton beam. The distribution of gamma rays resulting from neutron capture are indicated. The proton dose distributions are shown for comparison. A good correlation between the Bragg peak and the edge of the prompt gamma distribution is seen. The contrast of the edge of the prompt gamma distribution substantially decreases with increasing proton energy. Bottom: measured depth profile of positron emission activity from target fragmentation (solid line) for 110 MeV protons stopping in a PMMA target. Reprinted from [10]. The dotted line shows the corresponding calculated dose distributions

weapon State parties had yet to bring into force such a comprehensive safeguards agreement.

In-field inspections at nuclear facilities or locations outside facilities are an important activity to ensure that the nuclear material which the State declared to the

IAEA, and thus being under safeguards control, remains part of peaceful activities or is otherwise adequately accounted for. This nuclear material accountancy compares information on the nuclear material available at the facility with what was declared and, essentially, confirms that the material is present at the facility.

An especially interesting nuclide for safeguards is ^{239}Pu, due to its potential use as fissile material in nuclear weapons. ^{239}Pu is produced via the capture of a neutron by ^{238}U, leading to the formation of ^{239}U, which is followed by beta decay to ^{239}Np that beta decays to ^{239}Pu. Nuclear fuel used in nuclear power plants contains a large amount of ^{238}U. Spent nuclear fuel (SNF), i.e. fuel removed from the nuclear reactor when it is no longer efficient in contributing to the power production, therefore contains a fair amount of ^{239}Pu. In order to rule out the diversion of SNF from legitimate declared uses, for example, the extraction of ^{239}Pu, safeguards inspections put special emphasis on verifying that all SNF is accounted for.

For use in a nuclear reactor, nuclear fuel is arranged in so-called fuel assemblies, containing, depending on reactor type, from about 50 to 300 fuel rods. Each rod, 2–4 m long, is made up of fuel pellets (typically 1 cm diameter and 1 cm long) inside a zirconium alloy cladding tube. As an example, Fig. 7.3 shows a model ATRIUM 10 fuel assembly. After their useful time in the nuclear reactor, fuel assemblies are not disassembled. The complete spent fuel assemblies (SFAs) are typically stored in water pools until the fuel is reprocessed or moved to some form of final disposal. Across the world, geological repositories are planned and constructed for the final disposal of SFAs. Finland is scheduled to be the first country to start with the final disposal of SNF, 400 m underground at the Onkalo facility in Eurajoki, Finland [14]. The time between removal from the reactor and disposal in a geological repository is several decades. During this time, safeguards inspections need to be able to verify the declaration of completeness of the SFAs stored under water.

IAEA Safeguards distinguishes between a gross defect, i.e. a complete fuel assembly missing, and partial defects, i.e. part of a fuel assembly missing. It only seems logical to measure radiation emitted by the SFAs, so-called passive measurements, for verification purposes. The traditional instruments passively measuring radiation in order to detect partial defects are the Improved and the Digital Cerenkov Viewing Devices (ICVD, DCVD) and the Fork detector (FDET). The former are optimized to detect the ultraviolet Cerenkov radiation in the water surrounding spent fuel which is emitted by high-energy recoil electrons created in the interaction of gamma rays emitted in the radioactive decay of fission fragments inside the fuel. The latter contains detectors measuring the total gamma and neutron intensities emitted by the SNF assembly. The CVD devices are looking at the top of a SFA in its storage position from outside the storage pool; the FDET is installed inside the storage pool with SFAs moved very close to the FDET for measurement. Both the CVD and FDET devices perform crude, global measurements, basically confirming that the object under study is highly radioactive. They typically can only detect half or more of the SNF missing from an assembly.

With the goal of more precise partial defect verification, the IAEA, in collaboration with some of its Member States, has developed since the 1980s Passive Gamma ray Emission Tomography (PGET) for imaging SFAs [15–18]. This effort

Fig. 7.3 Close-up picture of a model ATRIUM 10 nuclear fuel assembly. For clarity, some fuel rods are removed. Three partial fuel rods as well as the 3x3 water channel can be seen. PGET images from an ATRIUM 10 fuel assembly are shown in Fig. 7.16. Reprinted from www.framatome.com with permission from Framatome

has culminated at the end of 2017 in the approval by the IAEA to use the PGET device for inspections of SFAs [19, 20].

7.3 Principles of Gamma Ray Emission Imaging

The most essential requirement for gamma ray emission imaging is to know the direction in which the gamma rays travel. We discuss different gamma camera principles to obtain this information and deduce the demands this places on the camera detectors. In a tomographic imaging procedure, 1D or 2D gamma ray emission projections are measured at different angles around the object. In the image reconstruction process, 1D projections vs. angle are transformed into a 2D image, while 2D projections vs. angle are transformed into a 3D image. We limit the discussion to iterative image reconstruction methods as these are by far most common in the gamma ray emission imaging applications discussed in these notes.

As it will become evident in this section, there are two different types of gamma ray emission as far as imaging methods are concerned: gamma rays emitted in a transition between nuclear energy levels and positron annihilation photons. The main difference between these two types is that gamma rays from nuclear transitions are emitted "alone", whereas positron annihilation photons are emitted in pairs travelling in opposite directions.

In order to be useful for imaging, the energy of a gamma ray photon needs to be sufficiently high such that the probability of exiting the object being imaged is large enough. On the detector side, design and performance depends on the gamma ray energy. In nuclear medicine, the gamma ray energy can, to some extent, be chosen because different isotopes emit different energy gamma rays. There are however a number of other factors that determine the suitability of a certain isotope, e.g. half-life, ease of production, chemical properties enabling the synthesis of radiotracers and cost. In the other applications discussed in these notes, there is no or a very limited choice of isotope and thus gamma ray energy. Positron annihilation photons, obviously, always have an energy of 511 keV.

7.3.1   Basic Principles

Gamma ray emission imaging looks at gamma rays emitted from within an object to provide an image of the location of emission of the gamma rays inside the object. The general idea is illustrated in Fig. 7.4. The imaging device is most often called a "camera" or "scanner".

The most essential requirement to obtain imaging information is to know the direction of propagation of the gamma rays. The gamma rays used in emission imaging are emitted isotropically, and thus the camera has to provide information on the gamma ray direction. There are two approaches to this: (1) the camera uses a collimator to select a certain direction or directions; (2) the camera measures directional information for each gamma ray detected. In the former approach, often referred to as "physical collimation", various collimator types are used depending on the application and the specific imaging situation. In contrast, the latter approach is often referred to as "electronic collimation": knowledge on the gamma ray direction

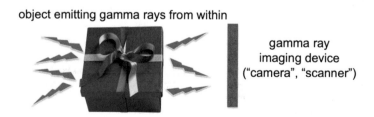

Fig. 7.4 Illustration of the general idea of gamma emission imaging. Gamma rays emitted from within an object are detected by an imaging device outside of the object, providing information on the spatial distribution of the emission of the gamma rays inside the object

is gained from measuring certain details of the interaction of individual gamma rays in the camera. The most common examples of electronic collimation, Compton imaging and positron emission imaging, are discussed in these notes.

Knowing only the direction of a gamma ray does not provide information on the location of the source of the gamma ray along its direction. If a 3D object is imaged using gamma rays travelling, from each point of the object, in only one direction (which is not necessarily the same direction for each point), a 2D projection (often referred to as a radiograph) will result. Obtaining a 3D image requires that from each point in the object gamma rays under multiple directions are seen by the camera. The practical implementation of this principle is to have the camera look at the object from different directions, ideally distributed over 360 degrees: such a procedure is called tomography. Tomographic imaging using a collimated gamma camera is usually referred to as single-photon emission computed tomography (SPECT). The tomographic versions of Compton imaging and positron emission imaging are called Compton scatter tomography (CST) and positron emission tomography (PET).

7.3.2   Essential Design Considerations for Gamma Ray Imaging Instruments

7.3.2.1 Gamma Cameras Based on Physical Collimation
The principle of a gamma camera using physical collimation is illustrated in Fig. 7.5 for a parallel-hole and a pinhole collimator. A parallel-hole collimator is what it

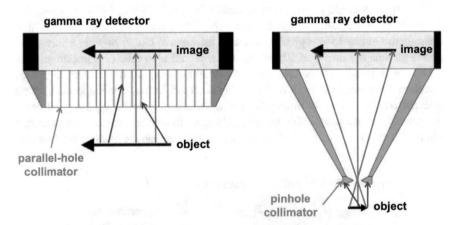

Fig. 7.5 The principle of a gamma camera using physical collimation; on the left a parallel-hole collimator, on the right a pinhole collimator. A gamma ray detector is placed behind the collimator to detect the gamma rays passing through the collimator. The object and its image in the gamma ray detector are indicated by the black arrows. The red arrows represent gamma rays blocked by the collimator, while the green arrows represent gamma rays passing through the collimator and detected by the detector. The geometry of the systems results in a mirrored and magnified image with the pinhole collimator and a non-mirrored and real-size image with the parallel-hole collimator. Figure adapted from [21]

says: a tight array of long and narrow holes with the purpose of allowing only gamma rays travelling along the length of the holes to pass through the collimator and hit the gamma ray detector behind it. A pinhole collimator is basically a gamma ray version of the well-known camera obscura used with visible light. Figure 7.5 shows that the gamma ray detector needs to determine at which location a gamma ray enters it in order to create an image; a position-sensitive gamma ray detector is thus needed. In most applications, the gamma camera is position-sensitive across two dimensions, making a 2D image of a 3D object. Position sensitivity in three dimensions, so also in the detector depth dimension, is in practice usually not needed because the thickness of the detector needed to provide a good detection efficiency is much smaller than the distance of the detector to the gamma ray origin inside the object.

7.3.2.2 Compton Cameras

Compton scattering is the process in which a photon scatters off an electron, creating a scattered photon of lower energy and a scattered electron; see Fig. 7.6. It is one of the main interaction mechanisms of gamma rays with matter and thus of the detection of gamma rays. Compton imaging makes use of the kinematics of Compton scattering: the relationships between incoming gamma ray energy and the energy and angles of the scattered gamma ray and electron. Generally speaking, the energy of the scattered electron is deposited very close to the point of interaction, while the scattered gamma ray interacts at another location in the detector(s) of the camera or leaves the camera without interaction. In the simplest implementation of a Compton camera, the location of two interactions and the energy deposited at those interaction points is measured; see Fig. 7.7. Obtaining the correct Compton cone, and thus valid imaging information, requires knowledge of two out of the three energies involved (E_γ, E'_γ, E_e). Thus, if the incoming gamma ray photon energy is not a priori known, the absorber detector must measure the full energy of the scattered gamma ray photon. This presents a difficulty as it is not possible to know for sure whether this is the case. A more complicated Compton camera design in which three interactions are measured does not require the energy of incoming or scattered photon to be known: the Compton cone can be reproduced from the

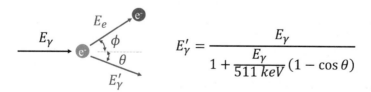

$$E'_\gamma = \frac{E_\gamma}{1 + \dfrac{E_\gamma}{511\ keV}(1 - \cos\theta)}$$

Fig. 7.6 In Compton scattering, an incoming gamma ray photon with energy E_γ scatters off an electron, resulting in a scattered gamma ray photon and electron with energies E'_γ and E_e and scattering angles θ and ϕ. The relationship between incoming and scattered gamma ray energies and gamma scattering angle, which follows from conservation of energy and momentum, is shown

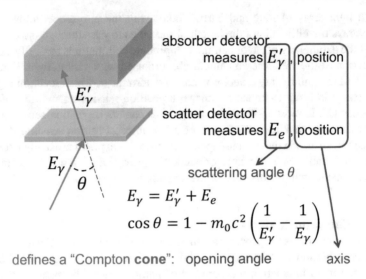

absorber detector
measures E'_γ, position

scatter detector
measures E_e, position

scattering angle θ

$$E_\gamma = E'_\gamma + E_e$$

$$\cos\theta = 1 - m_0 c^2 \left(\frac{1}{E'_\gamma} - \frac{1}{E_\gamma}\right)$$

defines a "Compton **cone**":   opening angle   axis

Fig. 7.7 Illustration of a Compton camera measuring two interaction points. Shown is a situation in which a scatter detector measures the location and scattered electron energy of the first, Compton scattering, interaction and an absorber detector measures the interaction location and the full energy of the scattered photon. The line connecting the two interaction locations defines the axis of the Compton cone, which is the surface from which the gamma ray was emitted. The opening angle of the Compton cone, i.e. the scattering angle of the first interaction, can be calculated using the equation shown. If the absorber detector does not measure the full energy of the scattered photon, the incoming photon energy needs to be known to determine the Compton cone. This is not required in a three-interaction Compton event

energy of two scattered electrons and the location of the three interactions (see, e.g. [22]). If the track of the scattered electron(s) is measured, the origin of the gamma ray emission can be limited to an angular section of the Compton cone, a so-called Compton arc (see Fig. 7.8).

General conclusions from the various Compton camera implementations discussed above are that a Compton camera needs position-sensitive detectors and that good energy resolution is needed to obtain good imaging information. A general issue with Compton cameras is that the time resolution of the detectors is usually too poor to unambiguously determine the time sequence of interactions. The resulting ambiguity leads to wrong imaging information for part of the detected events and thus degraded image quality. In most practical situations however, an "educated guess" enables to establish the correct time sequence in a majority of detected events (see, e.g. the gamma ray tracking algorithm in [24]).

7.3.2.3 Positron Emission Imaging

When a positron emitting radioactive nucleus decays inside an object, the emitted positron will readily annihilate with an electron in the object. Annihilation is most probable after the positron has stopped (i.e. thermalized), typically within less than

Fig. 7.8 Illustration of the
electron tracking Compton
camera. Reprinted from [23].
The track of the scattered
electron is measured by a set
of scatter detectors. The
electron track information
narrows the origin of the
gamma rays to a Compton
arc. Note that in the absence
of depth information, the
Compton arc represents an
angular section of the
Compton cone surface

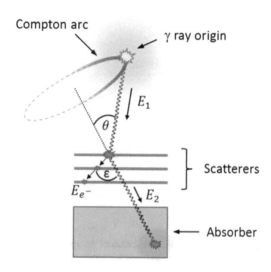

Fig. 7.8 Illustration of the electron tracking Compton camera. Reprinted from [23]. The track of the scattered electron is measured by a set of scatter detectors. The electron track information narrows the origin of the gamma rays to a Compton arc. Note that in the absence of depth information, the Compton arc represents an angular section of the Compton cone surface

a few mm from the place of radioactive decay. Due to conservation of energy and momentum, two 511 keV annihilation photons are emitted in opposite directions (with a small spread of 0.5 degree full-width-at-half-maximum (FWHM) due to thermal motion of the annihilating electron and positron). Two 511 keV photons detected in coincidence (i.e. within a very short time) will have originated from the same positron annihilation. The annihilation photons were thus emitted somewhere on the line connecting the locations of detection, the so-called line-of-response (LoR); see Fig. 7.9. The LoR thus provides the direction in which the photons travel and imaging is possible without physical collimation. The more accurately the LoR can be established, the higher the quality of the imaging information it contains. Detectors with good spatial resolution are thus required. An accurate measurement of the time difference between the coincident detection of positron annihilation photons enables to narrow down along the LoR where annihilation took place, providing depth information at the level of a single LoR. This so-called time-of-flight PET (TOF-PET) principle is illustrated in Fig. 7.10. Its introduction and continuous development have substantially improved PET image quality and diagnostic performance; see, e.g. [25].

Note that imaging using positron annihilation photons is obviously also possible without using coincident detection, thus treating the annihilation photons as single photons. For the purpose of these notes, however, we refer to the coincident detection of annihilation photons when talking about positron emission imaging/tomography.

7.3.2.4 Object and Detector Scatter

Gamma rays emitted from within an object may interact inside the object on their way to the camera. Complete absorption of a gamma ray obviously prevents it from being detected, and thus the number of counts detected in the imaging procedure

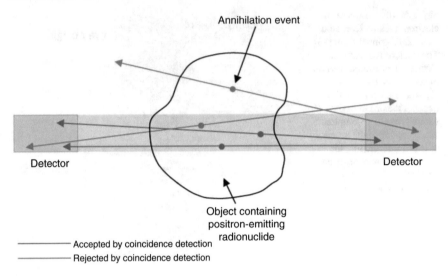

Fig. 7.9 The coincident detection of two back-to-back 511 keV annihilation photons defines a line-of-response that determines the direction in which the photons travel. This directional information provides the imaging information in positron emission imaging. Reprinted from [21]

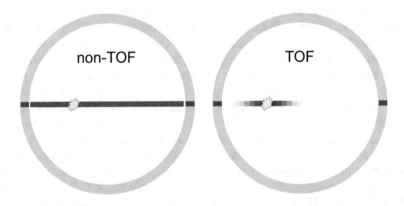

Fig. 7.10 The principle of time-of-flight positron emission tomography (TOF-PET). Assume that a positron annihilates at the location of the yellow "explosion" and that the 511 keV annihilation photons, emitted back-to-back, are detected by opposite detectors (dark blue) in a PET detector ring (light blue). Without TOF information (left figure), the contribution to the image is distributed uniformly along the LoR connecting the two detectors. With TOF information (right figure), the contribution to the image can be restricted to part of the LoR. The better the time resolution of the detectors, the narrower this part is and the larger the gain in image quality

will be reduced. This is relevant, because in many applications, the image quality is significantly determined by counting statistics and/or the number of emitted gamma rays is limited due to other considerations, e.g. radiation dose to a patient. In case a gamma ray Compton scatters in the object and the scattered gamma ray

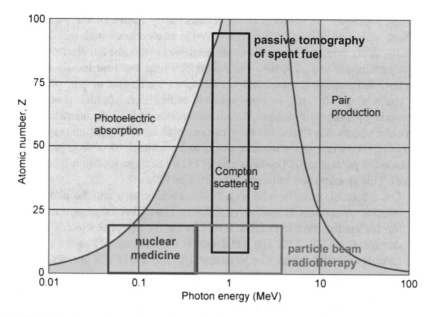

Fig. 7.11 Dominant photon interaction mechanism as function of photon energy and atomic number of the material with which a photon interacts. The boxes encompass the regions relevant for the objects imaged and gamma energies used in the applications considered in these notes. Related to nuclear safeguards, only the passive tomography of spent nuclear fuel is considered. One can conclude that Compton scattering is by far the most common interaction mechanism in the objects. Modified from [21]

leaves the object, detection of this scattered gamma ray will give wrong imaging information: the direction determined is that of the scattered gamma ray, not that of the emitted gamma ray. The solution to avoid that scattered gamma rays contribute to an image is good energy resolution in the detector: scattered gamma rays have a lower energy than the emitted gamma ray and can thus be rejected based on their detected energy. A better energy resolution in the detector will lead to better rejection. Figure 7.11 illustrates that Compton scattering is the most common photon interaction mechanism for the objects and gamma ray energies in the fields discussed in these notes.

7.3.2.5 General Demands to Gamma Camera Detectors

In this section, we summarize the general demands to gamma camera detectors as they follow from the principles and designs discussed in the previous sections.

- *Detector spatial resolution*
  In most imaging applications, image spatial resolution is a key performance characteristic. Image spatial resolution is to a large extent derived from the spatial resolution with which the gamma rays are detected in the camera.

When a gamma ray interacts inside a detector by means of the photoelectric effect, almost all of its energy is transferred to an electron, which will deposit its energy fairly locally. The small energy needed to overcome the electron binding energy is re-emitted as X-rays and Auger electrons and also locally absorbed. In case the gamma ray interacts via Compton interaction or pair production, a scattered gamma ray or two positron annihilation photons result. These secondary photons will interact, if at all, some distance from the interaction point. In order to correctly determine the direction of an incoming gamma ray, the first point of interaction in the detector needs to be determined. This is problematic in case the gamma ray deposits its energy in two or more locations relatively far apart. This problem can be (partially) solved as follows:

- Use a detector material with high atomic number such that the photoelectric effect dominates the interaction probability. The higher the gamma ray energy, the higher the atomic number needs to be for the photoelectric effect to dominate; see Fig. 7.12. The NaI detectors most commonly used for imaging gamma rays with energies between about 80 and 300 keV in nuclear medicine provide a majority of photoelectric interactions. For positron annihilation tomography, lutetium-based scintillation detectors are mostly used, providing a sizeable fraction of photoelectric interactions. Spent fuel tomography uses CdZnTe (CZT) detectors, making Compton interactions by far the most likely. For gamma ray energies above about 1 MeV, the photoelectric effect is not dominant even for uranium, the material with the highest atomic number found in nature. Obviously, a scatter detector of a Compton camera needs

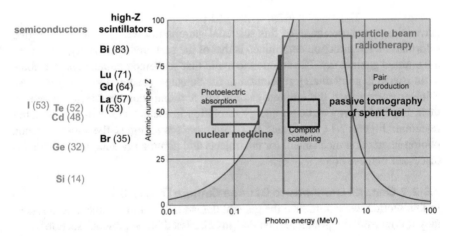

Fig. 7.12 Dominant photon interaction mechanism as function of photon energy and atomic number of the material with which a photon interacts. The chemical elements, and their atomic numbers, present in relevant semiconductor and scintillation detectors are given. The boxes encompass the regions relevant for the gamma ray energies used in the applications considered in these notes and the gamma ray detector materials used. Related to nuclear safeguards, only the passive tomography of spent nuclear fuel is considered. Modified from [21]

to be made from a material for which Compton scattering is dominant at the relevant gamma ray energy.

– Use a detector that is finely pixelated (physically or electronically) in three dimensions such that individual energy depositions originating from the same primary gamma ray can be disentangled. Most detectors are not fast enough to determine the order of interaction points from their order in time. However, based on the known probabilities and kinematics of the interaction mechanisms, an educated guess can be made, such that in a majority of cases, the correct point of first interaction is determined.

The usefulness of (a combination of) the solutions above depends on the properties of each imaging application.

- *Detector energy resolution*
  A good detector energy resolution allows to reject gamma rays scattered in the object and improves the imaging information obtained with Compton cameras. The best energy resolution is obtained with semiconductor detectors.
- *Detector efficiency*
  Because in many cases image quality is determined by the counting statistics, a high detector efficiency is needed. In nuclear medicine, the activity of the gamma ray source, i.e. the activity injected in the patient, is limited by the radiation dose to the patient. In imaging for the verification of particle beam radiotherapy, the number of gamma rays emitted is relatively low. It is determined by the radiation dose prescribed by the radiation oncologist and, thus, cannot be increased for the sake of imaging. The activity of spent fuel rods is determined by their properties, their reactor history and cooling time after removal from the reactor, but is generally very high.
- *Fast detector signals*
  In many cases, it is important to use detectors that provide fast signals. This minimizes detection dead time in case of high count rates, which are, e.g. typical in nuclear imaging with very short-lived isotopes such as ^{82}Rb (half-life 76 s). There is in principle no one-to-one relationship between fast signals and good timing resolution, but in practice this is most often the case. Good timing is essential for TOF-PET, with the image quality improving linearly with better timing resolution. Detectors with very good timing might even enable to unambiguously determine the interaction sequence in Compton cameras.

A general comment on gamma ray detector technology is relevant at this point. There are two major principles used in gamma ray detection: scintillation and ionization. In a scintillation detector, energy deposited by a gamma ray is transformed into optical scintillation photons which are turned into an electrical signal by a photosensor attached to the scintillation material. In an ionization detector, energy deposited by a gamma ray is transformed into electrical charges (electrons and ions in a gas detector, electron-hole pairs in a semiconductor detector). Scintillation photons are emitted isotropically, and it is hard to control the path they follow inside the scintillator towards the photosensor. In an ionization detector, the charges follow the field lines of the electric field applied across the detector towards the anode/cathode, during which they induce an electric signal. As such, an ionization detector enables better control and more flexibility and can provide a better position resolution than a scintillation detector. In situations where very good position resolution is needed, ionization detectors (often semiconductor detectors) are chosen. A drawback, however, is the typically lower maximum count rate because electrical charges move much slower than optical photons. The relatively slow signals also result in a poorer timing resolution compared to fast scintillation detectors. Semiconductor detectors on the other hand have far superior energy resolution.

Selecting the optimum detector technology and design for an imaging application can be a complicated problem in which various pros and cons need to be considered.

7.3.3   Iterative Image Reconstruction

In the previous sections, we discussed the principles of different gamma ray emission imaging systems. These systems measure projection data which provide indirect information about the source (or object) of the detected radiation. Such information can be used for recovering an image of the unknown object by means of, for example, reconstruction algorithms.

In general, reconstruction approaches can be either analytical or iterative. With an analytical approach, the solution is formulated in a closed-form, single equation. Instead, with iterative reconstruction algorithms, the final result is cast as the solution of a system of equations or the solution of an optimization problem, which is then solved with an iterative algorithm; see Fig. 7.13. Here, we focus on the latter case, namely, formulating the solution of the imaging problem as iterative solution of an appropriate optimization problem.

Being able to formulate the optimization problem requires, first and foremost, a (sufficiently accurate) mathematical model of the imaging process. This is, in general, called the *forward problem*, as opposed to the fact that the reconstruction of the unknown objects requires the inversion of such an operator. That is, in the imaging community, the reconstruction task is regarded as the solution of an *inverse problem*. The modelling of the forward problem includes also the data acquisition process, which amounts to determining how the detection system handles data sampling and data noise. Going into the details of the forward model would require a section of its own. Here, we report the main ideas and refer the reader to the literature (see, for instance, [26, chapter 3]) for all the details.

In emission tomography, sufficient modelling for the forward operator is provided by the well-known *Radon transform* [27]:

$$y(s, \theta) = (\mathcal{R}(f))(s, \theta) = \int_{\mathbb{R}} f(s\theta + t\theta^{\perp}) \, dt, \tag{7.1}$$

where (s, t) are coordinates of a point in the object; $\theta^{\perp}$ is perpendicular to $\theta = (\cos(\phi), \sin(\phi))$, with ϕ the rotation angle; and f represents the density of the

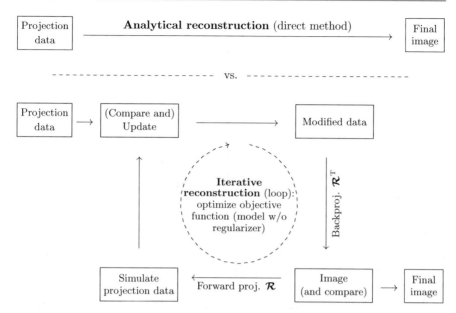

Fig. 7.13 Analytical vs. iterative approaches in image reconstruction. With analytical methods, the reconstruction is computed using a mathematical inverse of the forward transform (closed-form, single equation). In iterative reconstructions, the solution is computed numerically by iterating on a feedback loop: having defined a model or *objective function* for the problem, by starting from an initial guess for the solution, the current estimate is updated until it converges to an image that well explains the measured data. Notice that once a model or objective function is chosen, there can be different iterative algorithms that lead to the *same* solution

incoming photons. The underlying geometric setup is sketched in Fig. 7.14. In (7.1), f is the parameter (the object) we want to reconstruct: in the forward problem, we are given f, and we compute the measurements y, i.e. the projection data. In the inverse problem (i.e. the reconstruction task we are interested in), data and unknown are exchanged: that is, given the measurements y and the forward operator $\mathcal{R}$, we seek to determine (an approximation of) f.

In (7.1), phenomena like gamma ray scatter and attenuation are neglected. In emission tomography problems, however, it is important to compensate at least for the attenuation. This is easily achieved by using as forward operator the *attenuated Radon transform*:

$$y_\mu(s, \theta) = (\mathcal{R}_\mu(f))(s, \theta) = \int_{\mathbb{R}} e^{\int_t^{+\infty} \mu(s\theta + t'\theta^\perp)\,dt'} f(s\theta + t\theta^\perp)\,dt \qquad (7.2)$$

where μ is the linear attenuation coefficient and all the other parameters are defined as in (7.1). In the following, we will assume y_μ and $\mathcal{R}_\mu$ are given and the goal will be to recover f (and possibly μ). A direct or analytical inversion formula for the attenuated Radon transform exists and relies on the well-known Fourier slice

Fig. 7.14 Geometric setup
of the Radon transform
defined in Eq. (7.1). The
rotation angle ϕ determines
the direction
$\theta = (\cos(\phi), \sin(\phi))$. The
grey blob represents the
object f we wish to recover

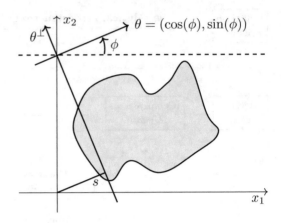

theorem. This is at the basis of, for instance, the *filtered back-projection* (FBP), a well-known algorithm for the inversion of 2D data, whose adaptation for 3D data is still largely employed in commercial CT machines [28].

With real-life measurements, the analytical inversion of the (attenuated) Radon transform (and, therefore, FBP-like algorithms) is often unreliable due to the *ill-posedness* of the Radon transform. A problem is defined to be well-posed when its solution is unique; it exists for any data function in a suitable space and depends continuously on the data with respect to the metric of the data space. The last condition (namely, the continuous dependence on the data) is the one usually violated by tomographic problems. In practice, it means that even if for certain data y the solution exists, a small perturbation on the data can result in data $\tilde{y}$ for which a solution does no longer exist. When it comes to real-life tomography, small perturbations in terms of *noise* are inherently present in the detection system. Indeed, the number of detected photons is subject to fluctuations due to the emission-transmission process, quantum conversion process and intrinsic properties (or defects) of the detection system. Being corrupted by noise, the tomographic measurements no longer satisfy the mathematical requirements for the inversion of the Radon transform (see, for instance, [29]), and alternative solutions need to be sought to provide reliable reconstructions.

To this end, the next step is to derive a discrete counterpart to the forward model (7.2). Since a detector counts the incoming photons by means of the imaging system, the measurement data are inherently discrete. Let's use $i = 1, \ldots, km$ to index the components y_i of the vector $\mathbf{y} \in \mathbb{R}^{km}$ denoting the acquired data. Here, km is the product between the number of projection angles k and the number of detector elements m. To discretize the unknown object, we use voxels:[2] the discrete

---

[2] Pixels and voxels are a traditional choice as basis functions to represent 2D and 3D objects, respectively. Given that, after discretization, the object $\mathbf{f}$ is reorganized by column-wise stacking it into a column vector, i.e. $\mathbf{f} \in \mathbb{R}^n$, with a slight abuse of terminology, in these notes we use the term "voxel" without distinguishing between 2D and 3D objects.

object $\mathbf{f} \in \mathbb{R}^n$ has components f_j, with $j = 1, \ldots, n$ and n total number of voxels, ordered lexicographically. Whenever μ is known—this is the case of PET, SPECT and CST as discussed in these notes—the forward model can be discretized by a linear operator represented by a matrix $\mathcal{R} \in \mathbb{R}^{km \times n}$. In this case, each element $\mathcal{R}_{i,j}$ of the matrix $\mathcal{R}$ is a "weight" of how much the ray connecting the source to the ith detector cell gets attenuated while travelling in the unknown jth voxel f_j. In the simplest case, $\mathcal{R}$ can contain only zeros and ones, namely, $\mathcal{R}_{i,j}$ is either 1 (when the ray connecting the source to the ith detector cell intersects the voxel f_j) or 0 (there is no intersection). More details on how to build $\mathcal{R}$ can be found, for instance, in [30, chapter 4] for PET and SPECT and [31, section 2.3] for CST. In both cases, the discrete linear model of the forward problem is given by:

$$\mathbf{y} = \mathcal{R}\mathbf{f}. \tag{7.3}$$

For PGET, the attenuation cannot be always assumed to be known. Therefore, the model is nonlinear and cannot be represented by a matrix. Nonetheless, the discrete model can still be written as:

$$\mathbf{y} = \mathcal{R}_\mu(\mathbf{f}) \tag{7.4}$$

where $\mathcal{R}_\mu$ is a discrete nonlinear operator depending on the (also unknown) attenuation vector $\mu \in \mathbb{R}^n$. More details on how to build $\mathcal{R}_\mu$ can be found in [32, section 2.2]. In any case, since $\mathcal{R}$, or $\mathcal{R}_\mu$ in the nonlinear case, is assumed to be given, the solution of the inverse problem is reduced to compute $\mathbf{f}$, or $(\mathbf{f}, \mu)$ in the nonlinear case, given $\mathbf{y}$. Given the physical quantities they represent, all components of $\mathcal{R}$, or $\mathcal{R}_\mu$, $\mathbf{y}$, $\mathbf{f}$ and μ are nonnegative.

Naturally, one might wonder whether having discretized the model improves our chances of getting an accurate reconstruction by directly "inverting" the discrete operators in (7.3)–(7.4). Unfortunately, the ill-posedness of the continuous model (7.2) translates into ill-conditioning of the discrete counterpart. Therefore, directly inverting $\mathcal{R}$ or $\mathcal{R}_\mu$ leads to very poor reconstructions. The reason for this is that the data $\mathbf{y}$ appearing in (7.3)–(7.4) are, in practice, not exact data but can only be a perturbation of the ideal data. The perturbation is due to several kinds of errors, the most important being noise. The principal source of noise is the so-called photon noise: this refers to the inherent natural variation of the incident photon flux (i.e. number of detected photons). Its effect appears when the number of photons captured by the detector elements in a given time interval is not very large. Under certain conditions (see, e.g. [26]), y_i is the realization of an integer valued *Poisson random variable* Y_i with a probability distribution given by:

$$\mathcal{P}^P_{Y_i}(y_i) = e^{-z_i} \frac{z_i^{y_i}}{y_i!}$$

where y_i is the number of photons detected by the ith detector element with average value z_i. This is usually the error model assumed for PET, SPECT and CST, where the number of emitted photons is usually small. For PGET, since the number of emitted photons y_i and their averages z_i are in general large, the error model can be approximated by *additive Gaussian noise*:

$$e^{-z_i} \frac{z_i^{y_i}}{y_i!} \approx \frac{1}{\sqrt{2\pi z_i}} e^{\frac{(y_i - z_i)^2}{2z_i}} = \mathcal{P}_{Y_i}^G(y_i).$$

While this is commonly regarded as a reasonable assumption, the implications are quite relevant. Indeed, additive Gaussian noise is independent of the signal; therefore it spreads uniformly across the voxels. Instead, Poisson noise is signal dependent: this means that the noise variance is not uniform over the image domain, but it is voxel dependent.

Even more, having the information about the statistical properties of the data at disposal, it would be natural to look for statistical approaches to solve the inverse problem. Indeed, the assumption we make on the noise statistics for the data can also be understood as a priori information we can use to compensate for the loss of information due to the perturbed data. More generally, a way to "cure" ill-posedness is to *regularize* the problem, that is, combining the (statistics behind the) forward model by some prior information which might be available on the solution. This is the rationale behind *maximum likelihood* (ML) estimation and *maximum a posteriori* (MAP) estimation,[3] which will lead us to iterative optimization approaches to face our reconstruction tasks.

Let's start by introducing a function that in statistics is generally used to measure the goodness of fit of a model to sampled data for given values of the unknown parameters. This is the so-called likelihood function, defined as the joint probability distribution of the sample, but viewed and used as a function of the parameters only:

$$\mathcal{L}_\mathbf{y}(\mathbf{z}) = \prod_{i=1}^{m} \mathcal{P}_{Y_i}(y_i). \tag{7.5}$$

where m is the number of detector elements and $\mathcal{P}_{Y_i}(y_i)$ depends on the noise assumptions (e.g. Poisson or Gaussian noise). Clearly, this is only a function of $\mathbf{z}$ since the data $\mathbf{y}$ are given. Provided the likelihood function has a maximum point, the ML estimate of $\mathbf{z}$ is defined to be any nonnegative object $\mathbf{z}^*$ that maximizes the likelihood function:

$$\mathbf{z}^* = \underset{\mathbf{z} \in \mathbb{R}^n}{\mathrm{argmax}} \, \mathcal{L}_\mathbf{y}(\mathbf{z}).$$

---

[3] The reader interested in deepening this topic can refer to [33]. A presentation of ML and MAP estimation tailored to iterative image reconstruction (including tomographic imaging) which follows the same structure used in these notes can be found in [26, 34].

This problem can be transformed into a minimization problem by considering the negative logarithm of the likelihood function:

$$\mathbf{z}^* = \underset{\mathbf{z} \in \mathbb{R}^n}{\operatorname{argmax}} \; \Gamma_0(\mathbf{z}; \mathbf{y}) \quad \text{with} \quad \Gamma_0(\mathbf{z}; \mathbf{y}) = -c_1 \log(\mathcal{L}_\mathbf{y}(\mathbf{z})) + c_2 \qquad (7.6)$$

where $c_1, c_2 \in \mathbb{R}$ are constants to simplify the expression of the functional[4] $\Gamma_0(\mathbf{z}; \mathbf{y})$. Clearly, depending on the probability distribution in (7.5), we have different expressions for $\Gamma_0(\mathbf{z}; \mathbf{y})$. Let us consider the two error models we discussed above for the particular case of the linear model (7.3). This means that $\mathbf{z} = \mathcal{R}\mathbf{f}$ with $\mathcal{R}$ given while the source object $\mathbf{f}$ is unknown. Therefore, Γ_0 is now a function $\Gamma_0(\mathbf{f}; \mathbf{y})$ of $\mathbf{f}$.

- *Poisson noise.* In this case, we have:

$$\mathcal{P}_Y(\mathbf{y}; \mathbf{f}) = \prod_{i=1}^{m} \mathcal{P}_{Y_i}^{P}(y_i) = \prod_{i=1}^{m} \frac{e^{-(\mathcal{R}\mathbf{f})_i} (\mathcal{R}\mathbf{f})_i^{y_i}}{y_i!}.$$

After suitable approximation and choice of the constants c_1 and c_2, we reach:

$$\Gamma_0(\mathbf{f}; \mathbf{y}) = \sum_{i=1}^{m} \left\{ y_i \ln\left(\frac{y_i}{(\mathcal{R}\mathbf{f})_i}\right) + (\mathcal{R}\mathbf{f})_i - y_i \right\} \qquad (7.7)$$

and therefore the ML approach coincides with the so-called Kullback-Leibler (KL) divergence. It is well-known that KL is convex,[5] strictly convex if and only if the equation $\mathcal{R}\mathbf{f} = 0$ has the single trivial solution $\mathbf{f} = 0$. The properties of the continuous version of the KL divergence and its minimization are investigated in a series of papers (see [35–37]). In particular, in [35] an example is given where the functional does not have a minimum in the classical sense, hence proving the ill-posedness of this minimization problem.

---

[4] In calculus of variations (a research field in mathematical analysis), the term *functional* refers to a mapping from a space X into the real (or complex) numbers for the purpose of establishing a calculus-like structure on X. For example, when the space X is a space of functions, a functional can be thought as a "function of a function". Throughout these notes, the term functional is used with this meaning.

[5] We recall that a function g is called *convex* if the line segment between any two points on the graph of the function lies above the graph between the two points, that is, $g(t\xi_1 + (1 - t)\xi_2) \leq tg(\xi_1) + (1 - t)g(\xi_2)$ for all $0 \leq t \leq 1$. The function g is called *strictly convex* if the inequality is strict. In optimization theory, (strict) convexity is a highly desirable property for functions in order to ensure convergence of optimization algorithms.

- *Gaussian noise.* In this case, assuming expected value 0 and variance σ^2, we have:

$$\mathcal{P}_Y(\mathbf{y}; \mathbf{f}) = \prod_{i=1}^{m} \mathcal{P}_{Y_i}^G(y_i) = \left(\frac{1}{\sqrt{2\pi\sigma^2}}\right)^m e^{-\frac{1}{2\sigma^2}\|\mathbf{y}-\mathcal{R}\mathbf{f}\|_2^2}$$

where $\|\cdot\|_2$ denotes the usual 2-norm (or Euclidean norm). After a suitable choice of the constants c_1 and c_2, we obtain:

$$\Gamma_0(\mathbf{f}; \mathbf{y}) = \frac{1}{2}\|\mathcal{R}\mathbf{f} - \mathbf{y}\|_2^2 \tag{7.8}$$

and therefore the ML approach coincides with the well-known least squares (LS) approach [38]. The LS functional (7.8) is well-known to be convex, strictly convex if and only if the equation $\mathcal{R}\mathbf{f} = 0$ has only the solution $\mathbf{f} = 0$. Moreover, the LS problem has always a solution. However, also in this case the ill-posedness of the continuous version of this minimization problem can be proven [39]. In particular, the ill-posedness of this problem is the starting point of the Tikhonov regularization theory [40], which we will briefly discuss below.

As we stressed already, in the case of an image reconstruction task, ML problems are generally ill-posed (or ill-conditioned). Therefore, in line of principle, one is not interested in computing the minimum points $\mathbf{f}^*$ of the functionals Γ_0 corresponding to the different noise models—usually referred to as *data mismatch functional*—because they do not provide sensible estimates of the unknown object. However, numerical experience shows that constraining the solution to the nonnegative orthant (i.e. requiring that the components of the solution are all nonnegative) and/or terminating the iterations of an iterative algorithm by *early stopping* already provides sufficient regularization to the problem resulting in an acceptable solution.

We conclude these mathematical preliminaries by presenting some other common ways of regularizing the problem which will be useful for the nonlinear case of PGET. The role of regularization is to introduce additional information on the solution—a so-called *prior*—by adding constraints in the variational formulation provided by ML. Since we are already considering a statistical approach, it is worth to formulate also this prior information in a statistical framework: this is usually referred to as the *Bayesian paradigm* [41].

Within the Bayesian approach, also the unknown object $\mathbf{f}$ is assumed to be the realization of a random variable F. The additional information on $\mathbf{f}$ is provided in the form of the so-called Gibbs prior:

$$\mathcal{P}_F(\mathbf{f}) = \frac{1}{Z} e^{-\alpha\Gamma_1(\mathbf{f})} \tag{7.9}$$

where Z is a normalization constant, α is a positive parameter, and $\Gamma_1(\mathbf{f})$ is a nonnegative functional. Now, if the probability distribution $\mathcal{P}_Y(\mathbf{y}; \mathbf{f})$ is viewed as a conditional probability of Y for a given value of F, namely, $\mathcal{P}_Y(\mathbf{y}; \mathbf{f}) = \mathcal{P}_Y(\mathbf{y}|F = \mathbf{f}) = \mathcal{P}_Y(\mathbf{y}|\mathbf{f})$, then Bayes' formula yields:

$$\mathcal{P}_F(\mathbf{f}|\mathbf{y}) = \frac{\mathcal{P}_Y(\mathbf{y}|\mathbf{f})\,\mathcal{P}_F(\mathbf{f})}{\mathcal{P}_Y(\mathbf{y})}. \tag{7.10}$$

We now substitute in (7.10) the detected value of $\mathbf{y}$, therefore obtaining the a posteriori probability density of F:

$$\mathcal{P}_F(\mathbf{f}|\mathbf{y}) = \mathcal{L}_Y(\mathbf{f})\,\frac{\mathcal{P}_F(\mathbf{f})}{\mathcal{P}_Y(\mathbf{y})}. \tag{7.11}$$

Then, a *maximum a posteriori* (MAP) estimate of the unknown object is defined as any object $\mathbf{f}^*$ that maximizes the a posteriori probability density, namely:

$$\mathbf{f}^* = \underset{\mathbf{f} \in \mathbb{R}^n}{\operatorname{argmax}}\ \mathcal{P}_F(\mathbf{f}|\mathbf{y}). \tag{7.12}$$

If we now take the negative logarithm as we did with ML estimation, and we neglect the terms depending only on $\mathbf{y}$, rearranging constants, we reach:

$$\mathbf{f}^* = \underset{\mathbf{f} \in \mathbb{R}^n}{\operatorname{argmin}}\ \Gamma(\mathbf{f}) \quad \text{with} \quad \Gamma(\mathbf{f}; \mathbf{y}) = \Gamma_0(\mathbf{f}; \mathbf{y}) + \alpha\Gamma_1(\mathbf{f}) \tag{7.13}$$

where we assumed a Gibbs prior for $\mathcal{P}_F(\mathbf{f})$. The functional $\Gamma(\mathbf{f}; \mathbf{y})$ is called *regularization functional*, and $\alpha > 0$ is the so-called regularization parameter. Notice that it is not obvious that a minimum point $\mathbf{f}^*$ of $\Gamma(\cdot; \mathbf{y})$ is a sensible estimate of the unknown object: this depends on the choice of Γ_1 and the value of the regularization parameter α, which balances the trade-off between the data mismatch and the regularization term. Studying the existence and uniqueness properties of regularized functionals is an active research field in inverse problems.

One of the prime examples of regularization functionals, for which existence and uniqueness of the solution have been widely investigated, is given by the family of *Tikhonov regularizers*. In its most classic form, Tikhonov regularization is given by the square of the 2-norm of $\mathbf{f}$:

$$\Gamma_1(\mathbf{f}) = \frac{1}{2}\|\mathbf{f}\|_2^2 = \frac{1}{2}\sum_j f_j^2.$$

This type of regularization promotes a solution whose 2-norm is small. In a similar fashion, one can consider the square of the 2-norm of some higher-order *derivative* of the solution: this leads to *generalized* Tikhonov regularization. For example, the *Generalized Tikhonov regularization of order 2* is given by the square of the 2-norm of the discrete Laplacian Δ of $\mathbf{f}$:

$$\Gamma_1(\mathbf{f}) = \frac{1}{2} \|\Delta \mathbf{f}\|_2^2 = \frac{1}{2} \sum_j (\Delta \mathbf{f})_j^2. \tag{7.14}$$

This type of regularization enforces the 2-norm of the Laplacian of the solution to be small, that is, the solution's derivatives cannot be too large. This amounts to penalizing rapid changes in the solution, therefore promoting "smoothness" of the solution. An example of how this regularizer can be relevant for tomographic reconstructions is presented below for the PGET case. Tikhonov regularizers are widely employed in diverse imaging applications because the 2-norm is differentiable (or smooth): like convexity, this is another highly desirable property that makes the study of (the existence of) minimizers much easier. There exists a whole class of nonsmooth regularizers where the 2-norm is replaced by the 1-norm:[6] we leave it out of these notes as it is not relevant in the following.

Finally, the nonnegativity constraint can be easily taken into account in the Bayesian framework if in place of (7.9) we consider:

$$\mathcal{P}_F(\mathbf{f}) = \frac{\chi_+(\mathbf{f})}{Z} e^{-\alpha \Gamma_1(\mathbf{f})} \tag{7.15}$$

where $\chi_+(\mathbf{f})$ is the characteristic function of the nonnegative orthant,[7] which is 1 on the orthant and 0 elsewhere. By taking the negative logarithm as we did above, (7.13) reads as:

$$\mathbf{f}^* = \underset{\mathbf{f} \in \mathbb{R}^n}{\text{argmin}} \; \Gamma(\mathbf{f}) \quad \text{with} \quad \Gamma(\mathbf{f}; \mathbf{y}) = \Gamma_0(\mathbf{f}; \mathbf{y}) + \alpha \Gamma_1(\mathbf{f}) + \iota_+(\mathbf{f}) \tag{7.16}$$

where $\iota_+(\mathbf{f})$ is the indicator function of the nonnegative orthant, i.e. the function which is 0 on the orthant and ∞ outside. In Fig. 7.15 we show examples of reconstructions from tomographic data with two different regularizers to demonstrate how different priors predispose the reconstruction towards different solutions.

We have now laid the foundation to present some iterative methods proposed for solving the minimization problems derived from the ML and Bayesian approaches for the specific cases of PET, SPECT, CST and PGET. First, we introduce the well-known *expectation maximization* method, and its acceleration *ordered subset expectation maximization*, which is extensively used in commercial PET and SPECT scanners [42]. This method is also the state-of-the-art for CST. Then, we

---

[6] The 1-norm of a vector $\mathbf{f} \in \mathbb{R}^n$ is given by the sum of the absolute value of its components, namely, $\|\mathbf{f}\|_1 = \sum_j |f_j|$. Such regularizers are used in imaging problems where we expect the solution (or its higher order derivatives) to have many components equal to zero, i.e. it is *sparse*. For this reason, 1-norm regularization is usually said to "enforce sparsity".

[7] In 2D, an orthant is a quadrant; in 3D, an orthant is an octant. The *nonnegative orthant* is the generalization of the first quadrant to n-dimensions with $n \geq 2$.

Ground truth   Tikhonov regularizer   Tikhonov regularizer &
                                                    nonnegativity constraint

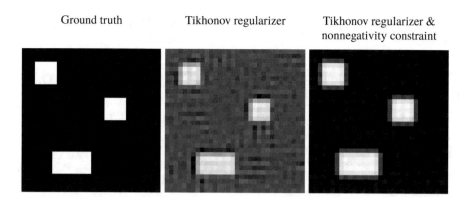

Fig. 7.15 An example of reconstructions from CT data (ground truth on the left) to showcase the role of regularization in reconstruction. We consider two different regularizers: classic Tikhonov regularization (middle) and classic Tikhonov regularization with additional nonnegativity constraint (right). For this object—a toy example with three rectangular regions of intensity 1 (white) on a uniform background of intensity 0 (black)—it is clear that nonnegativity constraints are essential to deliver a more sensible estimate of the (true) solution

present a MAP approach using generalized Tikhonov regularization of order 2 and another ad hoc hand-crafted 2-norm prior, coupled with a *Levenberg-Marquardt trust region* method for the PGET case. Additional considerations regarding iterative reconstruction methods in commercial devices are discussed in Sect. 7.4.

PET/SPECT and CST: Expectation Maximization Algorithm

The expectation maximization (EM) algorithm was first introduced by Shepp and Vardi for the case of emission tomography in [43]. Their derivation of the algorithm, starting from the ML estimate of KL, is not easy. Therefore we give here an alternative presentation, which provides an heuristic explanation of EM based on the Karush-Kuhn-Tucker (KKT) conditions and the fixed point method.[8] KKT are necessary and sufficient conditions for a point $\mathbf{f}^*$ to be a minimum of $\Gamma_0(\mathbf{f}; \mathbf{y})$:

$$\mathbf{f}^* \, \nabla\Gamma_0(\mathbf{f}^*; \mathbf{y}) = 0 \quad \text{with} \quad \mathbf{f}^* \geq 0, \ \nabla\Gamma_0(\mathbf{f}^*; \mathbf{y}) \geq 0. \tag{7.17}$$

It is easy to verify that when $\Gamma_0(\mathbf{f}; \mathbf{y})$ is the KL divergence (7.7), its gradient is given by:

$$\nabla\Gamma_0(\mathbf{f}^*; \mathbf{y}) = \mathcal{R}^\mathrm{T}\mathbf{1} - \mathcal{R}^\mathrm{T}\frac{\mathbf{y}}{\mathcal{R}\mathbf{f}}$$

where the quotient of vectors (here and in the following) is meant component-wise and $\mathbf{1}$ is the vector with all the components equal to 1. Now, starting from an initial

---

[8] The interested reader who is not familiar with these topics can refer, for example, to [44].

guess $\mathbf{f}^{(0)}$, the EM algorithm computes an approximation of $\mathbf{f}^*$ by successive updates given by:

$$\mathbf{f}^{(k+1)} = \frac{\mathbf{f}^{(k)}}{\mathbf{v}} \mathcal{R}^{\mathrm{T}} \frac{\mathbf{y}}{\mathcal{R}\mathbf{f}^{(k)}} \tag{7.18}$$

where $\mathbf{v} = \mathcal{R}^{\mathrm{T}}\mathbf{1}$. It is immediate to see that we can write the KKT condition (7.17) as the following fixed point equation:

$$\mathbf{f}^* = \frac{\mathbf{f}^*}{\mathbf{v}} \mathcal{R}^{\mathrm{T}} \frac{\mathbf{y}}{\mathcal{R}\mathbf{f}^*}.$$

In addition, (7.18) can be seen as a step of the scaled gradient descent method [45], with step size 1, the descent direction given by the negative gradient multiplied by a diagonal and positive scaling factor given by the current iteration:

$$\mathbf{f}^{(k+1)} = \mathbf{f}^{(k)} - \frac{\mathbf{f}^{(k)}}{\mathbf{v}} \nabla\Gamma_0(\mathbf{f}^{(k)}; \mathbf{y}).$$

Several convergence proofs of the algorithm to a ML solution are available (see, for instance, [37, 46, 47]). In addition, if the initial guess $\mathbf{f}^{(0)}$ is strictly positive, then it is easy to prove by induction that all the iterates are strictly positive, since $\mathcal{R}\mathbf{f}^{(0)}$ is also strictly positive. Therefore, with EM the nonnegativity constraint is automatically satisfied whenever $\mathbf{f}^{(0)}$ is strictly positive. Finally notice that an "implicit" regularization effect can be obtained by early stopping of the iterations. This is because of the empirical property of *semi-convergence*, that is, the iterations first approach a sensible solution and then diverge. There is no formal proof of this, but it is usually observed in practice. This is a useful property because it provides, at the very least, a first approximation of the object, which in some cases can turn out to be quite accurate.

One of the main drawbacks of EM is that convergence is very slow. This means that even using early stopping, it can require a large number of iterations before reaching a sensible solution. Faster convergence can be gained with the *ordered subset expectation maximization* (OSEM) algorithm, first proposed by Hudson and Larkin in [48]. The idea consists of partitioning the data into ordered subsets and applying the EM iteration to each subset. An iteration of OSEM amounts to a cycle over the selected subsets. In details, we have that the set $\mathcal{I}$ is subdivided into (in general, non-overlapping) subsets $\mathcal{I}^{(\ell)}$ such that $\bigcup_{\ell=1}^{L} \mathcal{I}^{(\ell)} = \mathcal{I}$. Correspondingly, $\mathbf{y}^{(\ell)}$ denotes the data vector with components y_i such that $i \in \mathcal{I}^{(\ell)}$ and $\mathcal{R}^{(\ell)}$ is the block of the matrix $\mathcal{R}$ consisting of the rows with $i \in \mathcal{I}^{(\ell)}$. Similarly, $\mathbf{v}^{(\ell)}$ is the vector whose components are given by $\mathbf{v}_j^{(\ell)} = \sum_{i \in \mathcal{I}^{(\ell)}} \mathcal{R}_{ij}$. A detailed description of OSEM can be found in Algorithm 1. A proof of convergence for OSEM exists only in the so-called consistent case, namely, when the matrices $\mathcal{R}^{(\ell)}$ are *balanced*, i.e. $\mathbf{v}_j^{(\ell)}$ is independent of ℓ (see [48]). In the non-consistent case, a proof of convergence is still not available. The computational cost per iteration of EM and OSEM is

---

Algorithm 1: OSEM algorithm

Input: Initialize $\mathbf{f}^{(0)} > 0$ and choose a subdivision of $I = \bigcup_{\ell=1}^{L} I^{(\ell)}$.

while $k \leq K_{\max}$ *and until convergence* **do**

    Set $\mathbf{f}^{(k,0)} = \mathbf{f}^{(k)}$.

    for $\ell = 1, \ldots, L$ **do**

        $\mathbf{f}^{(k,\ell)} = \frac{\mathbf{f}^{(k,\ell-1)}}{\mathbf{v}^{(\ell)}} (\mathcal{R}^{(\ell)})^{\mathrm{T}} \frac{\mathbf{y}^{(\ell)}}{\mathcal{R}^{(\ell)} \mathbf{f}^{(k,\ell-1)}}$

    end

    Set $\mathbf{f}^{(k+1)} = \mathbf{f}^{(k,L)}$.

end

---

approximately the same, but a wise choice of the subsets (and of their order in the internal cycle) can result in a considerable gain in terms of convergence speed. This is a key point when it comes to commercial devices (see Sect. 7.4.1.3).

While we presented OSEM because of its widespread adoption as a standard method for iterative reconstruction in PET, SPECT and CST, many other iterative methods exist, and a nice review is given, for instance, in [42]. More recent approaches for PET within regularization theory and using sparsity-enforcing regularizers are available, for example, in [49, 50].

PGET: Simultaneous Reconstruction of Emission and Attenuation with Levenberg-Marquardt Algorithm

The iterative approach we present for PGET was very recently introduced in [32]. The baseline idea of the method is the simultaneous estimation of emission and attenuation, to overcome the limitation of analytical methods like FBP which require either that the attenuation map is known (from elsewhere) or that there is a post-processing step to correct for the attenuation. Since spent fuel is highly absorbing, attenuation cannot be measured in any practical way and therefore is not available when using the PGET device. Since both the (object's) emission $\mathbf{f}$ and attenuation $\boldsymbol{\mu}$ are unknown, the model (7.2) yields a nonlinear problem.

The reconstruction problem is formulated as a constrained minimization problem with a LS data mismatch term (i.e. it implicitly assumes a Gaussian distribution for the noise) and several regularization terms P_l:

$$(\mathbf{f}^*, \boldsymbol{\mu}^*) = \underset{(\mathbf{f},\boldsymbol{\mu})\in\mathbb{R}_+^{2n}}{\operatorname{argmin}} \left\{ \left\| \mathcal{R}_{\boldsymbol{\mu}}(\mathbf{f}) - \mathbf{y} \right\|_2^2 - \sum_l \alpha_l P_l(\mathbf{f}, \boldsymbol{\mu}) \right\} \tag{7.19}$$

$$\text{subject to} \quad \mathbf{A} \begin{bmatrix} \mathbf{f} \\ \boldsymbol{\mu} \end{bmatrix} \leq \mathbf{u}.$$

The constrained formulation (7.19) requires not only the nonnegativity of the unknowns $\mathbf{f}$ and $\boldsymbol{\mu}$, but it also forces a linear upper bound $\mathbf{u}$ on both. In particular, the bounds allow the three materials relevant for the application: water (no gamma ray emission, low attenuation), spent fuel rod (high gamma ray emission, high attenuation) and fresh fuel rod (no gamma ray emission, high attenuation), but they exclude the physically unlikely case of a material with high emission and low attenuation. In [32], two different choices for the regularization terms are presented. The first one is the generalized Tikhonov regularization of order 2 (7.14), in the following referred to as *smoothness prior*, applied separately to emission and attenuation:

$$\alpha_{\mathbf{f}}\|\Delta \mathbf{f}\|_2^2 + \alpha_\mu\|\Delta \mu\|_2^2.$$

In practice, the discrete Laplacian operator Δ is a 2D convolution with the kernel:

$$\ker_\Delta = \begin{bmatrix} 0 & 1 & 0 \\ 1 & -4 & 1 \\ 0 & 1 & 0 \end{bmatrix}.$$

The other penalty term is designed specifically for the PGET application. It assumes that the positions and the diameters of the possible fuel rods are known, although it requires no information about whether these rods are actually present or not. In practice, the algorithm is predisposed towards reconstructions that consist of rods of predefined sizes in certain predefined places, but it has no preference about the emission and attenuation values of the rods beyond uniformity within a single rod. The geometry of the fuel assembly can be inferred, for example, from an initial FBP reconstruction which allows to identify the fuel assembly type. For this reason it is referred to as *geometry aware prior*. Similarly to the Tikhonov prior, the geometry aware prior is applied separately to attenuation and emission:

$$\alpha_{\mathbf{f}}\|P_{\mathbf{f}} \mathbf{f}\|_2^2 + \alpha_\mu\|P_\mu \mu\|_2^2.$$

For the details of how to construct the matrices $P_{\mathbf{f}}$ and P_μ, we refer the reader to [32]. Both regularization terms are such that the minimization functional (7.19) can be naturally written as a nonlinear LS problem:

$$\|r(\mathbf{f}, \mu)\|_2^2 = \left\| \begin{matrix} \mathcal{R}_\mu(\mathbf{f}) - \mathbf{y} \\ \sqrt{\alpha_{\mathbf{f}}} \, M_{\mathbf{f}} \mathbf{f} \\ \sqrt{\alpha_\mu} \, M_\mu \mu \end{matrix} \right\|_2^2 \tag{7.20}$$

where the matrices $M_{\mathbf{f}}$ and M_μ depend on the choice of the penalty. To solve (7.20) we can use a classic algorithm for the solution of nonlinear LS problems: the Levenberg-Marquardt (LM) algorithm [51]. Denote by $\mathbf{x}$ the concatenation of the emission and attenuation vectors, i.e. $\mathbf{x} = [\mathbf{f}^T, \mu^T]^T$, and $r(\mathbf{x}) = r(\mathbf{f}, \mu)$ for the residual. At each iteration, LM minimizes, with regard to the next step $\mathbf{x}_{\text{step}}$, a linear LS term that results from linearizing the residual $r(\mathbf{x})$ at the current iterate $\mathbf{x}^{(k)}$ and from adding a regularization term:

$$\left\| \begin{bmatrix} J_r(\mathbf{x}^{(k)}) \\ \sqrt{\beta^{(k)}}\mathbb{1} \end{bmatrix} \mathbf{x}_{\text{step}} + \begin{bmatrix} r(\mathbf{x}^{(k)}) \\ \mathbb{0} \end{bmatrix} \right\|_2^2 , \qquad (7.21)$$

Here $J_r(\mathbf{x}^{(k)})$ is the Jacobian matrix of the residual $r(\mathbf{x})$, $\mathbb{1}$ is the identity matrix, and $\beta^{(k)}$ is the LM parameter modified at each step. Since the original formulation (7.19) includes also the box constraints on $\mathbf{x}$, the LM iteration (7.21) is modified by using linear constraints that keep the next iterate $\mathbf{x}^{(k+1)} = \mathbf{x}^{(k)} + \mathbf{x}_{\text{step}}$ feasible:

$$\mathbf{A}\mathbf{x}_{\text{step}} \leq \mathbf{u} - \mathbf{A}\mathbf{x}^{(k)}. \qquad (7.22)$$

This minimization is done using the scaled gradient projection (SGP) method [45]. Therefore the SGP iterative step is "nested" into the outer LM iteration: this is because a direct (analytical) solution of (7.21) would be computationally unfeasible due to the large-scale nature of the problem. In Fig. 7.16 can be seen an example of reconstructions, with the two different regularization terms, from PGET data of an ATRIUM 10 fuel assembly (see Fig. 7.3), measured at the Olkiluoto nuclear power plant (Finland) in 2019 (see Fig. 7.22).

Some variations of LM can be used as a regularization method by themselves, but for this specific application of PGET, the regularizations terms and the bounds

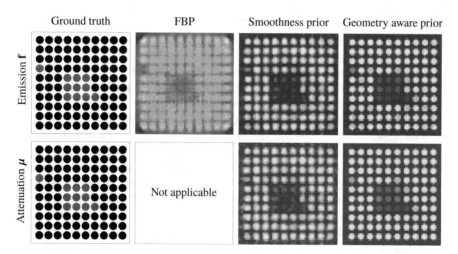

Fig. 7.16 The ground truth and the reconstruction images for the ATRIUM 10 fuel assembly (see also Fig. 7.3), with a 3×3 water channel in the centre and two missing rods (in red). In the top row there are the gamma ray emission images and in the bottom row the attenuation images. In the columns from left to right: the ground truth, the FBP reconstruction, the iterative reconstruction using the smoothness prior and the same using the geometry aware prior. The FBP reconstruction clearly provides a poorer reconstruction compared to the method proposed in [32], suffering from the fact that there is no attenuation correction

play a crucial role to deliver sensible solutions. Finally, we observe that the strategy introduced in [32] was the first one to propose a nonlinear regularized approach for PGET. Recently, other iterative approaches have been proposed. In [52], the authors couple a linear forward model with a sparsity-enforcing regularization term. In [53] (and their previous papers) the authors explore Bayesian iterative approaches, such as MLEM and OSEM.

7.4 Implementation of Gamma Ray Emission Imaging

We present some examples of state-of-the art imaging devices as implemented in the various fields of application considered. These examples illustrate how the purpose and requirements discussed in Sects. 7.2 and 7.3 are implemented in practice. We also discuss some additional considerations on iterative reconstruction methods when employed in commercial devices.

7.4.1 Nuclear Medicine

7.4.1.1 Gamma Camera and SPECT

In nuclear medicine, "gamma camera" refers to the device used for imaging single gamma rays (called "single" to distinguish them from the pairs of 511 keV positron annihilation photons). Figure 7.5, left, illustrates the basic design of the parallel-hole gamma camera. As this is by far the most often used gamma camera, we limit the discussion to this type. The parallel-hole collimator ensures that only gamma rays travelling in one direction, along the length of the holes, can be detected by the gamma ray detector that is located right behind the collimator.

The sensitivity and spatial resolution of the gamma camera as a whole are determined more by the collimator than by the detector: the collimator is the weak link in the gamma camera performance. Careful design of the collimator is thus needed. Sensitivity and spatial resolution are the main performance characteristics of a collimator. Sensitivity, also called efficiency, is the fraction of gamma rays arriving at the collimator that pass through one of the holes. The spatial resolution is the uncertainty in determining at which location in the camera a gamma ray passed through a collimator hole. There is a trade-off between collimator sensitivity and spatial resolution: long, narrow holes have poor sensitivity and good spatial resolution, whereas short, wide holes have good sensitivity but poor spatial resolution (see Fig. 7.17, left). As collimators are meant to block gamma rays not travelling in a particular direction, they are best made out of a material with a high probability to absorb gamma rays: most often collimators are made out of lead. When going from the lowest to the highest energy used in single-photon imaging (from about 80 to 350 keV), the gamma ray interaction probability in lead decreases by about a factor of 10. Therefore, collimators for higher energy gamma rays need to have thicker septa and/or be thicker (so they have longer holes), reducing the collimator sensitivity (see Fig. 7.17, right).

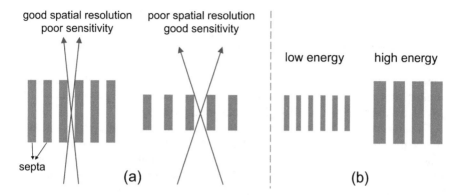

Fig. 7.17 Parallel-hole design. The walls of a collimator are called septa. (**a**) Trade-off between spatial resolution and sensitivity. The arrows illustrate the angular range of gamma rays being able to pass through the collimator holes. The collimator sensitivity scales with the fraction of hole area to total area. (**b**) Collimators optimized for low and high-energy gamma rays, with the same hole diameter

Table 7.1 Typical parallel-hole collimator parameters. The values shown are for the Siemens Symbia TruePoint SPECT/CT scanner, but are very similar to those of other manufacturers. An all-purpose collimator has an intermediate sensitivity and resolution

Collimator type	Low energy high sensitivity	Low energy all purpose	Low energy high resolution	High energy
Hole length (mm)	24.05	24.05	24.05	59.7
Hole diameter (mm)	2.54	1.45	1.11	4
Septal thickness (mm)	0.36	0.20	0.16	2
Sensitivity	4.6×10^{-4}	1.5×10^{-4}	0.92×10^{-4}	0.67×10^{-4}
Spatial resolution (mm)	14.6	8.3	6.4	13.2

    The trade-off between collimator efficiency and spatial resolution, and the energy dependence of the gamma ray interaction probability, mean that there is no one-size-fits all parallel-hole collimator. For optimum performance in a wide range of single-photon imaging procedures, a gamma camera has several interchangeable collimators. Table 7.1 gives typical parameters for various collimators.

    Gamma rays that pass through the collimator should be detected with high efficiency and with a reasonably good energy resolution (in order to identify and reject gamma photons that Compton-scattered in the patient). Since the first development of the gamma camera, NaI has been the scintillator material of choice: it is inexpensive, and its properties in terms of gamma ray detection efficiency, scintillation light yield and energy resolution are in general adequate. Better scintillation materials on each of these properties are available, but do not essentially improve the gamma camera image quality (partly because the collimator is the weak link) and have thus not been implemented in commercial cameras.

Cameras based on semi-conductor detectors, notably CZT, have been developed [54–56]. They exhibit better spatial and energy resolution than NaI-based cameras. A major purpose of these systems is dual-energy imaging: simultaneous imaging of two different radiotracers. The excellent energy resolution of CZT, and thus energy selection of the data, is needed for dual-energy imaging of the interesting combination of ^{99m}Tc and ^{123}I tracers as their gamma ray energies (140 and 159 keV) are close together; see, e.g. [57] for an example study.

Gamma cameras are typically available with two different NaI scintillation crystal thicknesses: 9.5 mm and 15–20 mm. The thinner version is perfectly fine when imaging with a radiotracer containing ^{99m}Tc, the most used radioisotope: the 140 keV gamma rays emitted have a half-thickness in NaI of just under 3 mm (the half-thickness is the thickness at which half of the gamma rays have interacted). The thicker version is needed for good detection efficiency when imaging with gamma rays of about 300 keV or higher (the half-thickness at 300 keV is 12 mm).

A gamma camera contains a continuous slab of NaI scintillator of typically 60×40 cm^2. Position sensitivity is obtained by so-called Anger logic, named after Hal Anger who introduced the principle in the first gamma camera developed [58]. In Anger logic, a relatively large detector surface area is covered by multiple relatively small photosensors (photomultiplier tubes, PMTs, in the case of NaI-based gamma cameras). In such an arrangement, the scintillation light that is created when a gamma ray interacts with the scintillator is distributed over and detected by a number of photosensors. Analysing the spatial distribution of the photosensor signals allows to pinpoint the location of the interaction. In a typical gamma camera such as the SymbiaTM TruePoint SPECT/CT, a 59.1×44.5 cm^2 NaI crystal is covered by a hexagonal array of 59 PMTs (with a diameter of either 7.6 or 5.1 cm). For the 9.5-mm-thick NaI crystal, the spatial resolution of the NaI detector on its own (the so-called intrinsic gamma camera resolution) is 3.8 mm FWHM. It is interesting to note, and perhaps surprising to realize, that the spatial resolution is much better than the size of the individual photosensors. Achieving the same resolution with an array of individual detectors would require detectors with a diameter of a few mm; but a few 10,000 detectors would then be needed to cover the surface area of the NaI slab. The large simplification and cost savings by using Anger logic are thus obvious. Gamma cameras based on CZT detectors do not use Anger logic. Being semiconductor detectors, the signal is generated by the collection of electron-hole pairs created following the interaction of a gamma ray. Pixellation of a large area detector is easily obtained by creating a 2D array of small area anodes. The General Electric NM/CT 670 CZT and NM/CT 870 CZT scanners, e.g. use 4×4 cm CZT crystals with a 16×16 array of anodes, effectively creating an array of 2.5×2.5 mm^2 detector pixels.

As explained in Sect. 7.3.1, 3D images can be obtained by taking 2D projections (radiographs) from all angles around the patient, the so-called tomographic procedure. The gamma camera has to be sufficiently large to cover the full cross-sectional size of a patient, hence the typical size of 60×40 cm^2. A full ring scanner made out of gamma cameras and with an inner diameter of 80 cm (typical for patient scanners) can contain not more than about five cameras and thus measures simultaneously at only a small number of angles. Sufficient angular sampling for tomographic imaging thus requires the addition of rotational motion. In practice, SPECT scanners most often contain two gamma cameras that are rotated around the patient.

7.4.1.2 Positron Emission Tomography Scanner

The back-to-back emission of 511 keV positron annihilation photons that are detected in coincidence defines the direction of emission to be the LoR, the line connecting the detector locations where the photons are detected. Positron emission imaging by means of the coincident detection of annihilation photons therefore does not need a collimator to select a certain direction. As a result, a much higher sensitivity than with a gamma camera is obtained. Modern PET scanners have a sensitivity of about 1.5% for a line source of uniform radioactivity placed along the central axis of the scanner, roughly a hundred times higher than a gamma camera. Part of this higher sensitivity stems from the fact that positron emission imaging is almost always performed in tomographic mode with a full-ring scanner which has a substantially larger solid angle than a typical SPECT scanner with two gamma cameras. Such a geometry, without the need of rotation, is made possible due to the absence of a collimator.

Since about 1980, the performance of PET scanners has improved tremendously. Several detector technology innovations have made this possible. The introduction of bismuth germanate (BGO, $Bi_4Ge_3O_{12}$) scintillation crystals around 1980 enabled the much more efficient detection of 511 keV photons than the NaI detectors used up to then (the half-thickness of 511 keV photons in BGO is 7.3 mm, while it is 20.4 mm in NaI). A major breakthrough happened when BGO crystals were used in so-called block detectors [59]. A block detector contains a fairly large array of small scintillation crystals which is read out by just four PMTs. The block detector is designed such that scintillation light from any crystal is shared over the four PMTs. Comparing the relative sizes of the PMT signals (i.e. Anger logic) enables to determine in which crystal of the array the 511 keV interaction took place. As a typical example, Fig. 7.18, left, shows a block detector from the Siemens Biograph mCT Flow PET/CT scanner. It consists of a 13×13 array of $4 \times 4 \times 20$ mm^3 LSO scintillation crystals read out by a 2×2 array of cylindrical PMTs. A full performance evaluation of this scanner is given by [60]. Figure 7.18, right, shows a 2D histogram of the interaction positions, determined via Anger logic, when the block detector is uniformly illuminated by 511 keV photons, a so-called flood map. It demonstrates how Anger logic can identify all 169 scintillation crystals. The fact that the flood map pattern is not regular but shows a severe pincushion distortion is not relevant; identifying the correct crystal in which an interaction took place is the relevant purpose, with the crystal positions known from the detector/scanner design.

A few prototype time-of-flight (TOF) capable PET scanners were constructed in the 1980s using CsF [61] or BaF_2 [62, 63] scintillation crystals. These were however abandoned in favour of the much more efficient BGO-based scanners. Note that the timing properties of both NaI and BGO are not sufficient to provide useful TOF information. The advent of the lutetium-based scintillation crystals, lutetium oxyorthosilicate (LSO, Lu_2SiO_5) and lutetium-yttrium oxyorthosilicate

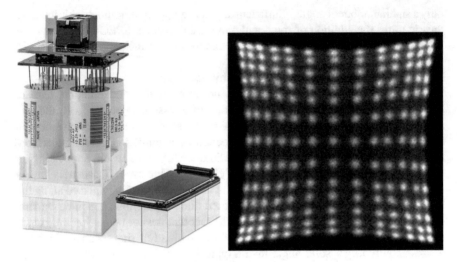

Fig. 7.18 Left: Comparison of PMT and SiPM-based PET detector modules. Left: block detector from the Siemens Biograph mCT Flow PET/CT scanner. Middle: detector module of the Siemens Biograph Vision PET/CT scanner. Right: flood-map of a Siemens Biograph mCT block detector. Left/middle picture reprinted from www.siemens-healthineers.com with permission from Siemens Healthineers

(LYSO, $Lu_{1.8}Y_{0.2}SiO_5$), which have a good efficiency (half-thickness of 7.9 mm) and excellent timing properties, as well as a better energy resolution than BGO, revived the TOF-PET scanner (see [64] for an update on the latest advances in TOF-PET). The first commercially available TOF-PET scanner was the Philips TOF PET/CT GEMINI-TF scanner in 2006, with a coincidence resolving time (CRT) of 600 ps FWHM. CRT is the parameter used to quantify the TOF performance; it is the FWHM of the distribution of the time differences between 511 keV photons detected in coincidence. For a Gaussian time difference distribution, the CRT is 3.33 times larger than the single detector standard deviation time resolution, the timing resolution measure usually used when discussing detector development. The most recent major innovation is the use of Geiger-mode avalanche photodiodes (G-APD, most commonly referred to as silicon photomultipliers, SiPM), which improved the timing performance considerably (see [65, 66] for some reviews of SiPMs and their timing properties). Nowadays, the best CRT of a commercial PET scanner is 210 ps [67], allowing to narrow down the location of positron annihilation along the LoR to a region of just over 3 cm FWHM, resulting in a huge reduction in noise compared to non-TOF PET. As a typical example, Fig. 7.18, middle, shows a SiPM-based TOF-PET detector module of the Siemens Biograph Vision PET/CT scanner. This module contains a 4×2 arrangement of mini blocks. A mini block consists of a 5×5 array of $3.2 \times 3.2 \times 20$ mm^3 LSO crystals coupled to a 4×4 SiPM array fully covering the crystals. A full performance evaluation of the Biograph Vision scanner is given by [67]. As the typical size of a SiPM is very similar to the size of individual scintillation crystals, the use of SiPMs allows to use one-to-one coupling of crystal to photosensor, removing the need for Anger logic, as in the Philips Vereos PET/CT system. The Vereos system uses so-called digital SiPMs and has a CRT of 310 ps [68].

Since about 2015, all major PET scanner manufacturers offer SiPM-based systems, which will, within the foreseeable future, replace the PMT-based systems. The newest innovation is the total-body PET scanner [69], which has a much larger axial length than the typical 20–25 cm so far. The first commercially available system, the μEXPLORER PET/CT system from United Imaging, has an axial length of 194 cm; a detailed performance evaluation is given in [70]. The main advantages of such systems stem from the very large sensitivity combined with the simultaneous coverage of the whole human body.

7.4.1.3 Additional Consideration on Iterative Reconstruction in Commercial PET and SPECT

The most widely used image reconstruction method in state-of-the-art SPECT and PET systems is the ordered subset expectation maximization (OSEM) algorithm (see Sect. 7.3.3). As we stressed already, the number of iterations and subsets used in OSEM are of key importance when deploying the algorithm, and, in practice, it varies between manufacturers and also depends on the specifics of the imaging task at hand. This is because, with increasing number of iterations, image noise (understood as pixel-to-pixel variations of a statistical nature) tends to grow in OSEM. This is a serious issue in nuclear medicine imaging because the number of counts is limited by the acceptable radiation dose delivered to the patient by the radiotracer. Often image quality is largely determined by the number of counts collected during the imaging procedure. Limiting noise requires to limit the number of iterations, which results in sub-optimal images. To counter this under-convergence to control image noise, and thus allow a larger number of iterations, noise reduction techniques are often applied. One example is implemented in the Q.Clear method developed by GE Healthcare [71], which uses a Bayesian penalized likelihood reconstruction algorithm [72, 73]. An evaluation of the Q.Clear method is presented by [74], even though details of the implementation are not publicly available (as is usually the case for commercial devices). Another key ingredient when it comes to practical implementation of image reconstruction algorithms in commercial devices is the correction of the so-called point spread function (PSF), i.e. the image of a point source. PSF varies throughout the field-of-view (FoV) of a scanner for different reasons. In SPECT, this is, to a large extent, due to the poorer spatial resolution further away from the collimator. In a full-ring PET scanner, the PSF becomes more and more asymmetric as one moves away from the centre of the FoV. Within the reconstruction algorithm, all manufacturers implement a sort of PSF correction, taking the variation of the 3D PSF throughout the FoV into account. As input information, the 3D PSF is modelled in a calibration procedure. In the case of PET, up to millions of accurately measured PSFs are incorporated in the reconstruction algorithms. As a result, the image spatial resolution is largely constant throughout the FoV and the overall image quality improves. An example of how PSF modelling is implemented is given by [75].

7.4.2   Particle Beam Radiotherapy

We are here selecting examples of imaging systems in the context of proton beam radiotherapy. Similar systems are in principle also applicable to ion beam therapy, although aspects such as the production of prompt gamma rays and positron emitting nuclides in the patient and radiation background issues will be different. For a review of prompt-gamma monitoring in particle beam radiotherapy and a quite complete overview of the instruments being developed for it, we refer to [76].

Modern proton beam therapy uses the pencil scanning technique: a narrow proton beam is scanned across the tumour using two dipole magnets moving the beam in two orthogonal directions perpendicular to the beam direction (the x- and y-directions), and the proton beam energy is varied to scan in the direction of the beam (the z-direction). Most often, spot dose delivery is used whereby the beam is static when delivering dose (the typical duration of one spot is a few milliseconds) and is switched off while changing the position and/or energy. At any time during a patient irradiation, the x-y position of the proton beam is known (it is monitored by detectors in the proton beam) and does not need to be determined by the gamma emission imaging system that is performing dose delivery verification. This simplifies the imaging system when imaging prompt gamma rays: the x-y position of the origin of each detected prompt gamma ray is known, and the imaging system only needs to determine the z-position. The imaging task is thus simplified to making a 1D image. The dose delivery verification system that is closest to be used in clinical routine, a knife-edge slit collimator camera, is designed to deliver such a 1D image.

Figure 7.19 illustrates the knife-edge slit camera. Prompt gamma rays produced along the proton beam path, and emitted isotropically, are projected by a slit collimator onto a linear array of LYSO scintillation crystals. The LYSO crystals are 4 mm wide in the direction of the proton beam, 31.5 mm thick and 100 mm high, with 2 rows of 20 crystals each mounted on top of each other, resulting in an 80-mm-wide and 200-mm-high detector array. The LYSO crystals are read out by SiPM photosensors. The slit camera, being a 1D pinhole camera (a camera obscura), enables magnification of the prompt gamma production profile, with the magnification factor (the image size to object size ratio) given by the ratio of the distance of the collimator slit to the detector and the distance of the proton beam to the collimator slit. A magnification close to 1 has so far been used in imaging during patient irradiations. Details on the design and data analysis are given in [78, 79]. As the camera measures a 1D profile, no image reconstruction as discussed in Sects. 7.3.3 and 7.4.1.3 is needed. The edge of the measured profiles, corresponding to the proton range, is determined. Measurements during the delivery of proton therapy for a brain cancer patient (Fig. 7.20) show that, when aggregating data from multiple beam spots, a beam range shift precision of 2 mm can be achieved [77]. The need for aggregation over multiple beam spots to achieve a clinically relevant precision is quite general in proton therapy verification because the number of secondary photons that is created is relatively small. It can obviously not be

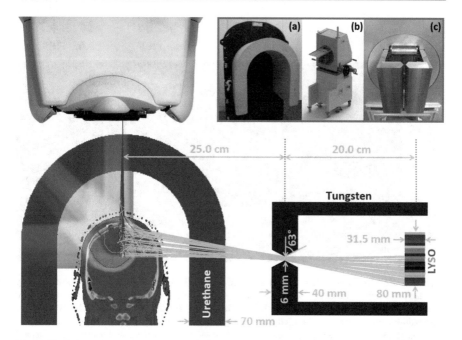

Fig. 7.19 Schematic setup and layout of the knife-edge slit collimator prompt gamma camera. Relevant dimensions are given. Protons (blue lines) travelling in the patient create isotropically emitted gamma rays. Some of these (green lines) pass through the knife-edge slit collimator and hit a linear array of LYSO scintillation crystals. The distribution of gamma rays detected along the LYSO array is an image of the production of prompt gamma rays along the proton beam path. Upper right pictures: (**a**) the table-mounted U-shaped urethane beam range shifter, (**b**) the trolley for camera positioning, (**c**) the knife-edge slit collimator. Reprinted from [77]

increased by a higher proton beam intensity as this is dictated by the prescribed radiation dose.

Several projects developing a Compton camera for imaging prompt gamma rays in particle beam radiotherapy have been and are being pursued around the world. A variety of detector technology is being used; see [76] for an overview. Most of these use one or more relatively thin scatter detectors and one relatively thick absorber detector, as illustrated in Figs. 7.7 and 7.8.

Instead of discussing in detail a particular Compton camera, we point out two detector technologies that have been driving the development of Compton cameras for in vivo range verification: monolithic scintillation detectors and 3D position-sensitive semiconductor detectors. We focus on how these detectors obtain their position sensitivity.

A monolithic scintillation detector is a scintillation crystal with a fairly large surface area that is read out by an array of photosensors, in modern systems most often SiPMs ([80] discusses the use of SiPMs in Compton cameras). Relatively thin monolithic scintillation detectors provide 2D position sensitivity and are used as

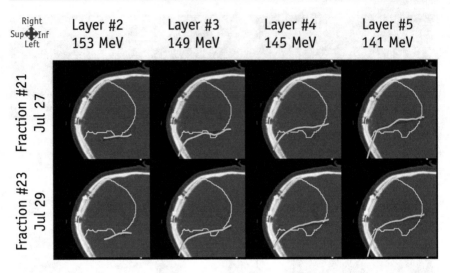

Fig. 7.20 Results from imaging with the knife-edge slit camera during the proton therapy treatment of a brain cancer patient. Horizontally, four different proton beam energy layers are shown. The rows show the results from the irradiations on two different days. The X-ray image of the irradiated region is overlaid with the following information: the yellow line depicts the volume that was planned to receive a high dose; the red and green lines show the planned and measured Bragg peak depths. Reprinted from [77]

scatter detectors. The position information is derived from the signal distribution of the photosensor array, similar to Anger logic in a gamma camera (see Sect. 7.4.1.1). For instance, the MACACO camera uses 5-mm and 10-mm-thick LaBr$_3$ crystals read out by 8×8 SiPM arrays [81]. For good detection sensitivity of the high-energy gamma rays emitted in particle beam radiotherapy, a fairly thick absorber detector is needed. For thick detectors, measuring the 2D interaction position is not sufficiently accurate to reconstruct the path of the scattered gamma ray and thus leads to poor imaging information. Better performance is obtained by having 3D position sensitivity, thus adding the so-called depth-of-interaction information. A good example is the 30-mm-thick LaBr$_3$ absorber in the electron tracking Compton camera described by [23]. Obtaining the 3D position information in such detectors requires an elaborate calibration procedure. Most recently, neural networks are used to determine the position of individual events [82].

Over the past decade, large semiconductor detectors (larger than about 1 cm in all three dimensions) with 3D position resolution on the mm scale have been under development. The position and energy resolution that can be obtained with such detectors are very appealing for Compton imaging. 2D position resolution in large semiconductor detectors is obtained by segmentation of the signal readout electrodes, either by having orthogonal strips on both the anode and cathode side (see, e.g. [24]) or a 2D readout grid on the anode and a planar cathode (see, e.g. [83]). Depth information is obtained by the relative timing of the anode and cathode signals. A major strength of such detectors is their capability to

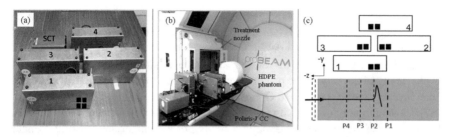

Fig. 7.21 A prototype Compton camera based on large CZT semiconductor detectors. (**a**) The four detection stages labelled 1–4 and the sychronization-coincidence timing module (SCT), with black squares showing the positioning of the CZT crystals in the stages. (**b**) The experimental setup for measurements at the Maryland Proton Treatment Center. The camera sits on the treatment couch next to a HDPE phantom. (**c**) Schematic (not to scale) of the experimental setup showing the proton beam (black arrow) incident on the phantom from the negative z-direction. Reprinted from [84] with permission from IOP Publishing

recognize multiple simultaneous interactions, such as Compton scattering and the subsequent photoelectric absorption of the scattered photon. This makes them very interesting for Compton imaging as the same detector functions as both scatter and absorber detector. However, the detector is too slow to determine the time order of multiple interactions. An "educated guess" is possible making use of the well-known Compton scattering kinematics and interaction probability and/or the known extent of the object being imaged, but some ambiguity will remain. Several commercial cameras are available (from, e.g. PHDS Co. and H3D Inc.), using germanium or CZT detectors, but these are optimized for gamma ray energies below about 2 MeV. During particle therapy, gamma rays up to about 6 MeV need to be imaged. Dedicated systems are thus being developed for particle therapy verification. The development of a Compton camera based on large CZT crystals is described by [84–86]. The system consists of four separate detector stages, each containing a 2×2 array of CZT crystals ($2 \times 1 \times 2$ cm^3 and $2 \times 1.5 \times 2$ cm^3 crystals are used); see Fig. 7.21. Images are produced using the stochastic origin ensemble (SOE) iterative reconstruction algorithm [22]. The SOE method does not reconstruct an image using the full Compton cone,[9] but selects a single, so-called representative, point on the cone (assuring that this point lies within the object being imaged). In consecutive iterations, new points are chosen in an attempt to improve the reconstruction by exchanging the current representative points with points where the probability of a gamma originating is higher. The iterative process is stopped when a steady-state situation, in which the image no longer changes from iteration to iteration, is reached.

Positron emission imaging for in vivo range verification has an advantage in that it can basically use the technology that is continuously being developed for PET

---

[9] Recall that each event in a Compton camera determines a cone-of-response (the Compton cone; see Sect. 7.3.2.2), not a line-of-response.

in the nuclear medicine context (see Sect. 7.4.1.2). A full ring PET scanner can be used to image the positron emitters after irradiation, once the patient has been moved from the treatment position to the PET scanner. A standard full ring scanner cannot be used to measure during an irradiation, a so-called in situ measurement, because it is physically in the way of the proton beam. A solution to this geometrical issue has been investigated: the so-called OpenPET system using a slanted full-ring scanner [87]. Most common for an in situ PET system is however a dual-panel configuration. Some of these systems are essentially a section of a commercial full-ring scanner (see, e.g. [88, 89]), while others are dedicated systems, but using technology very similar to nuclear medicine PET scanners (see, e.g. [90–94]). As the PET signal is delayed on a time scale of the same order as the duration of an irradiation, and an irradiation typically consist of several fields (irradiation from different directions) with overlapping dose delivery, 3D imaging is needed for good quality verification. Recently, the usefulness of positron emission imaging using the very short-lived nuclide ^{12}N (half-life = 11 ms) is under investigation [89,95]. It potentially provides a "prompt" PET signal for which instantaneous correlation with the beam delivery is possible and thus 1D imaging, as in the case of prompt gamma rays, suffices.

7.4.3 Nuclear Safeguards

A nuclear fuel assembly is an array of fuel rods several meters long; see an example in Fig. 7.3. Many different designs have been and are being used. The fuel rod diameter is approximately 1 cm, and a fuel assembly has a diameter of typically 15–25 cm. Fuel rods are assembled from fuel pellets (about 1 cm high) encased in a zirconium alloy tube. After use in a nuclear reactor, the radioactive decay of fission products in the spent fuel results in a tremendous gamma ray activity. For example, pressurized water reactor (PWR) spent fuel assemblies (SFAs) have a ^{137}Cs activity of the order of 10^{15} Bq [96]. However, most of the fission products are relatively short-lived, and after a few years of cooling time (time since removal from the nuclear reactor), only a handful of nuclides useful for gamma ray emission imaging remains; see Table 7.2.

The main purpose of gamma ray imaging in a nuclear safeguards context is the detection of missing fuel rods. Obtaining a 2D cross-sectional image is sufficient for this task, and thus the camera should measure 1D projections in a tomographic

Table 7.2 Decay properties of the most relevant nuclides in gamma ray emission imaging of spent nuclear fuel, for fuel with a cooling time of a few years

Nuclide	Half-life (y)	Main gamma ray energies and branching ratios
^{134}Cs	2.1	569 keV (15%), 605 keV (98%),796 keV (86%)
^{154}Eu	8.6	723 keV (20%), 1004 keV (18%), 1274 keV (35%)
^{137}Cs	30.1	662 keV (85%)

Fig. 7.22 A spent fuel assembly being lowered into the PGET device. The device sits 15 metres underwater at the bottom of the spent fuel pool at the Olkiluoto nuclear power plant in Finland. (Reprinted with permission from TVO)

imaging procedure (see Sect. 7.3.1). Information on the third dimension can of course be obtained by measuring 2D images at multiple places along the length of a fuel assembly.

We describe here the so-called PGET device developed under the auspices of the IAEA and in use by inspectors of the nuclear regulatory agencies since 2018. Figure 7.22 shows the PGET device in operation at the bottom of a spent nuclear fuel pool. Figure 7.23 illustrates the design and realization of the PGET device. The PGET device contains two detector heads, on opposite sides of a central channel into which a SFA is lowered for imaging. The detector heads are rotated over 360 degrees for tomographic imaging. Each detector head contains a 10-cm-thick slit collimator made out of tungsten with a small CZT detector behind each collimator slit. The slits are 1.5 mm wide and have a 4 mm pitch. The two detector heads are shifted by 2 mm with respect to each other, resulting, after interleaving the data from the two heads, in 1D projections with 2 mm bin size. Typically, measurements are

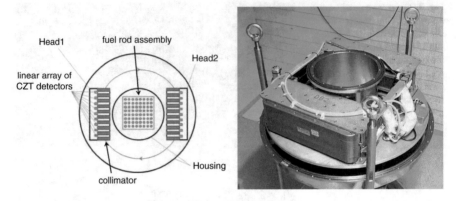

Fig. 7.23 Left: conceptual design of the PGET device. A cross-sectional view is shown. Modified from [15]. Right: picture of the inner parts of the PGET device, with the water-tight cover removed

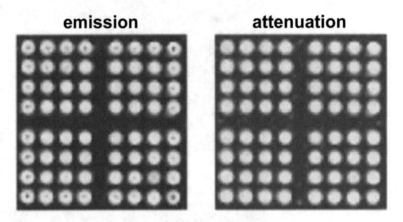

Fig. 7.24 Gamma ray emission (left) and attenuation (right) reconstructions for a SVEA-64 fuel assembly, obtained as discussed in Sect. 7.3.3. The outer rods show intra-rod emission variation. Reprinted from [98]

taken during continuous rotations and usually sorted into 360 one-degree steps. The typical measurement time is 5 minutes. During data acquisition, the detector count rates in user-defined preset energy intervals are recorded. The most relevant energy bin mostly used so far is 600–700 keV, containing the ^{137}Cs gamma rays.

Results from the PGET device are described in [20, 97, 98]. We limit ourselves here to a specific result showing the imaging power of the device. Figure 7.24 shows the results of imaging a SVEA-64 SFA [98]. The images are reconstructed using the method described in Sect. 7.3.3. This assembly has a relatively large rod diameter of 10.44 mm. The observed intra-rod variation in gamma ray emission can be understood from the fact that at the high temperatures inside a nuclear reactor, fission products diffuse from the hotter central region of the rod to the colder periphery. Especially isotopes of Cs, being a volatile metal, diffuse easily. Intra-rod Cs and Eu isotopic distributions were also seen and studied by [99]. This explanation is supported by the attenuation reconstruction, which does not show intra-rod differences. The intra-rod emission variations are best seen for the outer rods because the image quality for these rods is best. In this context, it is relevant to note that the regularization terms used (see Sect. 7.3.3) prefer solutions where the intra-rod activity and attenuation coefficient are uniform. This means that some of the intra-rod differences are smoothed out during image reconstruction.

7.5 Conclusions

The design of hardware and software for a gamma ray emission imaging application is a multifaceted task. The main considerations are:

- the requirements of the application
  - What does the image need to tell?
  - What accuracy and precision is needed?
  - What is the measurement environment?
  - What is the required/allowed duration of the imaging procedure?
- the properties of the object to be imaged
  - size and structure
  - material composition
  - gamma ray absorption and scattering
- the properties of the gamma ray source
  - spatial intensity distribution
  - gamma ray energies

To arrive at a practical instrument, cost will be an essential factor to consider as well. All these requirements are intertwined (e.g. absorption/scattering depends on both gamma ray energy and object size/structure), with often a trade-off to be made (e.g. between collimator spatial resolution and sensitivity or between performance and cost). The requirements of the application should be leading in taking decisions. These notes will have achieved their purpose if the reader is able to use the information presented as a starting point to make informed decisions on the design and use of gamma ray emission imaging systems. When doing so, remember that it is not about the detectors, not even about the images, but about what the images can do for your application.

Acknowledgments PD was partially supported by Business Finland under Grant 1845/31/2014. TAB was partially supported by the Royal Society through the Newton International Fellowship grant n. NIF\R1\201695 and was by the Academy of Finland through the postdoctoral grant, decision number 330522.

References

1. G. De Heversy, *Nobel Lecture* (1944). https://www.nobelprize.org/prizes/chemistry/1943/hevesy/lecture/
2. H.D. Kogelnik, Radiother. Oncol. **42**(3), 203 (1997)
3. R.R. Wilson, Radiology **47**(5), 487 (1946)
4. J.A. Langendijk, in *National Proton Therapy Conference*, vol. 29 (Delft, the Netherlands January 2009)
5. E.W. Korevaar, S.J.M. Habraken, D. Scandurra, R.G.J. Kierkels, M. Unipan, M.G.C. Eenink, R.J.H.M. Steenbakkers, S.G. Peeters, J.D. Zindler, M. Hoogeman, J.A. Langendijk, Radiother. Oncol. **141**, 267 (2019)
6. J. Unkelbach, T.C.Y. Chan, T. Bortfeld, Phys. Med. Biol. **52**(10), 2755 (2007)
7. A.C. Knopf, A. Lomax, Phys. Med. Biol. **58**(15), R131 (2013)
8. K. Parodi, J.C. Polf, Medical Physics **45**(11), e1036 (2018)
9. C.H. Min, H.R. Lee, C.H. Kim, S.B. Lee, Medical Physics **39**(4), 2100 (2012)
10. K. Parodi, Nucl. Med. Rev. **15**(C), 37 (2012)
11. https://www.iaea.org/sites/default/files/publications/documents/infcircs/1970/infcirc140.pdf
12. https://www.iaea.org/topics/basics-of-iaea-safeguards
13. *IAEA, Safeguards Techniques and Equipment: 2011 Edition (IAEA/NVS/1/2011)* (IAEA, Vienna, Austria, 2011). https://www.iaea.org/publications/8695/safeguards-techniques-and-equipment
14. https://www.posiva.fi/en/
15. T. Honkamaa, F. Lévai, A. Turunen, R. Berndt, S. Vaccaro, P. Schwalbach, in *Proc. Symp. Int. Safeguards, Vienna, Austria, Paper IAEA-CN-220* (2014)
16. S. Jacobsson Svärd, L.E. Smith, T.A. White, V. Mozin, P. Jansson, P. Andersson, A. Davour, S. Grape, H. Trellue, N. Deshmukh, E. Miller, R. Wittmam, T. Honkamaa, S. Vaccaro, J. Ely, ESARDA Bulletin **55**, 10 (2017)
17. F. Lévai, S. Dési, S. Czifrus, Technical Report STUK-YTO-TR 189, STUK, Helsinki, Finland (2002). https://www.julkari.fi/bitstream/handle/10024/123527/stuk-yto-tr189.pdf
18. F. Lévai, S. Dési, M. Tarvainen, R. Arlt, Technical Report STUK-YTO-TR 56, STUK, Helsinki, Finland (1993)
19. M. Mayorov, T. White, A. Lebrun, J. Brutscher, J. Keubler, A. Birnbaum, V. Ivanov, T. Honkamaa, P. Peura, J. Dahlberg, in *Proceedings 2017 IEEE Nucl. Sci. Symp. and Med. Imag. Conf., Atlanta, GA, USA* (2017)
20. T. White, M. Mayorov, A. Lebrun, P. Peura, T. Honkamaa, J. Dahlberg, J. Keubler, V. Ivanov, A. Turunen, in *Proceedings 59th Annual Meeting of the Institute of Nuclear Materials Management, Baltimore, MD, USA* (2017)
21. S.R. Cherry, J.A. Sorenson, M.E. Phelps, *Physics in Nuclear Medicine* (Elsevier Saunders, 2012)
22. D. Mackin, S. Peterson, S. Beddar, J. Polf, Phys. Med. Biol. **57**, 3537 (2012)
23. S. Aldawood, P.G. Thirolf, A. Miani, M. Böhmer, G. Dedes, R. Gernhäuser, C. Lang, S. Liprandi, L. Maier, T. Marinšek, M. Mayerhofer, D.R. Schaart, I.V. Lozano, K. Parodi, Radiat. Phys. Chem. **140**, 190 (2017)
24. L. Mihailescu, K.M. Vetter, M.T. Burks, E.L. Hull, W.W. Craig, Nucl. Instrum. Methods Phys. Res. Sect. A **570**(1), 89 (2007)
25. S. Surti, J. Nucl. Med. **56**(1), 98 (2015)
26. M. Bertero, P. Boccacci, V. Ruggiero, *Inverse Imaging with Poisson Data* (IOP Publishing, Bristol, 2018)
27. J. Radon, Ber. vor Sächs. Akad. Wiss., Leipzig Math. Phys. **69**, 262 (1917)
28. X. Pan, E.Y. Sidky, M. Vannier, Inverse Problems **25**(12), 123009 (2009)
29. F. Natterer, *The Mathematics of Computerized Tomography* (SIAM: Society for Industrial and Applied Mathematics, Philadelphia, PA, USA, 2001). https://doi.org/10.1137/1.9780898719284

30. A.C. Kak, M. Slaney, *Principles of Computerized Tomographic Imaging* (Society for Industrial and Applied Mathematics, Philadelphia, 2001)
31. E. Muñoz, J. Barrio, A. Etxebeste, P.G. Ortega, C. Lacasta, J.F. Oliver, C. Solaz, G. Llosá, Phys. Med. Biol. **62**(18), 7321 (2017)
32. R. Backholm, T.A. Bubba, C. Bélanger-Champagne, T. Helin, P. Dendooven, S. Siltanen, Inverse Probl. Imag. **14**(2), 317 (2020)
33. J. Kaipio, E. Somersalo, *Statistical and Computational Inverse Problems*, vol. 160 (Springer Science & Business Media, 2006)
34. M. Bertero, H. Lantéri, L. Zanni, Math. Methods Biomed. Imag. Intensity Modul. Radiat. Therapy (IMRT) **7**, 37 (2008)
35. H. Mülthei, Math. Methods Appl. Sci. **15**(4), 275 (1992)
36. H. Mülthei, B. Schorr, W. Törnig, Math. Methods Appl. Sci. **9**(1), 137 (1987)
37. H. Mülthei, B. Schorr, W. Törnig, Math. Methods Appl. Sci. **11**(3), 331 (1989)
38. Å. Björck, *Numerical Methods for Least Squares Problems* (SIAM, 1996)
39. H.W. Engl, M. Hanke, A. Neubauer, *Regularization of Inverse Problems*, vol. 375 (Springer Science & Business Media, 1996)
40. A.N. Tikhonov, V.Y. Arsenin, New York **1**(30), 487 (1977)
41. S. Geman, D. Geman, IEEE Trans. Pattern Anal. Mach. Intell. (6), 721 (1984)
42. J. Qi, R.M. Leahy, Phys. Med. Biol. **51**(15), R541 (2006)
43. L.A. Shepp, Y. Vardi, IEEE Trans. Med. Imag. **1**(2), 113 (1982)
44. S. Boyd, L. Vandenberghe, *Convex Optimization* (Cambridge University Press, 2004)
45. S. Bonettini, R. Zanella, L. Zanni, Inverse Problems **25**, 015002 (2009)
46. K. Lange, R. Carson, J. Comput. Assist. Tomogr. **8**(2), 306 (1984)
47. Y. Vardi, L. Shepp, L. Kaufman, J. Am. Stat. Assoc. **80**(389), 8 (1985)
48. H.M. Hudson, R.S. Larkin, IEEE Trans. Med. Imag. **13**(4), 601 (1994)
49. M. Burger, J. Müller, E. Papoutsellis, C.B. Schönlieb, Inverse Problems **30**(10), 105003 (2014)
50. A. Sawatzky, C. Brune, F. Wubbeling, T. Kosters, K. Schafers, M. Burger, in *2008 IEEE Nuclear Science Symposium Conference Record* (IEEE, 2008), pp. 5133–5137
51. C.T. Kelley, *Iterative Methods for Optimization* (SIAM, 1999)
52. M. Fang, Y. Altmann, D. Della Latta, M. Salvatori, A. Di Fulvio, Scientific Reports **11**(1), 1 (2021)
53. S. Shiba, H. Sagara, Ann. Nucl. Energy **150**, 107823 (2021)
54. P. Fernandez, H. de Clermont-Gallerande, S. Grenereau, E. Bittard, N. Balamoutoff, Médecine Nucléaire **43**(2), 241 (2019)
55. T. Ito, Y. Matsusaka, M. Onoguchi, H. Ichikawa, K. Okuda, T. Shibutani, M. Shishido, K. Sato, J. Appl. Clin. Med. Phys. **22**(2), 165 (2021)
56. O. Zoccarato, D. Lizio, A. Savi, L. Indovina, C. Scabbio, L. Leva, A. del Sole, C. Marcassa, R. Matheoud, M. Lecchi, M. Brambilla, J. Nucl. Cardiol. **23**(4), 885 (2016)
57. T. Blaire, A. Bailliez, F.B. Bouallegue, D. Bellevre, D. Agostini, A. Manrique, EJNMMI Physics **3**(1), 27 (2016)
58. H.O. Anger, Rev. Sci. Instrum. **29**(1), 27 (1958)
59. M.E. Casey, R. Nutt, IEEE Trans. Nucl. Sci. **33**(1), 460 (1986)
60. I. Rausch, J. Cal-González, D. Dapra, H.J. Gallowitsch, P. Lind, T. Beyer, G. Minear, EJNMMI Physics **2**(1), 26 (2015)
61. M.M. Ter-Pogossian, D.C. Ficke, M. Yamamoto, J.T. Hood, IEEE Trans. Med. Imag. **1**(3), 179 (1982)
62. T. Lewellen, A. Bice, R. Harrison, M. Pencke, J. Link, IEEE Trans. Nucl. Sci. **35**(1), 665 (1988)
63. B. Mazoyer, R. Trebossen, C. Schoukroun, B. Verrey, A. Syrota, J. Vacher, P. Lemasson, O. Monnet, A. Bouvier, J. Lecomte, IEEE Trans. Nucl. Sci. **37**(2), 778 (1990)
64. S. Surti, J.S. Karp, Phys. Medica Eur. J. Med. Phys. **80**, 251 (2020)
65. S. Gundacker, A. Heering, Phys. Med. Biol. **65**(17), 17TR01 (2020)
66. D.R. Schaart, Phys. Med. Biol. **66**(9), 09TR01 (2021)
67. J. van Sluis, J. de Jong, J. Schaar, W. Noordzij, P. van Snick, R. Dierckx, R. Borra, A. Willemsen, R. Boellaard, J. Nucl. Med. **60**(7), 1031 (2019)

68. I. Rausch, A. Ruiz, I. Valverde-Pascual, J. Cal-González, T. Beyer, I. Carrio, J. Nucl. Med. **60**(4), 561 (2019)
69. S.R. Cherry, T. Jones, J.S. Karp, J. Qi, W.W. Moses, R.D. Badawi, J. Nucl. Med. **59**(1), 3 (2018)
70. B.A. Spencer, E. Berg, J.P. Schmall, N. Omidvari, E.K. Leung, Y.G. Abdelhafez, S. Tang, Z. Deng, Y. Dong, Y. Lv, J. Bao, W. Liu, H. Li, T. Jones, R.D. Badawi, S.R. Cherry, J. Nucl. Med. **62**(6), 861 (2021)
71. https://www.gehealthcare.com/-/media/739d885baa59485aaef5ac0e0eeb44a4.pdf
72. C.T. Chen, V. Johnson, W. Wong, X. Hu, C. Metz, IEEE Trans. Nucl. Sci. **37**(2), 636 (1990)
73. E. Mumcuoglu, R. Leahy, S. Cherry, Z. Zhou, IEEE Trans. Med. Imag. **13**(4), 687 (1994)
74. E.J. Teoh, D.R. McGowan, R.E. Macpherson, K.M. Bradley, F.V. Gleeson, J. Nucl. Med. **56**(9), 1447 (2015)
75. A.M. Alessio, C.W. Stearns, S. Tong, S.G. Ross, S. Kohlmyer, A. Ganin, P.E. Kinahan, IEEE Trans. Med. Imag. **29**(3), 938 (2010)
76. J. Krimmer, D. Dauvergne, J.M. Létang, É. Testa, Nucl. Instrum. Methods Phys. Res. Sect. A **878**(Supplement), 58 (2018)
77. Y. Xie, E.H. Bentefour, G. Janssens, J. Smeets, F. Vander Stappen, L. Hotoiu, L. Yin, D. Dolney, S. Avery, F. O'Grady, D. Prieels, J. McDonough, T.D. Solberg, R.A. Lustig, A. Lin, B.K.K. Teo, Int. J. Radiat. Oncol. *Biol. *Phys. **99**(1), 210 (2017)
78. M. Priegnitz, S. Helmbrecht, G. Janssens, I. Perali, J. Smeets, F. Vander Stappen, E. Sterpin, F. Fiedler, Phys. Med. **61**(2), 855 (2016)
79. J. Smeets, F. Roellinghoff, D. Prieels, F. Stichelbaut, A. Benilov, P. Busca, C. Fiorini, R. Peloso, M. Basilavecchia, T. Frizzi, J.C. Dehaes, A. Dubus, Phys. Med. Biol. **57**(11), 3371 (2012)
80. G. Llosá, Nucl. Instrum. Methods Phys. Res. Sect. A **926**, 148 (2019)
81. P. Solevi, E. Muñoz, C. Solaz, M. Trovato, P. Dendooven, J.E. Gillam, C. Lacasta, J.F. Oliver, M. Rafecas, I. Torres-Espallardo, G. Llosá, Phys. Med. Biol. **61**(14), 5149 (2016)
82. M. Kawula, T.M. Binder, S. Liprandi, R. Viegas, K. Parodi, P.G. Thirolf, Phys. Med. Biol. **66**(13), 135017 (2021)
83. F. Zhang, Z. He, C.E. Seifert, IEEE Trans. Nucl. Sci. **54**(4), 843 (2007)
84. E. Draeger, D. Mackin, S. Peterson, H. Chen, S. Avery, S. Beddar, J.C. Polf, Phys. Med. Biol. **63**(3), 035019 (2018)
85. P. Maggi, S. Peterson, R. Panthi, D. Mackin, H. Yang, Z. He, S. Beddar, J. Polf, Phys. Med. Biol. **65**(12), 125004 (2020)
86. J.C. Polf, S. Avery, D.S. Mackin, S. Beddar, Phys. Med. Biol. **60**(18), 7085 (2015)
87. H. Tashima, E. Yoshida, N. Inadama, F. Nishikido, Y. Nakajima, H. Wakizaka, T. Shinaji, M. Nitta, S. Kinouchi, M. Suga, H. Haneishi, T. Inaniwa, T. Yamaya, Phys. Med. Biol. **61**(4), 1795 (2016)
88. W. Enghardt, P. Crespo, F. Fiedler, R. Hinz, K. Parodi, J. Pawelke, F. Pönisch, Nucl. Instrum. Methods Phys. Res. Sect. A **525**, 284 (2004)
89. I. Ozoemelam, E. van der Graaf, M.J. van Goethem, M. Kapusta, N. Zhang, S. Brandenburg, P. Dendooven, Phys. Med. Biol. **65**(24), 245013 (2020)
90. M.G. Bisogni, A. Attili, G. Battistoni, N. Belcari, N. Camarlinghi, P. Cerello, S. Coli, A.D. Guerra, A. Ferrari, V. Ferrero, E. Fiorina, G. Giraudo, E. Kostara, M. Morrocchi, F. Pennazio, C. Peroni, M.A. Piliero, G. Pirrone, A. Rivetti, M.D. Rolo, V. Rosso, P. Sala, G. Sportelli, R. Wheadon, J. Med. Imag. **4**(1), 1 (2017)
91. E. Fiorina, V. Ferrero, G. Baroni, G. Battistoni, N. Belcari, N. Camarlinghi, P. Cerello, M. Ciocca, M. De Simoni, M. Donetti, et al., Front. Phys. **8**, 660 (2021)
92. T. Nishio, A. Miyatake, T. Ogino, K. Nakagawa, N. Saijo, H. Esumi, Int. J. Radiat. Oncol. Biol. Phys. **76**, 277 (2010)
93. Y. Shao, X. Sun, K. Lou, X.R. Zhu, D. Mirkovic, F. Poenisch, D. Grosshans, Phys. Med. Biol. **59**(13), 3373 (2014)
94. H. Uchida, T. Okamoto, T. Ohmura, K. Shimizu, N. Satoh, T. Koike, T. Yamashita, Nucl. Instrum. Methods Phys. Res. Sect. A **516**(2–3), 564 (2004)

95. H.J.T. Buitenhuis, F. Diblen, K.W. Brzezinski, S. Brandenburg, P. Dendooven, Phys. Med. Biol. **62**(12), 4654 (2017)
96. M. Bengtsson, P. Jansson, U. Bäckström, F. Johansson, A. Sjöland, Nuclear Technology **0**(0), 1 (2021)
97. C. Bélanger-Champagne, P. Peura, P. Eerola, T. Honkamaa, T. White, M. Mayorov, P. Dendooven, IEEE Trans. Nucl. Sci. **66**(1), 487 (2019)
98. R. Virta, R. Backholm, T.A. Bubba, T. Helin, M. Moring, S. Siltanen, P. Dendooven, H. Tapani, ESARDA Bulletin **61**, 10 (2020)
99. S. Caruso, M.F. Murphy, F. Jatuff, R. Chawla, Nucl. Eng. Des. **239**(7), 1220 (2009)

Printed in the United States
by Baker & Taylor Publisher Services